NONMEASURABLE SETS AND FUNCTIONS

NORTH-HOLLAND MATHEMATICS STUDIES 195

(Continuation of the Notas de Matemática)

Editor: Jan van Mill
Faculteit der Exacte Wetenschappen
Amsterdam, The Netherlands

ELSEVIER

2004

Amsterdam – Boston – Heidelberg – London – New York – Oxford
Paris – San Diego – San Francisco – Singapore – Sydney – Tokyo

NONMEASURABLE SETS AND FUNCTIONS

A.B. KHARAZISHVILI
*I. Vekua Institute of Applied Mathematics
Tbilisi State University, Republic of Georgia*

ELSEVIER

2004

Amsterdam – Boston – Heidelberg – London – New York – Oxford
Paris – San Diego – San Francisco – Singapore – Sydney – Tokyo

ELSEVIER B.V.
Sara Burgerhartstraat 25
P.O. Box 211, 1000 AE Amsterdam
The Netherlands

ELSEVIER Inc.
525 B Street, Suite 1900
San Diego, CA 92101-4495
USA

ELSEVIER Ltd
The Boulevard, Langford Lane
Kidlington, Oxford OX5 1GB
UK

ELSEVIER Ltd
84 Theobalds Road
London WC1X 8RR
UK

First edition 2004

Library of Congress Cataloging in Publication Data
A catalog record is available from the Library of Congress.

British Library Cataloguing in Publication Data
A catalogue record is available from the British Library.

ISBN: 0 444 51626 3
ISSN: 0304-0208

∞ The paper used in this publication meets the requirements of ANSI/NISO Z39.48-1992 (Permanence of Paper).
Printed in The Netherlands.

Contents

Preface

This book contains a course of lectures devoted to the nonmeasurability property of certain subsets of an abstract space E which is equipped with a group G of its transformations and with a nonzero σ-finite G-invariant (more generally, G-quasi-invariant) measure μ. In particular, we will be dealing with various and unusual features of μ-nonmeasurable sets in such a space.

However, the main attention will be paid to the classical case where E coincides with the real line $\mathbf{R}$, G coincides with the group of all translations of $\mathbf{R}$, and μ is the standard Lebesgue measure on $\mathbf{R}$. We discuss the question of the existence of non Lebesgue-measurable sets in $\mathbf{R}$, their strange "geometric" properties, their connections with the measure extension problem and so on. At first sight, this topic seems to be narrow in scope. Nevertheless, we will try to demonstrate in due course that there are many interesting results concerning nonmeasurable sets (or nonmeasurable functions). Furthermore, it will be shown in the book that those results play an essential role in various domains of modern mathematics, such as set theory, real analysis, probability theory, and general topology. Deep relationships between the questions considered below and related areas of contemporary mathematics confirm, in our opinion, the importance of this topic. It should also be pointed out that there are many attractive unsolved problems about nonmeasurable sets and functions. Some of them will be formulated in subsequent chapters of the book.

Our starting point is the classical Vitali theorem [224] stating the existence of a subset of the real line $\mathbf{R}$, which is nonmeasurable in the Lebesgue sense and does not possess the Baire property (compare [125], [155], [165]).

Almost 100 years have passed since this remarkable theorem was proved, but it remains of living interest for real analysis and Lebesgue measure theory. Moreover, it has stimulated the emergence and further development of the following three fascinating branches of mathematics:

(1) The paradoxical decompositions of sets in finite-dimensional euclidean spaces $\mathbf{R}^1, \mathbf{R}^2, \mathbf{R}^3, \ldots$ and in other (more general) spaces endowed with various transformation groups;

(2) The theory of large cardinals;

(3) The theory of invariant (respectively, quasi-invariant) extensions of invariant (respectively, quasi-invariant) measures.

Each of the above-mentioned theories can be characterized as a beautiful field of mathematics and together they constitute a wide potential area for deep mathematical investigations. Besides, each of these theories vividly demonstrates how far modern set-theoretical methods lead in their sophisticated constructions, and how, by using these constructions, extraordinary results can be obtained which contrast with our practical intuition.

There are numerous works devoted to the Vitali theorem or its analogues. During the past century, this theorem was generalized and extended in several directions. For instance, in [68] the reader can find similar constructions of nonmeasurable sets in locally compact topological groups. In this context, see also [25]. Note, in addition, that in the monograph by Morgan [155] a large list of works is given, which are closely connected with the classical Vitali construction.

In the present book, we touch upon various aspects of the Vitali theorem and show its stimulating role for further investigations in this area. We also consider some essentially different constructions and results about nonmeasurable sets (functions) and compare them to each other. First of all, we mean here those sets which appear as a result of the classical Bernstein construction [10], nonmeasurable sets associated with a Hamel basis of the real line [63], sets participating in the Banach-Tarski paradox [9] and many others.

We maximally try to give the material on nonmeasurable sets and functions in a form accessible for a wide audience of potential readers (in particular, for graduate and post-graduate students whose interests lie in the above-mentioned domains of mathematics), and we focus our attention on the fundamental ideas and concepts which naturally play a dominant role in further studies. We also discuss some logical and set-theoretical aspects of measurability, which lead to a deeper and more profound understanding of the subject.

The present book is based on the course of lectures given by the author at I.Vekua Institute of Applied Mathematics (Tbilisi State University) several years ago. By virtue of our experience, we hope that the subject of these lectures will be interesting for a wide group of mathematicians with "a good mathematical taste". Moreover, since the concept of measurability is important for many mathematical disciplines, such as: real analysis, probability theory, optimization, and functional analysis, we can assert that various kinds of extraordinary sets (from the measure-theoretical viewpoint) deserve to be investigated more thoroughly. A similar situation we have in classical mathematical analysis (the advanced calculus). The differentiation operation for nice real-valued functions is extensively studied in analysis but, at the same time, it is always underlined, by presenting widely known examples, that there exist continuous nowhere differentiable functions (see [55], [165], [102]) and that those functions are typical in the sense of category.

As mentioned above, the first chapter of the book begins with discussion of the classical Vitali theorem for the real line $\mathbf{R}$ and the Lebesgue measure λ on it. We consider this theorem in detail and indicate some of its immediate consequences. In the same chapter, we point out close connections of the Vitali theorem with uncountable forms of the Axiom of Choice. This topic will be developed in subsequent sections of the book, especially in Chapter 15 and Appendix 1.

The second chapter is devoted to the classical Bernstein construction which plays a significant role not only in real analysis and measure theory, but also in general topology. Some exercises for that chapter illustrate this fact.

The third chapter of the book deals with Hamel bases of $\mathbf{R}$ and their applications to measurability properties of additive functionals. Note that this topic is widely presented in numerous works of mathematicians whose interests lie in the theory of functional equations (see, for instance, [118]).

In the next chapter, some nonmeasurable sets are under consideration whose existence is stated by using the classical Fubini theorem. Here the main role is played by a Sierpiński partition of the euclidean plane $\mathbf{R}^2$, which is possible only under the Continuum Hypothesis (see [200]). The ideas concentrated around this partition enable us to establish several deep results concerning nonmeasurability of sets. For example, the theorem of Kunen is presented saying that the real-valued measurability of the cardinality continuum $\mathbf{c}$ implies the existence of a non Lebesgue-measurable subset of $\mathbf{R}$ whose cardinality is strictly less than $\mathbf{c}$.

Chapter 5 is devoted to well-known classes of small subsets of $\mathbf{R}$. The smallness here means that these subsets are members of a fixed proper σ-

ideal in the power set Boolean algebra $\mathcal{P}(\mathbf{R})$. Typical examples of such classes are: the σ-ideal generated by Luzin sets, the σ-ideal generated by Sierpiński sets, and the σ-ideal of Marczewski sets. Various relationships between small sets and nonmeasurable sets (nonmeasurable functions) are observed.

In Chapter 6 some strange subsets of the euclidean plane $\mathbf{R}^2$ are considered and, in particular, further examples of non Lebesgue-measurable sets are obtained by starting with properties of such subsets. The classical Luzin problem on the existence of a function $f : \mathbf{R} \to \mathbf{R}$ whose graph $\Gamma_f \subset \mathbf{R}^2$ covers $\mathbf{R}^2$ by using countably many motions of $\mathbf{R}^2$ is discussed with its final solution due to Davies [31].

Chapter 7 contains special constructions of nonmeasurable sets, which essentially differ from the classical ones due to Vitali, Bernstein, and Hamel. Here we give several constructions in infinite-dimensional spaces, present purely combinatorial approach due to Ulam [222], and consider the method based on the existence of a nontrivial ultrafilter in the power set Boolean algebra $\mathcal{P}(\mathbf{N})$, where $\mathbf{N}$ stands for the set of all natural numbers. This method was first suggested by Sierpiński.

The next four chapters of the book are primarily devoted to some analogues and extensions of the Vitali theorem for abstract groups (or, more generally, for spaces endowed with various transformation groups). Namely, we consider an analogue of the Vitali theorem for a group of motions of a finite-dimensional euclidean space, examine measurability properties of selectors associated with countable and uncountable subgroups of a given transformation group, and introduce and investigate absolutely nonmeasurable sets in uncountable commutative groups. Most results presented in these chapters are due to the author.

In Chapter 12 we consider σ-ideals of sets producing nonmeasurable unions of their members. In this connection, see also [20] and [49].

In Chapter 13 we discuss measurability properties of subgroups of an uncountable group. Among other results, it is demonstrated that if finitely many subgroups of a standard group G are taken, then all of them can be made measurable with respect to an appropriate quasi-invariant extension of a given quasi-invariant Borel probability measure on G.

Chapter 14 is devoted to the group O_n^+ of all rotations of the euclidean n-dimensional space $\mathbf{R}^n$ $(n \geq 3)$ and to nonmeasurable sets in this space closely connected with special algebraic properties of O_n^+. We briefly touch upon the famous Banach-Tarski paradox for $\mathbf{R}^n$ (indicating the importance of various free subgroups of O_n^+), consider absolutely nonmeasurable sets on the euclidean n-dimensional unit sphere $\mathbf{S}_n$ ($n \geq 2$), and compare measurability properties of sets for the following two classical groups which essentially differ from each other: the group O_n^+ and the group T_n of all translations of $\mathbf{R}^n$, where $n \geq 3$.

Note also that many interesting facts concerning the group of rotations and the Banach-Tarski paradox can be found in the well-known monograph by Wagon [226].

In the last chapter we focus on deep connections between nonmeasurable sets and the corresponding properties of nontrivial filters in the Boolean algebra $\mathcal{P}(\mathbf{N})$. The central results here are Talagrand's theorem [218] on the nonmeasurability of so-called rapid filters and the theorem of Shelah and Raisonnier [176].

Each chapter ends with exercises. Some of them are rather difficult and are accompanied by a hint or a more detailed explanation.

In Appendix 1 several forms of the Axiom of Choice are briefly discussed with the Continuum Hypothesis and Martin's Axiom, and their influence on the measurability (in the Lebesgue sense) is pointed out.

In Appendix 2 general facts from the theory of infinite commutative groups are given. Some of them play an essential role in constructions of nonmeasurable sets, presented in this book.

A.B. Kharazishvili

Chapter 1
The Vitali theorem

In this introductory chapter we wish to recall the classical construction of Vitali [224] which yields the existence of a subset X of the real line $\mathbf{R}$, such that X is nonmeasurable in the Lebesgue sense and, simultaneously, does not have the Baire property (with respect to the standard euclidean topology on $\mathbf{R}$). Later, we will show some interesting connections of such subsets of $\mathbf{R}$ with fundamental problems in real analysis and measure theory.

Let $\mathbf{Q}$ denote the set of all rational numbers. Clearly, $\mathbf{Q}$ is a subgroup of the additive group $\mathbf{R}$. Let us introduce a binary relation $G \subset \mathbf{R} \times \mathbf{R}$ defined by the formula

$$(x, y) \in G \Leftrightarrow x - y \in \mathbf{Q}.$$

It is easy to see that G is an equivalence relation on $\mathbf{R}$. The graph of this relation is a simple subset of the euclidean plane $\mathbf{R}^2$. Namely, it can be represented as the union of a countable family of straight lines lying in $\mathbf{R}^2$ and parallel to the line

$$l = \{(x, y) \in \mathbf{R}^2 \ : \ x - y = 0\}.$$

Let us denote by $\{V_i \ : \ i \in I\}$ the partition of $\mathbf{R}$ canonically associated with G. For this partition, we have the equalities

$$card(V_i) = \omega \qquad (i \in I),$$

$$card(I) = \mathbf{c},$$

where ω is the first infinite cardinal number and $\mathbf{c}$ is the cardinality of the continuum. The family $\{V_i \ : \ i \in I\}$ is usually called the Vitali partition of the real line.

Let X be an arbitrary selector of the Vitali partition. In other words, let X be a subset of $\mathbf{R}$ satisfying the relation

$$card(X \cap V_i) = 1$$

for all $i \in I$. Obviously, the existence of such a selector follows directly from the Axiom of Choice (**AC**). In the sequel, X will usually be called a Vitali subset of the real line.

Vitali was the first mathematician to prove, in 1905, that X is not measurable with respect to the classical Lebesgue measure on **R**.

In order to establish this fact, let us first observe that the following two relations are true:

1) $\cup\{X + q \ : \ q \in \mathbf{Q}\} = \mathbf{R}$;

2) if $q \in \mathbf{Q}$, $r \in \mathbf{Q}$ and $q \neq r$, then $(X + q) \cap (X + r) = \emptyset$.

Let $\lambda = \lambda_1$ denote the Lebesgue measure on the real line. Suppose, for a moment, that X is a λ–measurable set. Then, since λ is invariant under the group $\mathbf{Q}$ and relation 1) holds, we get $\lambda(X) > 0$. Evidently, there exists a natural number n such that

$$\lambda(X \cap [-n, n]) > 0.$$

Let us put $Y = X \cap [-n, n]$ and consider the set

$$Z = \cup\{Y + q \ : \ q \in \mathbf{Q}, \ |q| < 1\}.$$

Taking into account relation 2) and the invariance of λ with respect to $\mathbf{Q}$, we see that $\lambda(Z) = +\infty$. On the other hand, it is clear that Z is a bounded subset of the real line, so we must have the inequality $\lambda(Z) < +\infty$. Thus, we obtained a contradiction which gives us the desired result.

Notice that the argument presented above also proves a more general statement. In order to formulate it, we need the notion of a set of Vitali type. Let Γ be a subgroup of the additive group **R**. Consider the partition of **R** canonically associated with the equivalence relation

$$x \in \mathbf{R} \ \& \ y \in \mathbf{R} \ \& \ x - y \in \Gamma.$$

Let X be any selector of this partition. We shall say in our further considerations that X is a Γ–selector (or that X is a set of Vitali type with respect to the group Γ).

It can easily be seen that the preceding argument enables us to establish the following result.

Theorem 1. *Let Γ be a countable dense subgroup of the additive group* **R** *and let μ be a measure defined on some σ–algebra of subsets of* **R**. *Suppose also that these three conditions are satisfied:*

1) μ is an invariant measure with respect to Γ; in other words, for all $g \in \Gamma$ and for all $Z \in dom(\mu)$, we have $g + Z \in dom(\mu)$ and $\mu(g+Z) = \mu(Z)$;
2) $[0,1] \in dom(\mu)$;
3) $0 < \mu([0,1]) < +\infty$.
Then every Γ-selector is nonmeasurable with respect to μ.

The proof of this result is left to the reader.

We also want to remark that conditions 1), 2), and 3) of Theorem 1 imply the following fact: the completion of the given measure μ is an extension of some measure on $\mathbf{R}$ which is proportional to λ. In other words, the completion of μ is an extension of the measure $t\lambda$ where $t = \mu([0,1])$ (compare Exercise 4 of this chapter).

A similar argument (however, slightly more complicated) shows us that no Vitali subset of $\mathbf{R}$ has the Baire property with respect to the standard topology of $\mathbf{R}$. Namely, we can formulate and prove the following more general result.

Theorem 2. *Let Γ be a countable subgroup of the additive group $\mathbf{R}$ and let $\mathcal{T}$ be a topology on $\mathbf{R}$, such that:*
1) the pair $(\mathbf{R}, \mathcal{T})$ is a topological group;
2) $(\mathbf{R}, \mathcal{T})$ is a second category topological space;
3) Γ is a nondiscrete subgroup of $(\mathbf{R}, \mathcal{T})$.
Then no Γ-selector has the Baire property in $(\mathbf{R}, \mathcal{T})$.

Proof. Let us take any Γ-selector X and let us show that X does not have the Baire property in the space $(\mathbf{R}, \mathcal{T})$. Suppose to the contrary that X possesses the Baire property in $(\mathbf{R}, \mathcal{T})$. Since the equality

$$\cup\{g + X : g \in \Gamma\} = \mathbf{R}$$

holds true and $card(\Gamma) \leq \omega$, we obtain, in view of relation 2), that X is not a first category set. Consequently, applying the Banach-Kuratowski-Pettis theorem (see Exercise 6 of this chapter), we can find a neighbourhood V of zero in $(\mathbf{R}, \mathcal{T})$ such that

$$V \subset X - X = \{y - z : y \in X,\ z \in X\}.$$

Further, according to relation 3), we have

$$(\Gamma \cap V) \setminus \{0\} \neq \emptyset.$$

Let $h \neq 0$ belong to $\Gamma \cap V$. Then we may write

$$(X + h) \cap X = \emptyset$$

since X is a Γ-selector. On the other hand, we must have

$$h \in V \subset X - X$$

from which it follows that there are some elements $y \in X$ and $z \in X$ such that $h = y - z$ or, equivalently, $h + z = y$. This immediately yields

$$(X + h) \cap X \neq \emptyset.$$

The contradiction obtained finishes the proof of Theorem 2.

Remark 1. As mentioned above, the proof of Vitali's classical result is based on the Axiom of Choice. Moreover, this result essentially needs an uncountable form of the Axiom of Choice. Indeed, Solovay demonstrated in his famous paper [210] that the existence of non Lebesgue-measurable subsets of the real line (or the existence of subsets of the real line without the Baire property) cannot be proved in the theory

$$\mathbf{ZF} \ \& \ (the\ countable\ form\ of\ the\ Axiom\ of\ Choice).$$

The precise formulation of the countable form of the Axiom of Choice and some of its direct consequences are given in Appendix 1. Actually, Solovay established in [210] a more general result stating that the existence of non Lebesgue-measurable subsets of $\mathbf{R}$ is unprovable in the theory

$$\mathbf{ZF} \ \& \ \mathbf{DC},$$

where $\mathbf{DC}$ denotes the so-called Axiom of Dependent Choice. Some information about this axiom is presented in Appendix 1 where it is also shown the equivalence of $\mathbf{DC}$ to the classical Baire theorem on category (for complete metric spaces).

Remark 2. It can be proved (see, for instance, [82] or [96]) that there exists a measure ν on $\mathbf{R}$ satisfying the following conditions:

1) ν is nonzero, σ-finite and nonatomic;

2) ν is invariant with respect to the group of all isometric transformations of $\mathbf{R}$ and, in particular, ν is invariant with respect to the group of all translations of $\mathbf{R}$;

3) $dom(\nu)$ contains the class of all Lebesgue measurable subsets of $\mathbf{R}$;

4) there is a Vitali subset X of $\mathbf{R}$ such that $X \in dom(\nu)$.

We thus see that some Vitali subsets of the real line $\mathbf{R}$ can be measurable with respect to certain nonzero σ-finite measures on $\mathbf{R}$ invariant under the group of all isometric transformations of $\mathbf{R}$.

Note that the construction of a measure ν satisfying conditions 1) - 4) is thoroughly considered in Example 2 of Chapter 11 of this book.

According to the definition introduced in the beginning of the present chapter, a Vitali set is an arbitrary selector of the classical Vitali partition $\{V_i : i \in I\}$ of $\mathbf{R}$. We already know that any such selector is λ-nonmeasurable (Theorem 1), but (see Remark 2 above) there are Vitali subsets of $\mathbf{R}$ measurable with respect to some nonzero σ-finite measures given on $\mathbf{R}$ and invariant under the group of all isometries of $\mathbf{R}$. In other words, there exist Vitali sets which are not absolutely nonmeasurable with respect to the class of all nonzero σ-finite measures given on the real line and invariant under the group of all isometric transformations of this line. The precise notion of an absolutely nonmeasurable set in a space equipped with a transformation group will be formulated and discussed in Chapter 11. This notion turns out to be helpful for obtaining new strong results about the existence of nonmeasurable sets and functions.

Now, we would like to discuss some connections of the classical Vitali construction with combinatorial properties of binary relations of a special type. We mean here the so-called $(n - n)$-correspondences, where n is a fixed natural number.

The notion of an $(n - n)$-correspondence is defined as follows. Let A and B be two arbitrary sets and let G be a binary relation between these sets; in other words, let

$$G \subset A \times B.$$

We say that G is an $(n - n)$-correspondence between A and B if for each element $a \in A$, the equality

$$card(\{b \in B : (a, b) \in G\}) = n$$

holds and, for each element $b \in B$, the equality

$$card(\{a \in A : (a, b) \in G\}) = n$$

holds, too. There are many interesting and important combinatorial facts concerning $(n - n)$-correspondences. We need one of these facts in our further considerations. First, we recall a very useful result from general set theory. It is due to Banach (see, for instance, [127] or [19]).

Theorem 3. *Let A and B be any two sets, let f be an injective function acting from A into B and let g be an injective function acting from B into A. Then there exist four sets A_1, A_2, B_1, B_2 satisfying the following conditions:*

1) $A_1 \cap A_2 = \emptyset$, $A_1 \cup A_2 = A$;

2) $B_1 \cap B_2 = \emptyset$, $B_1 \cup B_2 = B$;

3) $f|A_1$ is a bijection between the sets A_1 and B_1;

4) $g|B_2$ is a bijection between the sets B_2 and A_2.

In particular, we can define a bijection $h : A \to B$ by the following formula: $h(x) = f(x)$ if $x \in A_1$, and $h(x) = g^{-1}(x)$ if $x \in A_2$.

For the proof of Theorem 3, see [127] or [19].

The Banach theorem formulated above possesses many applications in various fields of mathematics. To illustrate this, it suffices to remember that:

1) the well known Cantor–Bernstein theorem from general set theory is a particular case of the Banach theorem;

2) the important result from classical descriptive set theory, stating that any two uncountable Borel subsets of a Polish topological space are Borel isomorphic, is essentially based on the Banach theorem;

3) the famous Banach–Tarski paradox, stating that any two bounded subsets of the euclidean space $\mathbf{R}^n$ ($n \geq 3$) with nonempty interiors are equivalent by finite decompositions, is essentially based on the Banach theorem.

Let us mention that the Banach-Tarski paradox with many related questions of equidecomposability theory are discussed in detail in the monograph by Wagon [226] (see also Chapter 14 of this book).

Remark 3. The proof of the Banach theorem can be carried out within the theory **ZF** (compare [127] or [19]).

Another useful combinatorial result which we need in our further considerations is due to Hall [60]. It is suitable to formulate this result in terms of set-valued mappings.

Theorem 4. *Let A and B be two sets, let $\mathcal{P}(B)$ denote the power set of B and let $F : A \to \mathcal{P}(B)$ be a set-valued mapping. Suppose also that the following two conditions hold:*

1) for each element $a \in A$, we have $card(F(a)) < \omega$;

2) for each finite subset X of A, we have

$$card(X) \leq card(\cup\{F(x) \ : \ x \in X\}).$$

Then there exists an injective mapping $f : A \to B$ such that

$$(\forall a \in A)(f(a) \in F(a)).$$

In other words, f is an injective selector of the given set-valued mapping F.

The proof of Theorem 4 is not difficult. Indeed, if $card(A) < \omega$, then the proof can be carried out by induction on $card(A)$, and this inductive process is a useful exercise from finite combinatorics. If $card(A) \geq \omega$, then the proof can be reduced to the previous case by the standard argument, using the classical Tychonoff theorem on the quasicompactness of products of quasicompact topological spaces. In fact, we need here only a very particular case of the Tychonoff theorem because, in our situation, all sets $F(a)$ $(a \in A)$ are finite and are assumed to be equipped with the discrete topology.

It is well known that in the theory **ZF**, the Tychonoff theorem on the quasicompactness of products of quasicompact topological spaces is equivalent to the Axiom of Choice. We recall that this result is due to Kelley [76]. The proof of the above-mentioned equivalence is outlined in Appendix 1.

Thus, we conclude that Theorem 4 heavily relies on the Axiom of Choice. Later, we will show that the Hall theorem cannot be proved in the theory **ZF & DC**.

The following two simple examples yield a typical application of the finite version of the Hall theorem.

Example 1. Let $(\Gamma, \cdot)$ be a finite group and let X and Y be some subgroups of Γ such that $card(X) = card(Y)$. Furthermore, let

$$\{z \cdot X \ : \ z \in \Gamma\}, \quad \{Y \cdot z \ : \ z \in \Gamma\}$$

be two partitions of Γ canonically associated with the subgroups X and Y, respectively. Then there exists a common selector of these two partitions.

Since our group Γ is not assumed to be commutative, a nontrivial particular case of Example 1 is the situation where $X = Y$.

Example 2. Let E be a set equipped with a probability measure μ, let $\{X_1, X_2, ..., X_m\}$ and $\{Y_1, Y_2, ..., Y_m\}$ be two finite partitions of E into μ-measurable sets, such that

$$\mu(X_i) = \mu(Y_j)$$

for all natural numbers $i \in [1, m]$ and $j \in [1, m]$. Then there exist pairwise distinct elements $z_1, z_2, ..., z_m$ from E, such that the set

$$Z = \{z_1, z_2, ..., z_m\}$$

turns out to be a common selector for these two partitions.

It is not difficult to see that Example 2 generalizes Example 1.

The next result follows immediately from Theorems 3 and 4.

Theorem 5. *Let $n > 0$ be a fixed natural number, let A and B be two sets and let G be an $(n-n)$-correspondence between these sets. Then there exists a bijection $g : A \to B$ such that the graph of g is contained in G.*

Now, we are going to show that the preceding theorem cannot be proved in the theory **ZF** & **DC**. For this purpose, let us return to the Vitali partition $\{V_i : i \in I\}$ of the real line **R**. First, let us observe that

$$\mathbf{Q} \in \{V_i : i \in I\},$$

where $\mathbf{Q}$ is the set of all rational numbers. Let us put

$$\{W_i : i \in I\} = \{V_i : i \in I\} \setminus \{\mathbf{Q}\}.$$

It is easy to check that for each index $j \in I$, we have the relations

$$-W_j \in \{W_i : i \in I\}, \quad -W_j \neq W_j.$$

Moreover, if $i \in I$, $j \in I$ and $-W_i \cup W_i = -W_j \cup W_j$, then we have the disjunction

$$W_i = W_j \quad \vee \quad -W_i = W_j.$$

Now, take the two-element set $\{0, 1\}$ and put

$$A = \{W_i : i \in I\},$$

$$B = \{-W_i \cup W_i \cup \{t\} : i \in I, \ t \in \{0, 1\}\}.$$

Furthermore, define a binary relation G between the sets A and B. Namely, for each element $W_i \in A$, put

$$G(W_i) = \{-W_i \cup W_i \cup \{t\} \ : \ t \in \{0,1\}\}.$$

Obviously, if $-W_i \cup W_i \cup \{t\}$ belongs to the set B, then

$$G^{-1}(-W_i \cup W_i \cup \{t\}) = \{-W_i, \ W_i\}.$$

So we claim that G is a $(2-2)$-correspondence between the sets A and B.

We shall show that the existence of a bijection $g \ : \ A \to B$ whose graph is contained in G cannot be established in the theory **ZF** & **DC**. The following argument is essentially due to Sierpiński (see [201]).

Suppose that a bijection $g \ : \ A \to B$ with the property mentioned above does exist. Then, for any index $i \in I$, we can write

$$g(W_i) = -W_i \cup W_i \cup \{t_i\}$$

where $t_i \in \{0,1\}$. Let us define a function

$$\phi \ : \ \mathbf{R} \setminus \mathbf{Q} \to \{0,1\}$$

as follows. Take an arbitrary element x from the set $\mathbf{R} \setminus \mathbf{Q}$. Then there exists a unique index $i \in I$ such that x belongs to the set W_i. Put

$$\phi(x) = t_i.$$

In this way, the required function ϕ is completely determined.

We assert that ϕ is not measurable with respect to the Lebesgue measure λ. Suppose to the contrary that ϕ is measurable in the Lebesgue sense. Then, starting with the definition of ϕ, it can directly be checked that the following two relations hold:

1) for each $x \in \mathbf{R} \setminus \mathbf{Q}$ and for each $q \in \mathbf{Q}$, the equality

$$\phi(x + q) = \phi(x)$$

is valid (in other words, our ϕ is a $\mathbf{Q}$-invariant function);

2) for each $x \in \mathbf{R} \setminus \mathbf{Q}$, we have the equality

$$\phi(-x) = 1 - \phi(x).$$

We now need to recall that the measure λ is metrically transitive with respect to any dense subgroup Γ of the additive group $\mathbf{R}$. In other words, λ possesses the following property: if a λ-measurable function $f : \mathbf{R} \to \mathbf{R}$ satisfies the relation $f(x + p) = f(x)$ for all $x \in \mathbf{R}$ and for all $p \in \Gamma$, then f is equivalent to a constant function (compare Exercise 7 from this chapter).

Since $\mathbf{Q}$ is a dense subgroup of $\mathbf{R}$, the Lebesgue measure is metrically transitive with respect to $\mathbf{Q}$. Taking into account this property of the Lebesgue measure and applying relation 1), we deduce that our function ϕ is constant almost everywhere. But we simultaneously have

$$ran(\phi) \subset \{0, 1\}.$$

Thus we get the disjunction: either $\phi = 0$ almost everywhere or $\phi = 1$ almost everywhere. But

$$dom(\phi) = \mathbf{R} \setminus \mathbf{Q}$$

and the set $\mathbf{R} \setminus \mathbf{Q}$ is symmetric with respect to the point 0. Now, relation 2) shows that if $\phi = 0$ almost everywhere, then $\phi = 1$ almost everywhere and, conversely, if $\phi = 1$ almost everywhere, then $\phi = 0$ almost everywhere. So we obtain a contradiction which gives us the nonmeasurability (in the Lebesgue sense) of our function ϕ.

Since the argument just presented belongs to the theory $\mathbf{ZF}$ & $\mathbf{DC}$, we can formulate the following result of Sierpiński.

Theorem 6. *The existence of a bijection $g : A \to B$ with the graph contained in the $(2 - 2)$-correspondence G (defined above) implies, in the theory $\mathbf{ZF}$ & $\mathbf{DC}$, the existence of a non Lebesgue-measurable function acting from $\mathbf{R}$ into $\mathbf{R}$.*

Taking into account the result of Solovay mentioned in Remark 1, we conclude that:

(1) the Hall theorem cannot be proved in the theory $\mathbf{ZF}$ & $\mathbf{DC}$;

(2) Theorem 5 cannot be proved in the same theory $\mathbf{ZF}$ & $\mathbf{DC}$.

We also have the next fact:

(3) it cannot be proved, within the theory $\mathbf{ZF}$ & $\mathbf{DC}$, that there exists a linear ordering of the Vitali partition $\{V_i : i \in I\}$.

Indeed, it is easy to see that, in the theory $\mathbf{ZF}$, the existence of a linear ordering of the family $\{V_i : i \in I\}$ implies the existence of a bijection

$$g : A \to B$$

whose graph is contained in the $(2-2)$-correspondence G.

We have already shown some nontrivial connections and interactions between nonmeasurable sets and infinite combinatorics. In fact, these connections are much deeper. In our further considerations, we will be able to underline many other relationships of this sort (see especially Chapter 7, Chapter 15 and Appendix 1). Various examples of nonmeasurable sets and sets without the Baire property will also be considered in the following sections of this book. Moreover, we will meet nonmeasurable sets and sets without the Baire property which have an additional algebraic structure. For example, they are groups or vector spaces with respect to the natural algebraic operations. Note that the constructions of such sets need methods essentially different from the ones described above. Also, it should be mentioned that there are some purely set-theoretical constructions and combinatorial methods leading to the existence of nonmeasurable sets with respect to nonzero σ-finite measures vanishing on singletons (such measures are usually called diffused or continuous). The main role in those constructions is played by an Ulam transfinite matrix [222]. Various properties of this matrix and its applications to measure theory and general topology are discussed in the well-known books [125], [155], [165] (see also Chapter 7 of the present book).

We have already indicated the role of the Axiom of Choice in problems concerning the existence of nonmeasurable sets or nonmeasurable functions. There are also constructions of nonmeasurable sets (functions) with some additional properties, which need much stronger set-theoretical assumptions. One of such assumptions is the famous Continuum Hypothesis (**CH**):

$$\mathbf{c} = \omega_1,$$

where ω_1 stands for the first uncountable cardinal. For instance, in Chapter 5 of this book we deal with the so-called Sierpiński sets which turn out to be extremely nonmeasurable: each uncountable subset of a Sierpiński set is not measurable in the Lebesgue sense. The existence of Sierpiński sets cannot be established without extra axioms, but easily follows from the Continuum Hypothesis.

The importance of the Continuum Hypothesis in numerous constructions of so-called "singular" subsets of the real line is well-known (see, for instance, [125], [147] and [165]). A lot of other extraordinary consequences of **CH** in measure theory, real analysis, and set-theoretical topology are also of interest from various points of view.

However, the Continuum Hypothesis is regarded as a very strong assumption because it maximally restricts the size of the continuum; namely, it states the identity between $\mathbf{c}$ and the first uncountable cardinal ω_1. Therefore, a natural question was posed to find an appropriate axiom (instead of $\mathbf{CH}$) which even in the absence of the Continuum Hypothesis could give some tools efficient enough for mathematical (primarily, set-theoretical) constructions. Martin's Axiom ($\mathbf{MA}$) turned out a good candidate to fill up this place. Extensive information about this axiom and its consequences can be found in [47], [64], [122] (see also Appendix 1). Sometimes, much weaker set-theoretical assumptions (formulated, for example, in terms of the σ-ideal of all Lebesgue measure zero sets or, respectively, in terms of the σ-ideal of all first category sets on $\mathbf{R}$) are sufficient to resolve the question of the existence of a non Lebesgue-measurable subset of $\mathbf{R}$ with some interesting additional properties (or, respectively, to resolve the question of the existence of a subset of $\mathbf{R}$ lacking the Baire property but possessing other important features). A number of examples of such subsets will be presented in this book. For an illustrative example, we can already mention here that Martin's Axiom enables us to state the existence of so-called generalized Sierpi'nski sets and generalized Luzin sets. In this connection, see Chapter 5 of the book.

EXERCISES

1. Let X be a Lebesgue measurable subset of $\mathbf{R}$. Show that for any compact set $K \subset \mathbf{R}$ with $\lambda(K) > 0$, there exists a compact set $K' \subset K$ with $\lambda(K') > 0$ satisfying the relation

$$K' \cap X = \emptyset \quad \vee \quad K' \cap (\mathbf{R} \setminus X) = \emptyset.$$

Conversely, let Y be a subset of $\mathbf{R}$ having the property that, for any compact set $K \subset \mathbf{R}$ with $\lambda(K) > 0$, there exists a compact set $K' \subset K$ with $\lambda(K') > 0$ such that

$$\lambda(K' \cap Y) = 0 \quad \vee \quad \lambda(K' \cap (\mathbf{R} \setminus Y)) = 0.$$

Show that Y is measurable in the Lebesgue sense.

Formulate and prove an analogous characterization of μ-measurable sets, where μ is the completion of some σ-finite Radon measure given on a Hausdorff topological space E.

For the definition of Radon measures and their properties, see [14], [62], [68] or [160].

2. We say that a partition of $\mathbf{R}$ is measurable (with respect to λ) if there exists a Lebesgue measurable function $f : \mathbf{R} \to \mathbf{R}$ such that this partition is canonically associated with f; in other words, elements x and y of $\mathbf{R}$ belong to the same class of the partition if and only if $f(x) = f(y)$.

Demonstrate that the Vitali partition of $\mathbf{R}$ is not measurable in the sense of this definition.

3. Give a detailed proof of Theorem 1.

4. Let Γ be a dense subgroup of $\mathbf{R}$ and let μ be a measure on $\mathbf{R}$ satisfying the following conditions:

(a) μ is Γ-invariant;

(b) $[0, 1] \in dom(\mu)$;

(c) $0 < \mu([0, 1]) < +\infty$.

Let X be an arbitrary μ-measurable set with $\mu(X) > 0$. By applying an argument similar to the Vitali construction, show that there exists a subset Y of X such that, for any measure ν on $\mathbf{R}$ extending μ and invariant under Γ, the relation $Y \notin dom(\nu)$ is valid. In particular, Y is nonmeasurable with respect to the original measure μ.

Demonstrate also that the restriction of μ to the Borel σ-algebra of $\mathbf{R}$ is proportional to the standard Borel measure on $\mathbf{R}$ (whose completion coincides with the Lebesgue measure λ).

In connection with the previous exercise, see also Chapter 11.

5. Let E be a topological space and let $(U_i)_{i \in I}$ be a family of open subsets of E, such that each U_i is of first category in E. Show that the open set $\cup\{U_i : i \in I\}$ is also of first category in E.

This classical result is due to Banach (see [125], [155] or [165]).

6. Let $(G, \cdot)$ be a topological group and let A be a subset of G having the Baire property. Demonstrate, by applying the preceding exercise, that if A is not of first category, then the set $A \cdot A^{-1}$ is a neighbourhood of the neutral element of G.

This result is sometimes called the Banach-Kuratowski-Pettis theorem (compare [77] or [125]). It may be regarded as a topological version of the Steinhaus property for invariant measures (see Exercise 8 below).

Deduce from the Banach-Kuratowski-Pettis theorem that if A and B are two subsets of G such that each of them possesses the Baire property and none of them is of first category, then the set $A \cdot B$ has nonempty interior.

7. Let E be a set and let G be a group of transformations of E. Suppose also that μ is a σ-finite G-invariant measure on E. As usual, the G-invariance of μ means that all transformations from G preserve μ.

According to the standard definition (see [27] or [62]), μ is metrically transitive (with respect to G) if for any μ-measurable set A with $\mu(A) > 0$, there exists a countable family $\{g_i : i \in I\}$ of elements from G, such that

$$\mu(E \setminus \cup\{g_i(A) : i \in I\}) = 0.$$

Note that the same definition can be introduced for G-quasi-invariant measures. See Chapter 9 where the notion of a G-quasi-invariant measure is formulated.

Show that the following two assertions are equivalent:

(a) μ is metrically transitive;

(b) if $f : E \to \mathbf{R}$ is a μ-measurable function such that all the functions

$$f \circ g \quad (g \in G)$$

coincide μ-almost everywhere with f, then f coincides μ-almost everywhere with a constant function.

Demonstrate also that if ν is a σ-finite (left) Haar measure on a locally compact topological group G, then ν is metrically transitive. (Apply the uniqueness property for a Haar measure.)

Moreover, demonstrate that if H is a dense subgroup of G and A is a ν-measurable set with $\nu(A) > 0$, then there exists a countable family $\{h_i : i \in I\}$ of elements from H, such that

$$\nu(G \setminus \cup\{h_i A : i \in I\}) = 0.$$

This equality shows that ν is metrically transitive with respect to H.

In particular, the classical Lebesgue measure λ (being the completion of the corresponding Haar measure on $\mathbf{R}$) is metrically transitive with respect to any dense subgroup of $\mathbf{R}$.

8. Let ν be a σ-finite (left) Haar measure on a locally compact topological group $(G, \cdot)$ and let A be a ν-measurable subset of G. Prove that

$$lim_{g \to e} \nu(A \cap gA) = \nu(A),$$

where e stands for the neutral element of G. Infer from this fact the following two assertions:

(a) if $\nu(A) > 0$, then there exists a neighbourhood $U(e)$ of e such that $A \cap gA \neq \emptyset$ for all elements $g \in U(e)$;

(b) if $\nu(A) > 0$ and B is another ν-measurable subset of G with $\nu(B) > 0$, then the set $A \cdot B$ has nonempty interior.

The assertion formulated in (a) is usually called the Steinhaus property of ν. In particular, the classical Lebesgue measure on $\mathbf{R}$ possesses this property.

Deduce from (a) that if H is a countable subgroup of G, the neutral element e is an accumulation point for H and A is a ν-measurable set with $\nu(A) > 0$, then any selector of $\{A \cap H_i : i \in I\}$ (where $\{H_i : i \in I\}$ denotes the family of all those left H-orbits in G whose intersections with A are nonempty) turns out to be nonmeasurable with respect to the completion of ν.

9. Let G again be a locally compact topological group and let ν be a σ-finite (left) Haar measure on G. Suppose also that G is not discrete. Let A be an arbitrary ν-measurable subset of G with $\nu(A) > 0$. Prove that there exists a set A' satisfying the following relations:

(a) $A' \subset A$;

(b) A' is nonmeasurable with respect to every (left) G-invariant measure on G extending ν.

Note that some essentially stronger versions of this exercise will be discussed later (compare Chapter 11).

10. Let E be a set and let μ be a measure defined on a σ-algebra of subsets of E. We say that μ satisfies the countable chain condition (or Suslin condition) if for any disjoint family of sets $\{X_i : i \in I\} \subset dom(\mu)$, the relation $(\forall i \in I)(\mu(X_i) > 0)$ implies the inequality $card(I) \leq \omega$. Verify that every σ-finite measure satisfies the countable chain condition. Give an example of a measure which satisfies this condition but is not σ-finite.

We would like to remark that the countable chain condition plays an important role in various constructions of nonmeasurable sets (see, for instance, Chapter 11).

11. Let G be a locally compact topological group and let ν denote a Haar measure on G. Verify that the following two assertions are equivalent:

(a) ν is a σ-finite measure;

(b) the group G is σ-compact; in other words, there exists a countable family $\{K_i : i \in I\}$ of compact subsets of G such that

$$G = \cup\{K_i \ : \ i \in I\}.$$

12. Let X be a Lebesgue measurable subset of $\mathbf{R}$. We recall that $x \in \mathbf{R}$ is a density point of X if

$$lim_{h \to 0+} \ \lambda([x - h, x + h] \cap X)/2h = 1.$$

It is well known that λ-almost all points of X are its density points (see [158], [165], [102]).

Now, for any set $X \in dom(\lambda)$, denote by $d(X)$ the set of all density points of X and put

$$\mathcal{T}_d = \{Y \in dom(\lambda) : Y \subset d(Y)\}.$$

Show that:

(a) $\mathcal{T}_d$ is a topology on $\mathbf{R}$ strictly extending the standard topology of $\mathbf{R}$;

(b) $(\mathbf{R}, \mathcal{T}_d)$ is a nonseparable Baire topological space;

(c) if $\{Y_i : i \in I\} \subset \mathcal{T}_d \setminus \{\emptyset\}$ is a disjoint family of sets, then $card(I) \leq \omega$ (in other words, the space $(\mathbf{R}, \mathcal{T}_d)$ satisfies the Suslin condition);

(d) every first category set in $(\mathbf{R}, \mathcal{T}_d)$ is nowhere dense;

(e) $X \in dom(\lambda)$ if and only if X has the Baire property in $(\mathbf{R}, \mathcal{T}_d)$;

(f) $\lambda(X) = 0$ if and only if X is of first category in $(\mathbf{R}, \mathcal{T}_d)$.

$\mathcal{T}_d$ is usually called the density topology on $\mathbf{R}$ (for more detailed information about $\mathcal{T}_d$, see especially [165] and [219]).

Notice that an abstract version of the density topology was introduced by von Neumann for any nonzero σ-finite complete measure μ (see [141], [165], [220]). However, the proof of the existence of a von Neumann topology for μ is not easy and needs uncountable forms of the Axiom of Choice.

Chapter 2
The Bernstein construction

The second widely known construction leading to nonmeasurable (in the Lebesgue sense) subsets of $\mathbf{R}$ is due to Bernstein [10] and was carried out by him in 1908. The same construction yields simultaneously examples of sets in $\mathbf{R}$ without the Baire property. Of course, Bernstein's argument is heavily based on the Axiom of Choice. Namely, Bernstein utilizes the fact that there exists a well ordering of the family of all uncountable closed subsets of $\mathbf{R}$.

The above-mentioned result of Bernstein is interesting in various respects. First of all, it admits generalizations to many other cases, where, for example, a topological space or a measure space are given and a Bernstein type subset of that space is required to be constructed (compare Exercise 5 of this chapter). In addition, Bernstein type subsets of an original space are frequently helpful as a tool for constructing counterexamples to some mathematical assertions which, at first sight, are expected to be valid in rather general situations (see, for instance, the remark in Exercise 12). At last, Bernstein sets are typical representatives from the large gallery of "singular" sets in analysis and topology (see [55], [125], [147]). They are often called freaks, monsters and so on but play an important role in delicate questions of analysis and topology (compare [55], [102]).

This chapter is devoted to the Bernstein construction and to properties of Bernstein sets. We begin with some standard definitions which will be useful in our further considerations.

Let E be a topological space and let X be a subset of E.

We say that X is totally imperfect in E if X contains no nonempty perfect subset of E (see [125]).

We say that X is a Bernstein subset of E if both X and $E \setminus X$ are totally imperfect in E. Equivalently, X is a Bernstein subset of E if, for each nonempty perfect set $P \subset E$, we have

$$P \cap X \neq \emptyset, \quad P \cap (E \setminus X) \neq \emptyset.$$

It immediately follows from this definition that $X \subset E$ is a Bernstein set if and only if $E \setminus X$ is Bernstein.

Also, if E' is another topological space, $h : E \to E'$ is a homeomorphism and X is a subset of E, then the following two relations are equivalent:

a) X is totally imperfect in E;

b) $h(X)$ is totally imperfect in E'.

In particular, X is a Bernstein subset of E if and only if $h(X)$ is a Bernstein subset of E'.

Remark 1. Let E be a complete metric space and let X be a subset of E. It is not hard to see that the following two statements are equivalent:

1) X is totally imperfect in E;

2) X does not contain a set homeomorphic to the Cantor discontinuum

$$C = 2^\omega = \{0, 1\}^\omega,$$

where $\{0, 1\}$ is equipped with the discrete topology.

Indeed, the equivalence of these two assertions is implied by the classical fact stating that any complete metric space without isolated points contains a topological copy of 2^ω (see, for instance, [125]).

In particular, assertions 1) and 2) are equivalent for every complete separable metric space E (in other words, for every Polish topological space E).

Remark 2. Obviously, each subset of the real line (more generally, of a Polish space) that has cardinality strictly less than the cardinality of the continuum $\mathbf{c}$ is totally imperfect. The question concerning the existence of totally imperfect subsets of the real line, having cardinality $\mathbf{c}$, turns out to be rather nontrivial. For its solution, we are forced to appeal to uncountable forms of the Axiom of Choice (in this connection, see Exercise 2 of the present chapter; compare also Theorem 1 from Chapter 12). Moreover, assuming the Axiom of Choice, one can obtain many interesting and important examples of totally imperfect subsets of the real line (or, equivalently, of an uncountable Polish space). A wide class of such sets was introduced and investigated by Marczewski (see [15] and Chapter 5 of this book).

Let us return to Bernstein sets. We now formulate and prove the classical Bernstein result on the existence of these sets (compare [125], [155], [165]).

Theorem 1. *There exists a Bernstein subset of the real line. All such subsets are nonmeasurable in the Lebesgue sense and do not possess the Baire property.*

Proof. Let α denote the least ordinal number for which $card(\alpha) = \mathbf{c}$. As known, the family of all nonempty perfect subsets of $\mathbf{R}$ is of cardinality $\mathbf{c}$. So we can denote this family by $\{P_\xi \ : \ \xi < \alpha\}$. Furthermore, we may assume without loss of generality that each of the partial families

$$\{P_\xi \ : \ \xi < \alpha, \ \xi \text{ is an even ordinal}\},$$

$$\{P_\xi \ : \ \xi < \alpha, \ \xi \text{ is an odd ordinal}\}$$

also consists of all nonempty perfect subsets of $\mathbf{R}$. Now, applying the method of transfinite recursion, we define an α-sequence of points

$$\{x_\xi \ : \ \xi < \alpha\} \subset \mathbf{R},$$

satisfying the following two conditions:
1) if $\xi < \zeta < \alpha$, then $x_\xi \neq x_\zeta$;
2) for each $\xi < \alpha$, we have $x_\xi \in P_\xi$.

Suppose that for $\beta < \alpha$, the partial β-sequence $\{x_\xi \ : \ \xi < \beta\}$ has already been defined. Take the set P_β. Clearly,

$$card(P_\beta) = \mathbf{c},$$

$$card(\{x_\xi \ : \ \xi < \beta\}) \leq card(\beta) < \mathbf{c}.$$

Hence, we can write

$$P_\beta \setminus \{x_\xi \ : \ \xi < \beta\} \neq \emptyset.$$

Choose any x from $P_\beta \setminus \{x_\xi : \xi < \beta\}$ and put $x_\beta = x$. Continuing in this manner, we are able to construct the α-sequence $\{x_\xi \ : \ \xi < \alpha\}$ of points of $\mathbf{R}$, satisfying conditions 1) and 2). Further, we put

$$X = \{x_\xi \ : \ \xi < \alpha, \ \xi \text{ is an even ordinal}\}.$$

It immediately follows from our construction that X is a Bernstein subset of $\mathbf{R}$ because both sets X and $\mathbf{R} \setminus X$ are totally imperfect in $\mathbf{R}$.

It remains to demonstrate that X is not Lebesgue measurable and does not possess the Baire property.

Suppose first that X is measurable in the Lebesgue sense, that is X belongs to the domain of the standard Lebesgue measure λ on $\mathbf{R}$. Then the set $\mathbf{R} \setminus X$ is Lebesgue measurable, too, and at least one of these two sets is of strictly positive measure. We may assume, without loss of generality, that $\lambda(X) > 0$. Then a well-known regularity property of λ implies that there exists a closed set $F \subset \mathbf{R}$ for which

$$F \subset X, \quad \lambda(F) > 0.$$

Since λ is a diffused (continuous) measure, that is λ vanishes on all one-element subsets of $\mathbf{R}$, we must have $card(F) > \omega$ and hence $card(F) = \mathbf{c}$. Denote by F_0 the set of all condensation points of F. Obviously, F_0 is a nonempty perfect subset of $\mathbf{R}$ included in X. But this contradicts the fact that X is a Bernstein set in $\mathbf{R}$.

Actually, using the same argument, we are able to establish that a Bernstein set $X \subset \mathbf{R}$ is nonmeasurable with respect to the completion of any nonzero σ-finite diffused Borel measure on $\mathbf{R}$ (compare Theorem 2 presented below and its consequences concerning measure and category).

Suppose now that X possesses the Baire property. Then the set $\mathbf{R} \setminus X$ possesses the Baire property, too, and at least one of these two sets is not of first category. We may assume, without loss of generality, that X is of second category. Hence, we have a representation of X in the form:

$$X = V \triangle Y = (V \setminus Y) \cup (Y \setminus V),$$

where V is a nonempty open subset of $\mathbf{R}$ and Y is a first category subset of $\mathbf{R}$. Applying the classical Baire theorem, we see that the set $V \setminus Y$ contains an uncountable G_δ-subset of $\mathbf{R}$ which can be regarded as an uncountable Polish space with respect to the induced topology. Since every uncountable Polish space contains in itself a topological copy of the Cantor discontinuum C, we immediately obtain that X contains a nonempty perfect subset of $\mathbf{R}$, which contradicts the fact that X is a Bernstein set in $\mathbf{R}$.

A result essentially more general than Theorem 1 is presented in Exercise 5 of this chapter (see also Theorem 2 below).

Let E be a topological space. As usual, we denote by $\mathcal{B}(E)$ the σ-algebra of all Borel subsets of E, in short the Borel σ-algebra of E. We recall that this σ-algebra is generated by the family of all open sets in E.

Let $\mathcal{I}$ be a σ–ideal of subsets of E.

We shall say that this σ–ideal has a Borel base if for each set $X \in \mathcal{I}$, there exists a set $Y \in \mathcal{I}$ such that $X \subset Y$ and $Y \in \mathcal{B}(E)$.

For example, if E is a second category space, then the σ–ideal $\mathcal{K}(E)$ of all first category subsets of E has a Borel base. Moreover, in this case, $\mathcal{K}(E)$ has a base consisting of some F_σ–subsets of E.

Similarly, if E is a metric space and μ is a nonzero σ–finite Borel measure on E, then the σ–ideal $\mathcal{I}(\mu)$ generated by the family of all μ–measure zero sets has a Borel base. Moreover, in this case $\mathcal{I}(\mu)$ has a base consisting of some G_δ–subsets of E.

The following result shows that all Bernstein sets in a Polish space E are very bad from the point of view of measurability with respect to the σ-algebra which is generated by the Borel σ-algebra of E and a σ-ideal in E with a Borel base.

Theorem 2. *Let E be an uncountable Polish topological space and let $\mathcal{I}$ be a σ–ideal of subsets of E, such that:*

1) all one–element subsets of E belong to $\mathcal{I}$;

2) $\mathcal{I}$ has a Borel base.

Further, let $\mathcal{S}$ denote the σ–algebra of subsets of E, generated by the family $\mathcal{B}(E) \cup \mathcal{I}$, and let Z be an arbitrary Bernstein set in E. Then Z does not belong to $\mathcal{S}$.

Proof. Suppose to the contrary that $Z \in \mathcal{S}$. Since the equality

$$Z \cup (E \setminus Z) = E$$

holds and $E \notin \mathcal{I}$, we get the disjunction

$$Z \notin \mathcal{I} \quad \vee \quad E \setminus Z \notin \mathcal{I}.$$

Without loss of generality, we may assume that $Z \notin \mathcal{I}$. Further, since the set Z belongs to $\mathcal{S}$, it can be represented in the form $Z = B \triangle X$, where B is a Borel subset of E and X is some set belonging to $\mathcal{I}$. According to condition 2) of the theorem, there exists a set Y such that

$$Y \in \mathcal{I}, \quad X \subset Y, \quad Y \in \mathcal{B}(E).$$

Consequently, we have the inclusions

$$B \setminus Y \subset B \setminus X \subset Z.$$

Taking into account the facts that Z is a Bernstein set and $B \setminus Y$ is a Borel set, we get the inequality

$$card(B \setminus Y) \leq \omega.$$

According to condition 1) of the theorem, we also have $B \setminus Y \in \mathcal{I}$. Therefore, we obtain the relation

$$Z \subset (Y \cup (B \setminus Y)) \in \mathcal{I},$$

which yields a contradiction. Thus, Theorem 2 is proved.

Finally, we have the following two important facts which are easy consequences of the above theorem:

1. If E is an arbitrary uncountable Polish topological space without isolated points, then no Bernstein set in E possesses the Baire property.

2. If E is an arbitrary Polish topological space and μ is the completion of a nonzero σ–finite diffused Borel measure on E, then each Bernstein set in E is nonmeasurable with respect to μ.

The next theorem shows that there exist subsets of the real line which are simultaneously Vitali sets and Bernstein sets.

Theorem 3. *There exists a subset X of $\mathbf{R}$ which is a Vitali set and a Bernstein set.*

Proof. Let α denote the first ordinal of cardinality continuum. Let $\{x_\xi : \xi < \alpha\}$ be an injective family of all points of $\mathbf{R}$ and let $\{F_\xi : \xi < \alpha\}$ denote an injective family of all uncountable closed subsets of $\mathbf{R}$. Similarly to the classical Bernstein construction, we define, by applying the method of transfinite recursion, an injective family $\{x_\xi : \xi < \alpha\}$ of points in $\mathbf{R}$. Suppose that, for an ordinal $\beta < \alpha$, the partial family of points $\{x_\xi : \xi < \beta\}$ has already been constructed. Consider the set

$$Z_\beta = \cup\{x_\xi + \mathbf{Q} : \xi < \beta\}.$$

Obviously,

$$card(Z_\beta) \leq card(\beta) \cdot \omega < \mathbf{c}.$$

Since $card(F_\beta) = \mathbf{c}$, we have

$$F_\beta \setminus Z_\beta \neq \emptyset.$$

Take any element $z \in F_\beta \setminus Z_\beta$ and put $x_\beta = z$. In this way, the required family of points $\{x_\xi : \xi < \alpha\}$ will be constructed. Now, we define

$$X' = \{x_\xi : \xi < \alpha\}.$$

Let us remark that in view of our construction any equivalence class of the Vitali partition $\{x + \mathbf{Q} : x \in \mathbf{R}\}$ contains at most one point from X'. Moreover, our construction implies at once that the set $\mathbf{R} \setminus X'$ is totally imperfect in $\mathbf{R}$. In other words, X' turns out to be a partial selector of the Vitali partition whose complement is totally imperfect. Evidently, we can extend X' to a selector of the same partition. We denote by X the selector obtained in this manner. According to the definition, X is a Vitali subset of $\mathbf{R}$. It remains to demonstrate that X is a Bernstein set, as well. To see this, first observe that the set $\mathbf{R} \setminus X$ is totally imperfect since it is contained in the set $\mathbf{R} \setminus X'$. Further, let us take any rational number $q \neq 0$ and consider the set $X + q$. Clearly, $(\mathbf{R} \setminus X) + q$ is totally imperfect in $\mathbf{R}$ and

$$(X + q) \cap X = \emptyset,$$

$$\mathbf{R} \setminus (X + q) = (\mathbf{R} \setminus X) + q,$$

from which it immediately follows that $X \subset (\mathbf{R} \setminus X) + q$ and, consequently, X turns out to be totally imperfect in $\mathbf{R}$. This argument completes the proof of the theorem.

We have already discussed the Vitali and Bernstein constructions which are concerned with certain sets of real numbers nonmeasurable in the sense of Lebesgue and lacking the Baire property. Evidently, the existence of such sets implies the existence of functions (acting from $\mathbf{R}$ into $\mathbf{R}$) nonmeasurable in the Lebesgue sense and without the Baire property. We now wish to consider a direct construction of a non Lebesgue-measurable function acting from $\mathbf{R}$ into $\mathbf{R}$. An analogous construction is applicable to the Baire property (see Exercise 9). The reader can easily observe that the construction presented below is rather similar to the Bernstein construction. Some generalizations of constructions of this type will be indicated in other chapters of the book; see, for instance, Chapter 5 where small (in a certain sense) nonmeasurable sets are discussed.

In our further considerations, the symbol λ_2 will denote the standard two-dimensional Lebesgue measure on the euclidean plane $\mathbf{R}^2$. Clearly, λ_2 is the completion of the product measure $\lambda \times \lambda$.

We recall that a subset X of $\mathbf{R}^2$ is λ_2-thick (or λ_2-massive) in $\mathbf{R}^2$ if, for each λ_2-measurable set $Z \subset \mathbf{R}^2$ with $\lambda_2(Z) > 0$, we have $X \cap Z \neq \emptyset$. In other words, X is λ_2-thick in $\mathbf{R}^2$ if and only if the equality

$$(\lambda_2)_*(\mathbf{R}^2 \setminus X) = 0$$

is fulfilled where the symbol $(\lambda_2)_*$ stands for the inner measure associated with λ_2.

Let us point out that if a subset X of $\mathbf{R}^2$ is λ_2-measurable and λ_2-massive simultaneously, then it is of full λ_2-measure; in other words, we may write $\lambda_2(\mathbf{R}^2 \setminus X) = 0$. Thus, if we already know that a set $X \subset \mathbf{R}^2$ is not of full λ_2-measure but is λ_2-thick, then we can immediately conclude that X is not λ_2-measurable.

The next classical statement (essentially due to Sierpiński) shows that there are functions acting from $\mathbf{R}$ into $\mathbf{R}$ whose graphs are λ_2-thick subsets of the plane (see, for instance, [55] and [64]).

Theorem 4. *There exists a function*

$$f \: : \: \mathbf{R} \to \mathbf{R}$$

whose graph is a λ_2-thick subset of $\mathbf{R}^2$. Consequently, for f, the following two assertions are true:

1) the graph of f is not a λ_2-measurable subset of $\mathbf{R}^2$;

2) f is not a λ-measurable function.

Proof. Let α be again the least ordinal number of cardinality continuum. Consider the family $\{B_\xi \: : \: \xi < \alpha\}$ consisting of all Borel subsets of $\mathbf{R}^2$ having strictly positive λ_2-measure. We are going to construct, by transfinite recursion, a family of points

$$\{(x_\xi, y_\xi) \: : \: \xi < \alpha\} \subset \mathbf{R}^2$$

satisfying these two conditions:

(1) if $\xi < \zeta < \alpha$, then $x_\xi \neq x_\zeta$;

(2) for each $\xi < \alpha$, the point (x_ξ, y_ξ) belongs to B_ξ.

Suppose that, for an ordinal $\beta < \alpha$, the partial family of points

$$\{(x_\xi, y_\xi) \: : \: \xi < \beta\} \subset \mathbf{R}^2$$

has already been defined. Let us take the set B_β. For each $x \in \mathbf{R}$, denote

$$B_\beta(x) = \{y \in \mathbf{R} : (x, y) \in B_\beta\}.$$

Applying the classical Fubini theorem, we see that the set

$$\{x \in \mathbf{R} \ : \ \lambda(B_\beta(x)) > 0\}$$

is λ-measurable and has strictly positive measure. Consequently, this set is of cardinality $\mathbf{c}$, and there exists a point $x \in \mathbf{R}$ such that

$$\lambda(B_\beta(x)) > 0, \quad (\forall \xi < \beta)(x \neq x_\xi).$$

We put $x_\beta = x$. Then we choose an arbitrary point y from the set $B_\beta(x_\beta)$ and put $y_\beta = y$. In this way, we obtain the point $(x_\beta, y_\beta) \in \mathbf{R}^2$.

Proceeding in this manner, we are able to construct the required family of points $\{(x_\xi, y_\xi) \ : \ \xi < \alpha\}$. Now, it easily follows from condition (1) that the set

$$\Gamma = \{(x_\xi, y_\xi) \ : \ \xi < \alpha\}$$

can be regarded as the graph of a partial function acting from $\mathbf{R}$ into $\mathbf{R}$. We extend arbitrarily this partial function to a function acting from $\mathbf{R}$ into $\mathbf{R}$ and denote the latter function by f. Then condition (2) implies that the graph of f (which contains Γ) is λ_2-thick in $\mathbf{R}^2$. Since there are uncountably many pairwise disjoint translates (in $\mathbf{R}^2$) of this graph, we conclude that it is not of full λ_2-measure and, hence, it is not a λ_2-measurable subset of $\mathbf{R}^2$.

Finally, the function f is not λ-measurable. Indeed, supposing that our f is Lebesgue measurable, we easily claim that the graph of f must be a λ_2-measure zero subset of the plane, which is impossible.

This ends the proof of the theorem. A much stronger result is formulated in Exercise 19.

Sometimes, Bernstein sets with additional algebraic properties are more useful for applications. For example, one can require to construct a Bernstein subset of $\mathbf{R}$ which simultaneously is a subgroup of the additive group $\mathbf{R}$. The construction of such a Bernstein set needs only slight modification of the classical Bernstein construction. Actually, the following statement is true.

Theorem 5. *There exist two subgroups G_1 and G_2 of the additive group* $\mathbf{R}$, *such that:*

1) $G_1 \cap G_2 = \{0\}$;

2) both G_1 and G_2 are Bernstein sets in $\mathbf{R}$.

Proof. For each subset X of $\mathbf{R}$, we put

$$[X] \ = \ the \ group \ generated \ by \ X.$$

Let α denote again the first ordinal number of cardinality continuum and let $\{P_\xi : \xi < \alpha\}$ be the family of all uncountable closed subsets of $\mathbf{R}$. We are going to define, by transfinite recursion, two α-sequences

$$\{G_{1,\xi} : \xi < \alpha\}, \quad \{G_{2,\xi} : \xi < \alpha\}$$

of subgroups of $\mathbf{R}$, satisfying the relations:

a) $G_{1,0} \subset G_{1,1} \subset ... \subset G_{1,\xi} \subset ...$;

b) $G_{2,0} \subset G_{2,1} \subset ... \subset G_{2,\xi} \subset ...$;

c) $G_{1,\xi} \cap G_{2,\xi} = \{0\}$ for each ordinal $\xi < \alpha$;

d) $G_{1,\xi} \cap P_\xi \neq \emptyset$ and $G_{2,\xi} \cap P_\xi \neq \emptyset$ for all ordinals $\xi < \alpha$;

e) $card(G_{1,\xi}) \leq card(\xi) + \omega$ and $card(G_{2,\xi}) \leq card(\xi) + \omega$ for each ordinal $\xi < \alpha$.

Suppose that, for a given ordinal $\xi < \alpha$, the partial families of groups

$$\{G_{1,\zeta} : \zeta < \xi\}, \quad \{G_{2,\zeta} : \zeta < \xi\}$$

have already been defined. Then, putting

$$G'_\xi = [\cup\{G_{1,\zeta} \cup G_{2,\zeta} : \zeta < \xi\}],$$

we obtain the group G'_ξ such that

$$card(G'_\xi) \leq card(\xi) + \omega < \mathbf{c}.$$

Since

$$card(P_\xi) = \mathbf{c} > card(G'_\xi),$$

we can choose two distinct points $u_\xi \in P_\xi$ and $v_\xi \in P_\xi$ having the property that for any two integers m and n, the relation

$$mu_\xi + nv_\xi \in G'_\xi$$

implies the relation

$$m = n = 0.$$

We now define

$$G_{1,\xi} = [\{u_\xi\} \cup (\cup\{G_{1,\zeta} : \zeta < \xi\})], \quad G_{2,\xi} = [\{v_\xi\} \cup (\cup\{G_{2,\zeta} : \zeta < \xi\})].$$

Continuing in this manner, we will be able to construct the required two α-sequences of subgroups of $\mathbf{R}$. Finally, putting

$$G_1 = \cup\{G_{1,\xi} : \xi < \alpha\}, \quad G_2 = \cup\{G_{2,\xi} : \xi < \alpha\},$$

we can easily verify (by virtue of our construction) that the groups G_1 and G_2 satisfy conditions 1) and 2) of the theorem. In particular, each of these groups is λ-thick (hence, everywhere dense) in $\mathbf{R}$.

Remark 3. In a similar manner, one can construct a subfield of $\mathbf{R}$ which simultaneously is a Bernstein set. However, it should be indicated that the construction carried out in the proof of the previous theorem utilizes some specific features of the additive group $\mathbf{R}$. It cannot be generalized even to the class of all commutative locally compact topological groups. Indeed, there is an example of an uncountable commutative σ-compact locally compact group Γ which admits no proper dense subgroup. In other words, each dense subgroup of Γ necessarily coincides with Γ. Since Γ is equipped with a nonzero σ-finite Haar measure μ, we readily conclude that Γ possesses no μ-thick proper subgroup and, consequently, there are no μ-thick groups $\Gamma_1 \subset \Gamma$ and $\Gamma_2 \subset \Gamma$ satisfying the relation

$$\Gamma_1 \cap \Gamma_2 = \{e\},$$

where e denotes the neutral element of Γ.

For more details about Γ, see [25]. The existence of such a Γ shows us that the classical Bernstein construction does not always work and if we want to find a nonmeasurable subgroup of a given commutative group, we must appeal to essentially different methods. One approach to the problem of finding nonmeasurable subgroups of uncountable commutative groups will be developed later in this book. Briefly speaking, that approach will be based on some combinatorial properties of Ulam transfinite matrices (for details, see Chapter 13).

EXERCISES

1. Let C denote the Cantor discontinuum. Starting with the existence of a continuous surjection

$$f : C \to [0, 1],$$

demonstrate that for any nonempty perfect set $P \subset \mathbf{R}$, there exists a disjoint family $\{P_j : j \in J\}$ of nonempty perfect subsets of $\mathbf{R}$ such that:

(a) $card(J) = \mathbf{c}$;

(b) $P_j \subset P$ for each $j \in J$.

2. Prove in the theory **ZF** & **DC** that if there exists a totally imperfect subset of **R** of cardinality **c**, then there exists a non Lebesgue-measurable subset of **R**.

Prove also an analogous fact for the Baire property.

In connection with the previous exercise, let us mention that a more general result can be obtained (compare Theorem 1 from Chapter 12).

3. Let n be a natural number greater than or equal to 2 and let X be a totally imperfect subset of the n-dimensional euclidean space $\mathbf{R}^n$. Show that the set $\mathbf{R}^n \setminus X$ is connected (in the usual topological sense). Infer from this fact that any Bernstein subset of $\mathbf{R}^n$ is connected.

4. Let E be an infinite set and let $\{X_j : j \in J\}$ be a family of subsets of E, such that:

1) $card(J) \leq card(E)$;

2) $(\forall j \in J)(card(X_j) = card(E))$.

Prove, by applying the method of transfinite recursion, that there exists a family $\{Y_j : j \in J\}$ of subsets of E, satisfying the relations:

(a) $(\forall j \in J)(\forall j' \in J)(j \neq j' \Rightarrow Y_j \cap Y_{j'} = \emptyset)$;

(b) $(\forall j \in J)(\forall j' \in J)(card(X_j \cap Y_{j'}) = card(E))$.

5. By starting with the result of the previous exercise, show that in every complete metric space E of cardinality continuum (consequently, in every uncountable Polish space) there exists a Bernstein set. Moreover, demonstrate that there exists a partition $\{Y_j : j \in J\}$ of E such that:

(a) $card(J) = \mathbf{c}$;

(b) for each $j \in J$, the set Y_j is a Bernstein subset of E.

Finally, show that if the space E has no isolated points, then all Bernstein subsets of E lack the Baire property.

6. Let E be an uncountable Polish space and let X be a subset of E. Demonstrate that the following two assertions are equivalent:

(a) X is a Bernstein subset of E;

(b) for each nonzero σ-finite diffused Borel measure μ given on E, the set X is nonmeasurable with respect to the completion of μ.

In other words, the equivalence of (a) and (b) yields some characterization of Bernstein subsets of a Polish space E in terms of topological measure theory.

Show also that in general (a) and (b) are not equivalent for a subset X of a nonseparable complete metric space E.

7. Let us consider the first uncountable ordinal number ω_1 equipped with its order topology, and let

$$\mathcal{I} = \{X \subset \omega_1 \ : \ (\exists F \subset \omega_1)(F \text{ is closed}, \ card(F) = \omega_1, \ F \cap X = \emptyset)\}.$$

Prove that $\mathcal{I}$ is a σ–ideal of subsets of ω_1. The elements of $\mathcal{I}$ are usually called nonstationary subsets of ω_1. Respectively, a set $Z \subset \omega_1$ is called a stationary subset of ω_1 if Z is not nonstationary. Let us put

$$\mathcal{S} = \mathcal{I} \cup \mathcal{I}^*,$$

where $\mathcal{I}^*$ is the δ-filter dual to $\mathcal{I}$. Observe that $\mathcal{S}$ is the σ–algebra generated by $\mathcal{I}$. Finally demonstrate that for any set $Y \subset \omega_1$, the following two relations are equivalent:

(a) the sets Y and $\omega_1 \setminus Y$ are stationary in ω_1;

(b) for every nonzero σ–finite diffused measure μ defined on $\mathcal{S}$, the set Y is not measurable with respect to μ.

Any set Y with the above-mentioned properties can be considered as an analogue for the topological space ω_1 of a Bernstein subset of **R**.

Note that there is a natural two-valued probability measure ν on $\mathcal{S}$ defined as follows: $\nu(X) = 0$ if $X \in \mathcal{I}$, and $\nu(X) = 1$ if $X \in \mathcal{I}^*$. This measure was introduced by Dieudonné and is called the Dieudonné measure.

For more information about the σ-ideal $\mathcal{I}$ and stationary subsets of ω_1, see [64], [112] and [122].

The next exercise assumes that the reader is familiar with the notion of a complete Boolean algebra. For the definition, see [127].

8. Let $\mathcal{P}(\mathbf{R})$ denote the complete Boolean algebra of all subsets of the real line **R**. Let $\mathcal{I}_0$ be the σ-ideal of all Lebesgue measure zero subsets of **R** and let $\mathcal{I}_1$ be the σ-ideal of all first category subsets of **R**. Consider the corresponding quotient algebras $\mathcal{P}(\mathbf{R})/\mathcal{I}_0$ and $\mathcal{P}(\mathbf{R})/\mathcal{I}_1$. Show that these Boolean algebras are not complete.

9. By applying the Kuratowski-Ulam theorem which is a topological analogue of the classical Fubini theorem (see [125] or [165]), prove for the Baire property a statement analogous to Theorem 4. Namely, show that there exists a function $f \ : \ \mathbf{R} \to \mathbf{R}$ such that its graph is thick in the topological sense, that is the graph of f intersects each second category subset of $\mathbf{R}^2$ having the Baire property.

Deduce from this fact that the graph of f does not have the Baire property in $\mathbf{R}^2$ and f does not have the Baire property as a function acting from $\mathbf{R}$ into $\mathbf{R}$.

10. Theorem 4 with Exercise 9 show us that there exist functions acting from $\mathbf{R}$ into $\mathbf{R}$ whose graphs are thick subsets of the plane; in particular, those graphs are nonmeasurable in the Lebesgue sense or do not possess the Baire property. On the other hand, prove that there exists a complete measure μ on $\mathbf{R}^2$ satisfying the following conditions:

(a) μ is an extension of the Lebesgue measure λ_2;

(b) μ is invariant under the group of all translations of $\mathbf{R}^2$ and under the central symmetry of $\mathbf{R}^2$ with respect to $(0,0)$;

(c) the graphs of all functions acting from $\mathbf{R}$ into $\mathbf{R}$ belong to $dom(\mu)$ and for any such graph Γ, we have $\mu(\Gamma) = 0$.

We thus see that the graphs of all functions acting from $\mathbf{R}$ into $\mathbf{R}$ are small (negligible) with respect to the measure μ.

This is a nontrivial example of a common property of all functions acting from $\mathbf{R}$ into $\mathbf{R}$.

In connection with the above-mentioned property, the following natural question arises:

Does there exist a measure on $\mathbf{R}^2$ extending λ_2, invariant under the group of all motions (i.e. isometric transformations) of $\mathbf{R}^2$ and such that the graphs of all functions acting from $\mathbf{R}$ into $\mathbf{R}$ are measurable with respect to this measure?

It turns out that the answer to the question is negative, and a detailed explanation will be given in Chapter 6.

11. Let X be a Bernstein subset of $\mathbf{R}$. Let Y be a λ-measurable set with $\lambda(Y) > 0$ and let Z be a second category subset of $\mathbf{R}$ having the Baire property. Show that the set $X \cap Y$ is not measurable in the Lebesgue sense and that the set $X \cap Z$ does not possess the Baire property.

12. Demonstrate that there exists a sequence $\{X_n : n < \omega\}$ of Bernstein subsets of the unit segment $[0, 1]$, such that:

(a) $X_{n+1} \subset X_n$ for all $n < \omega$;

(b) $\cap\{X_n : n < \omega\} = \emptyset$.

In particular, we have the relation

$$(\forall n < \omega)(\lambda^*(X_n) = 1),$$

where λ^* stands for the outer Lebesgue measure, but

$$\lambda(\cap\{X_n : n < \omega\}) = 0.$$

The existence of such a sequence of sets enables us to construct a counterexample to the famous Kolmogorov consistency theorem (or, in other words, Kolmogorov extension theorem) for those probability measures which are not assumed to be regular in an appropriate sense (see [160]).

13. Let E be a set, $\mathcal{S}$ be a σ-algebra of subsets of E, and let μ be a measure defined on $\mathcal{S}$. According to the standard terminology, we say that the triple $(E, \mathcal{S}, \mu)$ is a measure space (see [62]). Suppose that $\mu(E) = 1$; in other words, suppose that μ is a probability measure. This measure is called to be perfect if for every μ-measurable function $f : E \to \mathbf{R}$, there exists a set $B \in \mathcal{B}(\mathbf{R})$ such that

$$B \subset f(E), \quad \mu(f^{-1}(B)) = 1.$$

The notion of perfect probability measures was introduced by Gnedenko and Kolmogorov [56] and was investigated by numerous authors (see, for instance, [189]). These measures form a sufficiently wide class (which contains all Radon probability measures) and are good from the measure-theoretical and probabilistic viewpoint since they exclude various pathological situations. For example, the Kolmogorov extension theorem mentioned above (see Exercise 12) remains valid for perfect probability measures.

Let X be a Bernstein subset of $\mathbf{R}$ and let

$$E = X \cap [0, 1],$$

$$\mathcal{S} = \{E \cap Y : Y \in dom(\lambda) \cap \mathcal{P}([0, 1])\}.$$

Define a functional

$$\mu : \mathcal{S} \to [0, 1]$$

by putting

$$\mu(E \cap Y) = \lambda(Y) \quad (Y \in dom(\lambda) \cap \mathcal{P}([0, 1])).$$

Verify that this definition is correct and μ is a probability measure. Finally, demonstrate that μ is not perfect.

14. Let A be an arbitrary subset of $\mathbf{R}$. Assuming Martin's Axiom, prove that there exists a disjoint family $\{A_j : j \in J\}$ of subsets of A, satisfying the following relations:

$$card(J) = \mathbf{c}, \quad (\forall j \in J)(\lambda^*(A_j) = \lambda^*(A)).$$

Moreover, under suitable assumptions on a measure space $(E, \mathcal{S}, \mu)$, formulate and prove for μ a result analogous to the previous one.

15. Let E be a topological space. We shall say that E is resolvable if there exist two disjoint everywhere dense subsets of E.

Recall that this notion was introduced by Hewitt.

We shall say that a topological space E is isodyne if $card(U) = card(E)$ for any nonempty open set $U \subset E$.

Show that, for every topological space E, there exists a disjoint family of isodyne open subspaces of E, such that their union is everywhere dense in E.

Starting with this fact and using the argument similar to the one of Bernstein construction, demonstrate that any locally compact topological space (in particular, any locally compact topological group) without isolated points is resolvable.

16. Let E be an infinite set and let $\{X_i : i \in I\}$ be a family of subsets of E. We say that this family is almost disjoint (in E) if the following conditions hold:

(a) $(\forall i \in I)(card(X_i) = card(E))$;
(b) $(\forall i \in I)(\forall i' \in I)(i \neq i' \Rightarrow card(X_i \cap X_{i'}) < card(E))$.

Suppose that a family $\{Y_j : j \in J\}$ of subsets of E is given, such that:

(c) $card(J) \leq card(E)$;
(d) $(\forall j \in J)(card(Y_j) = card(E))$.

Show by applying the method of transfinite induction that there exists an almost disjoint family $\{X_i : i \in I\}$ of subsets of E, satisfying the relations:

(e) $card(I) > card(E)$;
(f) $(\forall i \in I)(\forall j \in J)(card(X_i \cap Y_j) = card(E))$.

In addition to this, for $E = \mathbf{N}$, demonstrate in the theory $\mathbf{ZF}$ that there exists an almost disjoint family $\{X_i : i \in I\}$ of subsets of E, such that

$$card(I) = \mathbf{c}.$$

Let us remark that the notion of an almost disjoint family of sets was first introduced and investigated by Sierpiński (see [194]).

17. Let E be an infinite topological space and let $\mathcal{K}(E)$ denote the family of all first category sets in E. Suppose that the following conditions are valid:

(a) E is not of first category;

(b) the σ-ideal $\mathcal{K}(E)$ possesses a base whose cardinality does not exceed $card(E)$;

(c) for each set $Z \subset E$ with $card(Z) < card(E)$, we have $Z \in \mathcal{K}(E)$.

Demonstrate, by using the result of the previous exercise, that there exists a subset of E without the Baire property. More generally, demonstrate that for every second category set $X \subset E$, there exists a subset of X lacking the Baire property.

Consider a particular case where $card(E) = \omega_1$. Namely, suppose that E is a second category topological space of cardinality ω_1, whose all one-element subsets are of first category. Suppose also that the σ-ideal $\mathcal{K}(E)$ has a base whose cardinality does not exceed ω_1. Then, in view of the said above, we can assert that E contains a subset without the Baire property.

In connection with the result presented in Exercise 17 and some of its generalizations, see [92].

18. According to the standard terminology (see [158] and [241]), a subset X of $\mathbf{R}$ is called a U-set if, for every trigonometric series

$$a_0 + \sum_{n \geq 1} (a_n cos(nx) + b_n sin(nx)),$$

the convergence to zero of this series at all points from $\mathbf{R} \setminus X$ implies the equalities

$$a_n = 0 \quad (n \geq 0),$$

$$b_n = 0 \quad (n \geq 1).$$

It immediately follows from this definition that each subset of a U-set is a U-set, too. Obviously, $\mathbf{R}$ is not a U-set.

It is well known that all countable subsets of $\mathbf{R}$ are U-sets.

There are also nonempty perfect U-sets. One of them is the Cantor discontinuum C. On the other hand, as proved by Menshov, there exist λ-measure zero subsets of $\mathbf{R}$ which are not U-sets (for more details, see [241]).

Show that any totally imperfect subset of $\mathbf{R}$ is a U-set. Infer from this fact that any Bernstein subset of $\mathbf{R}$ is a U-set and conclude that:

(a) the union of two U-sets is not, in general, a U-set;

(b) there exists a λ-thick U-set of second category in $\mathbf{R}$.

Note in connection with (b) that if $X \subset \mathbf{R}$ and $\lambda_*(X) > 0$, then X is not a U-set (see [158] or [241]).

19. Prove that there exists a function

$$f \ : \ \mathbf{R} \to \mathbf{R}$$

having the following property:

For any σ-finite diffused Borel measure μ on $\mathbf{R}$ and for any σ-finite measure ν on $\mathbf{R}$, the graph of f is a thick subset of $\mathbf{R}^2$ with respect to the product measure $\mu \times \nu$; in other words, this graph intersects every $(\mu \times \nu)$-measurable set of strictly positive measure.

20. Demonstrate that there exists an injective function

$$g \ : \ \mathbf{R} \to \mathbf{R}$$

whose graph Γ_g is a totally imperfect subset of the plane $\mathbf{R}^2$. (Define such a function by applying the method of transfinite induction.)

In addition, show that:

(a) g is not measurable in the Lebesgue sense;

(b) g does not possess the Baire property;

(c) $\mathbf{R}$ is the image of Γ_g under some bijective continuous mapping.

It is useful to compare the above-mentioned function g with a Sierpiński-Zygmund function which will be introduced later (see Exercise 12 from Chapter 3).

21. Let $\{X_i : i \in I\}$ be a partition of $\mathbf{R}$ such that

$$(\forall i \in I)(card(X_i) \geq 2 \ \& \ \lambda(X_i) = 0).$$

Assuming Martin's Axiom, prove that there exist two disjoint selectors X and Y of this partition which are λ-thick subsets of $\mathbf{R}$. In particular, both X and Y are nonmeasurable with respect to λ.

Chapter 3
Nonmeasurable sets associated with Hamel bases

In a famous article by Hamel (see [63]), for the first time, the real line $\mathbf{R}$ was considered as a vector space over $\mathbf{Q}$ and it was shown, by using transfinite methods, the existence of a basis in this space, that is the existence of a maximal rationally independent subset of $\mathbf{R}$. Any such subset is now called a Hamel basis of $\mathbf{R}$. In the above-mentioned article, the main goal of Hamel was to construct a nontrivial (equivalently, discontinuous) solution of the Cauchy functional equation

$$f(x + y) = f(x) + f(y) \qquad (x \in \mathbf{R}, \ y \in \mathbf{R}).$$

He was able to prove that there are functions $f : \mathbf{R} \to \mathbf{R}$ satisfying this equation but discontinuous at each point of $\mathbf{R}$. In other words, there are everywhere discontinuous homomorphisms of the additive topological group $\mathbf{R}$ into itself. This result stimulated further investigations of functional equations, and many works were written devoted to the theory of such equations. We especially refer the reader to the monograph by Kuczma [118] where a rich material from this area is presented.

Here we are mainly interested in various connections of Hamel bases with nonmeasurable (in the Lebesgue sense) subsets of $\mathbf{R}$. First of all, we wish to consider a purely logical aspect of the relationship between Hamel bases and those sets in $\mathbf{R}$ which are not Lebesgue measurable. The following simple statement is true.

Theorem 1. *In the theory* $\mathbf{ZF}$ *&* $\mathbf{DC}$, *the existence of a Hamel basis implies the existence of a subset of* $\mathbf{R}$ *nonmeasurable in the Lebesgue sense.*

Proof. Let $\{e_i : i \in I\}$ be a Hamel basis of $\mathbf{R}$. Fix an index i_0 from I and consider the vector subspace of $\mathbf{R}$ (over $\mathbf{Q}$, of course) generated by the partial family $\{e_i : i \in I \setminus \{i_0\}\}$. We denote this subspace by V. Actually, V is a vector hyperplane in $\mathbf{R}$ regarded as a vector space over $\mathbf{Q}$. Our purpose is to show that V cannot be Lebesgue measurable. Suppose to the contrary that $V \in dom(\lambda)$ where λ denotes, as usual, the standard Lebesgue measure on $\mathbf{R}$. Since

$$\cup\{V + qe_{i_0} : q \in \mathbf{Q}\} = \mathbf{R}, \quad \lambda(\mathbf{R}) > 0$$

and λ is $\mathbf{R}$-invariant, we must have the inequality $\lambda(V) > 0$. On the other hand, for any nonzero $q \in \mathbf{Q}$, we can write

$$V \cap (V + qe_{i_0}) = \emptyset.$$

Taking q arbitrarily small, we see that the set V does not have the Steinhaus property (see Exercise 8 of Chapter 1). This circumstance immediately yields a contradiction with $\lambda(V) > 0$. The contradiction obtained finishes the proof.

Remark 1. It can easily be observed that V turns out to be a Vitali type set for the group $\Gamma = \mathbf{Q}e_{i_0}$. This implies at once that the set V is not Lebesgue measurable (see Theorem 1 from Chapter 1).

By using the same Steinhaus property, it can be shown (in the theory **ZF** & **DC**) that any discontinuous solution of the Cauchy functional equation is a function nonmeasurable in the Lebesgue sense. This fact also leads to another proof of Theorem 1. In addition, the following statement is valid.

Theorem 2. *In the theory* **ZF** & **DC**, *the existence of a Hamel basis implies the existence of a subset of* **R** *without the Baire property.*

Proof. Since the vector space V (over $\mathbf{Q}$) from the proof of Theorem 1 is a Vitali type set, we conclude that V does not possess the Baire property (see Theorem 2 from Chapter 1). This simple remark ends the proof and also shows that, in the same theory, the existence of a Hamel basis implies the existence of sets in $\mathbf{R}$ which are not Lebesgue-measurable and, simultaneously, do not have the Baire property.

Now, a natural question arises whether a given Hamel basis is measurable in the Lebesgue sense. It turns out that the answer to this question depends on additional features of a Hamel basis. We are going to demonstrate that there are Lebesgue measurable Hamel bases and there are non Lebesgue-measurable ones.

Let us first note that if a Hamel basis is Lebesgue measurable, then it necessarily is of Lebesgue measure zero. This fact can be established by utilizing an argument similar to the proof of Theorem 1. Indeed, we easily infer that a Hamel basis cannot have the Steinhaus property. So, in the case of its Lebesgue measurability, it must be of Lebesgue measure zero. We now give a construction of such a Hamel basis in $\mathbf{R}$.

We will need the following auxiliary geometric proposition concerning one additive property of the classical Cantor discontinuum.

Lemma 1. *Let C denote the Cantor discontinuum on the unit segment* $[0,1]$. *Then the set*

$$C + C = \{x + y : x \in C,\ y \in C\}$$

coincides with the segment $[0,2]$.

Proof. We use the standard geometric argument presented, for example, in [155]. Namely, let us introduce a mapping

$$\phi : \mathbf{R} \times \mathbf{R} \to \mathbf{R}$$

defined by

$$\phi(x,y) = x + y \qquad (x \in \mathbf{R},\ y \in \mathbf{R}).$$

This mapping can be described in another way. Denoting

$$l = \{(x,y) \in \mathbf{R} \times \mathbf{R} : x + y = 0\},$$

we see that ϕ is identical with the projection

$$pr_l : \mathbf{R} \times \mathbf{R} \to \mathbf{R} \times \{0\}$$

whose direction is determined by the straight line l. Now, from the geometric viewpoint it is almost evident that

$$C \times C = \cap\{Z_n : n < \omega\},$$

where $\{Z_n : n < \omega\}$ is some decreasing (with respect to inclusion) sequence of compact subsets of the unit square $[0,1] \times [0,1]$ and, in addition, the equality

$$pr_l(Z_n) = [0,2]$$

holds for any natural number n. This fact readily implies, by virtue of the compactness of all sets Z_n, that the relation

$$pr_l(C \times C) = [0, 2]$$

holds true, too, which is equivalent to the relation $C + C = [0, 2]$. The lemma has thus been proved.

Lemma 2. *There exists a set $A \subset \mathbf{R}$ of Lebesgue measure zero and of first category, such that*

$$A + A = \{x + y : x \in A, \ y \in A\} = \mathbf{R}.$$

In particular, the σ-ideal of all Lebesgue measure zero subsets of $\mathbf{R}$ (respectively, the σ-ideal of all first category subsets of $\mathbf{R}$) is not closed under the operation of vector sum of its members.

Proof. Let us put

$$A = \cup\{nC \ : \ n \in \mathbf{Z}\},$$

where C stands again for the Cantor discontinuum on $\mathbf{R}$ and $\mathbf{Z}$ stands for the set of all integers. Since the set C is of Lebesgue measure zero (even of Jordan measure zero) and is nowhere dense in $\mathbf{R}$, the set A is of Lebesgue measure zero and of first category in $\mathbf{R}$. Now, we may write

$$A + A = \cup\{nC + mC \ : \ n \in \mathbf{Z}, \ m \in \mathbf{Z}\}$$

and, consequently,

$$\cup\{nC + nC : n \in \mathbf{Z}\} \subset A + A.$$

Taking into account Lemma 1, we conclude that

$$\mathbf{R} = \cup\{n(C + C) : n \in \mathbf{Z}\} \subset A + A \subset \mathbf{R}$$

and, therefore, $A + A = \mathbf{R}$. This completes the proof.

The following classical result is due to Sierpiński (see [196]).

Theorem 3. *There exists a Hamel basis of first category, whose Lebesgue measure is equal to zero.*

Proof. Let A be a subset of $\mathbf{R}$ described in Lemma 2. In virtue of the Kuratowski-Zorn lemma, there exists a maximal (with respect to the inclusion relation) rationally independent subset of A. We fix such a subset and denote it by H. Our goal is to show that H is a Hamel basis in $\mathbf{R}$. Suppose to the contrary that there is an element $r \in \mathbf{R}$ for which

$$r \notin lin_{\mathbf{Q}}(H),$$

where $lin_{\mathbf{Q}}(H)$ stands for the linear hull (over $\mathbf{Q}$) of the set H. In view of the equality

$$A + A = \mathbf{R},$$

there are two elements $a_1 \in A$ and $a_2 \in A$ such that $r = a_1 + a_2$. Obviously, at least one of these elements does not belong to $lin_{\mathbf{Q}}(H)$. Without loss of generality, we may assume that $a_1 \notin lin_{\mathbf{Q}}(H)$. Now, consider the set

$$H' = \{a_1\} \cup H \subset A.$$

Then H' is also rationally independent and contains H as a proper subset. This contradicts our assumption that H is a maximal rationally independent subset of A. The obtained contradiction shows that H is a Hamel basis in $\mathbf{R}$ and, since $H \subset A$, this basis is of first category and of Lebesgue measure zero.

Theorem 3 has thus been proved.

Now, let us demonstrate that a certain transfinite construction (very similar to the classical Bernstein construction considered in Chapter 2) enables us to obtain λ-nonmeasurable Hamel bases of $\mathbf{R}$ (compare [118], [196]).

Theorem 4. *There exists a Hamel basis in $\mathbf{R}$ which simultaneously is a Bernstein subset of $\mathbf{R}$.*

Proof. Let us denote by α the least ordinal of cardinality continuum, and let $\{F_\xi : \xi < \alpha\}$ be an enumeration of all uncountable closed subsets of $\mathbf{R}$. We are going to define, by the method of transfinite recursion, a family $\{e_\xi : \xi < \alpha\}$ of rationally independent elements of $\mathbf{R}$ with the additional property that

$$(\forall \xi < \alpha)(e_\xi \in F_\xi).$$

Suppose that, for an ordinal $\xi < \alpha$, the partial family $\{e_\zeta : \zeta < \xi\}$ has already been defined. Consider the set

$$T_\xi = lin_{\mathbf{Q}}(\{e_\zeta : \zeta < \xi\}).$$

Clearly, we have

$$card(T_\xi) \leq card(\xi) + \omega < \mathbf{c}.$$

Consequently, $F_\xi \setminus T_\xi \neq \emptyset$ and we can choose an element

$$e_\xi \in F_\xi \setminus T_\xi.$$

In this way, the required family $\{e_\xi : \xi < \alpha\}$ will be constructed. Now, extend this family to a Hamel basis of $\mathbf{R}$ and denote the obtained Hamel basis by H. We assert that H is a Bernstein subset of $\mathbf{R}$. Indeed, let us first observe that the complement of H is totally imperfect in view of our construction. Further, it is easy to see that there exists an element $h \in \mathbf{R}$ for which

$$(h + H) \cap H = \emptyset.$$

This implies at once that H is totally imperfect, too. Thus, we have established that H is a Bernstein set in $\mathbf{R}$.

In particular, the above-mentioned Hamel basis is not Lebesgue measurable and does not possess the Baire property (see Theorem 1 from Chapter 2).

Now, starting with Theorem 3, we are going to demonstrate that the vector sum of two Lebesgue measure zero sets (respectively, of two first category sets) can be nonmeasurable in the Lebesgue sense (respectively, can be without the Baire property). This result is also due to Sierpiński.

Theorem 5. *There exist two sets A and B of Lebesgue measure zero, such that the set*

$$A + B = \{a + b \ : \ a \in A, \ b \in B\}$$

is nonmeasurable in the Lebesgue sense.

Analogously, there exist two sets A' and B' of first category, such that the set $A' + B'$ does not possess the Baire property.

Proof. We shall establish only the validity of the first part of Theorem 5 concerning nonmeasurability in the Lebesgue sense. The second part of this theorem, concerning the Baire property, can be obtained similarly by applying a dual argument.

Let $H = \{e_i : i \in I\}$ be a Hamel basis in $\mathbf{R}$ whose Lebesgue measure is zero (see Theorem 3). For any natural number n, we denote by E_n the set of all those elements

$$e = \sum_{i \in I} q_i e_i \qquad (q_i \in \mathbf{Q})$$

which satisfy the relation

$$card(\{i \in I : q_i \neq 0\}) \leq n.$$

Obviously, we have

$$\{0\} = E_0 \subset E_1 \subset E_2 \subset ... \subset E_n \subset ...,$$

$$\cup\{E_n : n \in \mathbf{N}\} = \mathbf{R}.$$

Therefore, we can find the least natural number m for which $\lambda^*(E_m) > 0$ and, consequently,

$$(\forall n > m)(\lambda^*(E_n) > 0).$$

Observe also that $m \geq 2$ since

$$E_1 \subset \mathbf{Q}H, \quad \lambda^*(E_1) \leq \lambda(\mathbf{Q}H) = 0,$$

in view of the definition of H. Notice, in addition to this, that no set E_n possesses the Steinhaus property. Indeed, taking $\{e_0, e_1, ..., e_{2n}\} \subset H$ and denoting

$$e = e_0 + e_1 + ... + e_{2n},$$

we have $e \neq 0$ and

$$(E_n + qe) \cap E_n = \emptyset$$

for any nonzero rational number q. Since q can be arbitrarily small, we claim that the Steinhaus property does not hold for E_n. It immediately follows from this fact that all the sets

$$E_m, E_{m+1}, ..., E_n, ...$$

are nonmeasurable in the Lebesgue sense. To finish the proof, let us consider two possible cases.

1. m is an even natural number. Then for some natural $k < m$, we have $m = 2k$. In this case, we may write

$$\lambda(E_k) = 0, \quad E_k + E_k = E_m,$$

and the sets $A = E_k$ and $B = E_k$ are the required ones.

2. m is an odd natural number. Then for some natural $k < m$, we have $m = 2k + 1$. In this case, we may write

$$\lambda(E_1) = \lambda(E_{2k}) = 0, \qquad E_1 + E_{2k} = E_m,$$

and the sets $A = E_1$ and $B = E_{2k}$ are the required ones.

Thus, the proof of the theorem is completed.

A more general approach (by using an Ulam transfinite matrix) is developed in paper [95] where a much stronger result than Theorem 5 is obtained (compare also [101]).

Let us consider another example of a λ-nonmeasurable set produced by a Hamel basis of $\mathbf{R}$.

Let $H = \{e_i : i \in I\}$ be an arbitrary Hamel basis in $\mathbf{R}$. Obviously, we may identify the set I with the smallest ordinal α of cardinality continuum. Then, for any $x \in \mathbf{R}$, we have a unique representation

$$x = \sum_{\xi < \alpha} q_\xi e_\xi$$

where all q_ξ ($\xi < \alpha$) are rational numbers and

$$card(\{\xi < \alpha \; : \; q_\xi \neq 0\}) < \omega.$$

For each $x \in \mathbf{R} \setminus \{0\}$, denote by $\xi = \xi(x)$ the largest ordinal from the interval $[0, \alpha[$, satisfying the relation $q_\xi \neq 0$, and define

$$A = \{x \in \mathbf{R} \; : \; q_{\xi(x)} > 0\}, \quad B = \{x \in \mathbf{R} \; : \; q_{\xi(x)} < 0\}.$$

It is clear that

$$A \cap B = \emptyset, \quad A \cup B \cup \{0\} = \mathbf{R}, \quad -A = B.$$

The last equality means that the sets A and B are symmetric to each other (with respect to the origin of $\mathbf{R}$). Actually, both sets A and B are convex (more precisely, $\mathbf{Q}$-convex) cones in $\mathbf{R}$ considered as a vector space over $\mathbf{Q}$. Let us point out an interesting property of these sets. It is not hard to verify that for any $y \in \mathbf{R}$, the inequalities

$$card(A \triangle (A + y)) < \mathbf{c}, \quad card(B \triangle (B + y)) < \mathbf{c}$$

are valid. To see this, take the representation

$$y = \sum_{\xi < \alpha} q'_\xi e_\xi$$

of y (with respect to H). Then, for those $x \in \mathbf{R}$ which satisfy the relation $\xi(x) > \xi(y)$, we have

$$x \in A \Leftrightarrow x + y \in A, \quad x \in B \Leftrightarrow x + y \in B.$$

It remains to observe that

$$card(\{x \in \mathbf{R} : \xi(x) \leq \xi(y)\}) < \mathbf{c},$$

which yields at once the required inequalities. In other words, we obtain that the sets A and B are almost invariant under the group of all translations of $\mathbf{R}$.

We now assert that the sets A and B are nonmeasurable in the Lebesgue sense. Indeed, suppose otherwise. Then at least one of these sets is Lebesgue measurable and, in view of the relation $-A = B$, we claim that both these sets must be Lebesgue measurable. Since

$$\{0\} \cup A \cup B = \mathbf{R},$$

we derive that

$$\lambda(A) = \lambda(B) > 0.$$

On the other hand, the metrical transitivity of the Lebesgue measure (see Exercise 7 from Chapter 1) implies

$$\lambda(\mathbf{R} \setminus A) = 0 \quad \vee \quad \lambda(\mathbf{R} \setminus B) = 0$$

which leads to a contradiction. We thus conclude that each of the sets A and B is nonmeasurable in the Lebesgue sense. Moreover, an easy argument based on the same property of metrical transitivity of λ shows that both these sets are λ-thick in $\mathbf{R}$; in other words, we have

$$\lambda_*(A) = \lambda_*(B) = 0.$$

However, the last relation enables us to consider the sets A and B as measurable ones with respect to some measure on $\mathbf{R}$ which extends λ and is invariant under the group of all motions (isometric transformations) of $\mathbf{R}$. Indeed, let us denote:

S = the σ-algebra of subsets of $\mathbf{R}$, generated by $dom(\lambda) \cup \{A, B\}$;

J = the σ-ideal of all those subsets of $\mathbf{R}$ whose cardinalities are strictly less than $\mathbf{c}$ (that is $J = [\mathbf{R}]^{<\mathbf{c}}$);

S' = the σ-algebra of subsets of $\mathbf{R}$, generated by $S \cup J$.

Note that any set U from the σ-algebra S can be represented in the form

$$U = (A \cap X) \cup (B \cap Y) \cup (Z \cap \{0\}),$$

where $\{X, Y, Z\} \subset dom(\lambda)$, and such a representation is unique in the sense that any analogous equality

$$U = (A \cap X_1) \cup (B \cap Y_1) \cup (Z_1 \cap \{0\})$$

for U implies the relations

$$\lambda(X \triangle X_1) = 0, \quad \lambda(Y \triangle Y_1) = 0.$$

Now, define a functional

$$\lambda' : S \to \mathbf{R} \cup \{+\infty\}$$

by the formula

$$\lambda'(U) = (1/2)(\lambda(X) + \lambda(Y)).$$

The remark made above gives us the correctness of this definition. Moreover, an easy calculation shows that λ' turns out to be a measure on S extending λ. Obviously, this measure can be uniquely extended to a measure on the σ-algebra S', by putting $\lambda'(V) = 0$ for all $V \in J$. We preserve the same notation for the extended in this manner measure. Finally, we may assert that the obtained measure λ' on S' is invariant under all isometric transformations of $\mathbf{R}$. To see this, let us observe that the equality $-A = B$ and the definition of λ' immediately yield the invariance of λ' with respect to the symmetry s_0 of $\mathbf{R}$ defined by

$$s_0(x) = -x \quad (x \in \mathbf{R}).$$

Further, the almost invariance of A and B under the group of all translations of $\mathbf{R}$ implies at once the translation-invariance of λ'. It remains to utilize the elementary geometric fact saying that the group of all motions of $\mathbf{R}$ is generated by s_0 and all translations of $\mathbf{R}$.

We would like to mention one important property of sets belonging to $dom(\lambda')$. Suppose that μ is any measure on $\mathbf{R}$ extending λ and invariant under the group of all motions of $\mathbf{R}$. Suppose also that a set U belongs simultaneously to $dom(\mu)$ and $dom(\lambda')$. Then we can assert that

$$\mu(U) = \lambda'(U),$$

which means that the value $\mu(U)$ does not depend on μ. Indeed, without loss of generality, we may assume that U is representable in the form

$$U = (A \cap X) \cup (B \cap Y)$$

where X and Y are some Lebesgue measurable subsets of $\mathbf{R}$. If $\lambda(X \triangle Y) = 0$, then

$$\mu(U) = \lambda'(U) = \lambda(X) = \lambda(Y)$$

and there is nothing to prove. Consider now the case where $\lambda(X \triangle Y) > 0$. In this case, we have

$$\lambda(X \setminus Y) > 0 \quad \vee \quad \lambda(Y \setminus X) > 0$$

and we may assume, without loss of generality, that $\lambda(X \setminus Y) > 0$. Observe now that the set

$$A \cap (X \setminus Y) = U \cap (X \setminus Y)$$

belongs to the domain of μ. Applying the metrical transitivity of λ and the almost invariance of A, we readily deduce that $A \in dom(\mu)$ and, consequently, $B \in dom(\mu)$. Moreover,

$$A \cap T \in dom(\mu), \quad B \cap T \in dom(\mu)$$

for any Lebesgue measurable set $T \subset \mathbf{R}$. If $T = [t_1, t_2]$ is an arbitrary closed subinterval of $\mathbf{R}$, then, denoting by s_T the symmetry of $\mathbf{R}$ with respect to the point $(t_1 + t_2)/2$, we get

$$\mu(A \cap T) = \mu(s_T(A \cap T)) = \mu(B \cap T),$$

from which it follows

$$\mu(A \cap T) = \mu(B \cap T) = (1/2)\mu(T) = (1/2)\lambda(T).$$

This conclusion also implies that

$$\mu(A \cap W) = (1/2)\lambda(W)$$

for any open set $W \subset \mathbf{R}$, and we easily claim that

$$\mu(A \cap X) = (1/2)\lambda(X), \quad \mu(B \cap Y) = (1/2)\lambda(Y).$$

Thus, we finally obtain

$$\mu(U) = (1/2)(\lambda(X) + \lambda(Y)),$$

that is the value $\mu(U)$ does not depend on the choice of μ.

This property of all sets from $dom(\lambda')$ is closely connected with the so-called uniqueness property for invariant measures (see the precise definition below), which plays a significant role in many questions of analysis and measure theory.

Developing the idea presented above, let us give a more elaborated application of the sets A and B to the uniqueness property of invariant measures. First, we need some preliminary notions and auxiliary statements.

Let E be a set, G be a group of transformations of E and let μ be a probability measure on E invariant under all transformations from G. The triple (E, G, μ) will be called a space with an invariant measure.

We say that μ has the uniqueness property (on the domain of μ) if for any probability G-invariant measure ν on E, the relation $dom(\nu) = dom(\mu)$ implies the equality $\nu = \mu$.

Notice that if G' is a group of transformations of E, containing G, and μ is a probability G'-invariant measure on E having the uniqueness property as a probability G-invariant measure, then μ has the same property as a probability G'-invariant measure.

A much stronger version of the uniqueness property for a probability invariant measure can be introduced in the following manner.

We say that a probability G-invariant measure μ on E has the strong uniqueness property if for each set $X \in dom(\mu)$ and for any probability G-invariant measure ν on E, the relation $X \in dom(\nu)$ implies the equality $\nu(X) = \mu(X)$.

Obviously, if μ has the strong uniqueness property, then it has also the uniqueness property. The converse assertion is not true, as will be demonstrated below.

The following statement shows that the uniqueness property is preserved under products of invariant measures.

Theorem 6. *Let (E_1, G_1, μ_1) and (E_2, G_2, μ_2) be two spaces with invariant probability measures and let each of these measures possess the uniqueness property. Then the invariant product measure on the space*

$$(E, G, \mu) = (E_1 \times E_2, G_1 \times G_2, \mu_1 \times \mu_2)$$

possesses the uniqueness property, too.

Proof. Let ν be an arbitrary probability $(G_1 \times G_2)$-invariant measure defined on $dom(\mu_1 \times \mu_2)$. First, let us observe that a functional

$$\mu' : dom(\mu_1) \to [0, 1]$$

given by the formula

$$\mu'(X) = \nu(X \times E_2) \qquad (X \in dom(\mu_1))$$

is a probability G_1-invariant measure on $dom(\mu_1)$. Hence, in view of the uniqueness property of μ_1, we claim that μ' coincides with μ_1. So we may write

$$\mu_1(X) = \nu(X \times E_2) \qquad (X \in dom(\mu_1)).$$

Fix now a set $X \in dom(\mu_1)$ with $\mu_1(X) > 0$ and consider a functional

$$\mu'' : dom(\mu_2) \to [0, 1]$$

defined by the formula

$$\mu''(Y) = \frac{1}{\mu_1(X)} \nu(X \times Y) \qquad (Y \in dom(\mu_2)).$$

Taking into account the above stated, it is not hard to check that μ'' is a probability G_2-invariant measure on $dom(\mu_2)$. Applying the uniqueness property of μ_2, we deduce that μ'' coincides with μ_2. Consequently, we get

$$\mu_1(X)\mu_2(Y) = \nu(X \times Y)$$

for any $Y \in dom(\mu_2)$. Now, it is easy to verify that the above formula remains true for arbitrary sets $X \in dom(\mu_1)$ and $Y \in dom(\mu_2)$. This circumstance shows us that

$$\nu = \mu_1 \times \mu_2,$$

and the theorem is proved.

In particular, if we have arbitrary groups G_1 and G_2 with probability (left) G_1-invariant and (left) G_2-invariant measures μ_1 and μ_2, respectively, then according to the theorem just established, we may assert that the $(G_1 \times G_2)$-invariant product measure $\mu_1 \times \mu_2$ possesses the uniqueness property provided that both given measures have this property.

In many cases important from the point of view of applications, measures on groups which are invariant under translations turn out to be invariant with respect to the symmetry as well. For instance, it suffices to recall the invariance with respect to the symmetry of a Haar measure on a compact topological group (see [62], [68]). It is reasonable to call such measures symmetric ones. Evidently, if (G_1, μ_1) and (G_2, μ_2) are two symmetric probability measures on groups G_1 and G_2, respectively, then the product measure $\mu_1 \times \mu_2$ is symmetric on the product group $G_1 \times G_2$.

We shall consider below the uniqueness property for the product measure of two symmetric invariant probability measures. It will be shown that in this case, the uniqueness property is not preserved under products.

To give the corresponding example, we need some auxiliary constructions. Let us take the number π and let us extend the one-element set $\{\pi\}$ to a Hamel basis in $\mathbf{R}$. We denote the obtained Hamel basis by

$$H = \{e_\xi : \xi < \alpha\},$$

where α is the least ordinal of cardinality continuum, and we suppose in the sequel (without loss of generality) that $e_0 = \pi$. Let A and B be the two $\mathbf{Q}$-convex cones described earlier and associated with the Hamel basis H. Further, let

$$\phi : \mathbf{R} \to \mathbf{S}_1$$

be the canonical surjective group homomorphism defined by

$$\phi(x) = (cos(x), sin(x)) \qquad (x \in \mathbf{R}),$$

where $\mathbf{S}_1$ is the unit circumference in the plane $\mathbf{R}^2$, regarded as a commutative compact topological group with respect to the standard group operation and topology.

Then it is not hard to check that:

1) $s(\phi(A)) = \phi(B)$ where s stands for the symmetry in the group $\mathbf{S}_1$;

2) $card(\phi(A) \cap \phi(B)) \leq \omega$;

3) the sets $\phi(A)$ and $\phi(B)$ are almost invariant with respect to the group of all rotations of the circumference $\mathbf{S}_1$ about its centre;

4) $\phi(A) \cup \phi(B) \cup \{(1,0)\} = \mathbf{S}_1$.

For the sake of simplicity of notation, let us put

$$A' = \phi(A), \qquad B' = \phi(B).$$

Also, let us denote by λ_1 the standard Lebesgue probability measure on $\mathbf{S}_1$ invariant under the group of all isometries of $\mathbf{S}_1$. A direct verification shows that both sets A' and B' are λ_1-thick in $\mathbf{S}_1$. Keeping in mind the above-mentioned properties of the sets A' and B' and utilizing an argument similar to the one given earlier, we can easily construct a measure ν on $\mathbf{S}_1$ which satisfies the following properties:

(1) ν is a rotation-invariant measure on $\mathbf{S}_1$ extending λ_1;

(2) ν is invariant under the symmetry in $\mathbf{S}_1$;

(3) $\{A', B'\} \subset dom(\nu)$ and $\nu(A') = \nu(B') = 1/2$;

(4) ν has the strong uniqueness property.

The properties (1) - (3) are verified directly. To see the validity of property (4), it suffices to apply an argument analogous to the one utilized for the invariant extension λ' of λ. Briefly speaking, the construction of ν can be carried out by an argument similar to the construction of the extension λ' of λ, which forces the sets A and B to be measurable with respect to λ'. We leave to the reader the corresponding details of checking (1) - (4).

Starting with the measure ν indicated above, we are ready for proving the following statement.

Theorem 7. *The product measure $\nu \times \nu$ on $\mathbf{S}_1 \times \mathbf{S}_1$ considered as a symmetric $(\mathbf{S}_1 \times \mathbf{S}_1)$-invariant probability measure, does not possess the uniqueness property. More precisely, there exists a symmetric $(\mathbf{S}_1 \times \mathbf{S}_1)$-invariant probability measure on $dom(\nu \times \nu)$ which differs from $\nu \times \nu$.*

Proof. Obviously, we may write

$$\{A' \times A', A' \times B', B' \times A', B' \times B'\} \subset dom(\nu \times \nu).$$

At the same time, denoting $\lambda_2 = \lambda_1 \times \lambda_1$, we have

$$\lambda_2((\mathbf{S}_1 \times \mathbf{S}_1) \setminus (A' \times A' \cup A' \times B' \cup B' \times A' \cup B' \times B')) = 0.$$

Now, it is clear that the general form of an element W from $dom(\nu \times \nu)$ is the following:

$$W = ((A' \times A') \cap X) \cup ((A' \times B') \cap Y) \cup ((B' \times A') \cap Z) \cup ((B' \times B') \cap T)$$

where X, Y, Z, T are elements of $dom(\lambda_2)$. Let us mention that we omit in this representation some sets of λ_2-measure zero, which do not play any role here.

Now, we fix a real number $r \in \ ]0, 1[$ and define a measure μ on the domain of $\nu \times \nu$ by the formula

$$\mu(W) = (1/2)(r\lambda_2(X) + (1 - r)\lambda_2(Y) + r\lambda_2(T) + (1 - r)\lambda_2(Z)).$$

The correctness of this definition is implied by the fact that all the sets

$$A' \times A', \ \ A' \times B', \ \ B' \times A', \ \ B' \times B'$$

are thick in $\mathbf{S}_1 \times \mathbf{S}_1$ with respect to λ_2. Obviously, we have

$$\nu \times \nu \neq \mu$$

whenever $r \neq 1/2$. The reader can also easily verify that μ is a symmetric $(\mathbf{S}_1 \times \mathbf{S}_1)$-invariant probability measure on $dom(\nu \times \nu)$. This completes the proof of the theorem.

Remark 2. The measure $\nu \times \nu$ has the uniqueness property if it is regarded as a G'-invariant measure, where G' stands for the group generated by all pairs (g, s) and (s, h). Here g and h are arbitrary rotations of $\mathbf{S}_1$ and s is the symmetry in $\mathbf{S}_1$.

Indeed, it suffices to apply Theorem 6 to the product measure $\nu \times \nu$ and to the product group G'.

EXERCISES

1. Deduce from Theorem 5 that there exists a set $X \subset \mathbf{R}$ of Lebesgue measure zero, such that the set

$$X + X = \{y + z : y \in X, \ z \in X\}$$

is not measurable in the Lebesgue sense.

Formulate and prove an analogous fact in terms of the category and the Baire property.

In connection with this exercise, see also [23].

2. Let $h : \mathbf{R} \to \mathbf{R}$ be a homomorphism of the additive group $\mathbf{R}$ into itself. Show that the following assertions are equivalent:

(a) h is not continuous at some point of $\mathbf{R}$;

(b) h is everywhere discontinuous (in other words, there exists no point of $\mathbf{R}$ at which h is continuous);

(c) h is not Lebesgue measurable;

(d) h does not possess the Baire property;

(e) the graph of h is everywhere dense in the plane $\mathbf{R}^2$;

(f) h is not of the form

$$h(x) = rx \qquad (x \in \mathbf{R}),$$

where r is some real number (depending only on h).

Further, let $H = \{e_i : i \in I\}$ be a Hamel basis in $\mathbf{R}$. Fix an index $j \in I$ and consider a linear (over $\mathbf{Q}$) functional $h : \mathbf{R} \to \mathbf{R}$ defined as follows: $h(x) = q_j$ for each real $x = \sum_{i \in I} q_i e_i$.

Verify that h is a nontrivial solution of the Cauchy functional equation.

3. Let E be a set, n be a natural number and let $R \subset E^n$ be an n-ary relation on E. We shall say that a set $X \subset E$ is independent with respect to R if

$$(x_1, x_2, ..., x_n) \notin R$$

for all those $(x_1, x_2, ..., x_n) \in X^n$ whose coordinates are pairwise distinct.

Suppose that E is a complete metric space and denote by $Comp(E)$ the family of all nonempty compact subsets of E. Equip $Comp(E)$ with the standard Hausdorff metric (which induces the Vietoris topology on the same family). It is well known (see [37], [77]) that, in this manner, $Comp(E)$ becomes a complete metric space. Suppose, in addition, that a countable family $\{R_k : k < \omega\}$ of relations on E is given, such that each R_k has its own weight n_k (in other words, $R_k \subset E^{n_k}$) and the graph of R_k is a first category set in the product space E^{n_k}. Demonstrate that the family of those sets $X \in Comp(E)$ which are simultaneously independent with respect to all relations R_k $(k < \omega)$ is a residual (co-meager) subset of $Comp(E)$.

This result is due to Kuratowski (see [126] where some applications are presented as well; compare also [157] and [226]). Note that the above-mentioned result can be established within the theory **ZF** & **DC**.

4. Let R be the Vitali equivalence relation on **R**; in other words, let us put

$$R = \{(x,y) \in \mathbf{R} \times \mathbf{R} : x - y \in \mathbf{Q}\}.$$

We know that the graph of R is representable as the union of a countable family of straight lines lying in the plane $\mathbf{R} \times \mathbf{R}$, hence this graph is a first category set in $\mathbf{R} \times \mathbf{R}$.

Starting with this simple observation and applying the Kuratowski result formulated in Exercise 3, demonstrate the existence of an uncountable compact subset P of **R** which intersects every R-equivalence class in at most one point.

Infer from this fact that there exists a Vitali set in **R** containing P.

Conclude that there exist Vitali sets which are not totally imperfect and, hence, are not Bernstein sets (compare Theorem 3 from Chapter 2).

5. Let T stand for the family of all nonempty finite sequences of nonzero rational numbers. Clearly, $card(T) = \omega$. For any

$$\tau = (q_1, q_2, ..., q_n) \in T,$$

denote by R_τ the family of all those vectors $(x_1, x_2, ..., x_n) \in \mathbf{R}^n$ which satisfy the equality

$$q_1 x_1 + q_2 x_2 + ... + q_n x_n = 0.$$

In other words, R_τ coincides with the hyperplane in $\mathbf{R}^n$ determined by the above-mentioned equation, and this hyperplane is a closed nowhere dense subset of $\mathbf{R}^n$.

Starting with the family of relations $\{R_\tau : \tau \in T\}$ and applying again the Kuratowski result presented in Exercise 3, show that there exists an uncountable compact subset of **R** linearly independent over **Q**.

Deduce from the latter fact that there exists a Hamel basis of **R** containing a nonempty perfect set.

Conclude that there exist Hamel bases in **R** which are not totally imperfect and, hence, are not Bernstein sets (cf. Theorem 4 of this chapter).

In connection with the preceding exercise, let us remark that a much stronger result was obtained. Namely, as shown by von Neumann [159], there exists a nonempty perfect set of real numbers which are mutually algebraically independent (over $\mathbf{Q}$). The latter fact can also be derived from the Kuratowski result formulated in Exercise 3.

6. Let X be a subset of $\mathbf{R}$. Denote by $\mathcal{T}_X$ the family of all translates of X in $\mathbf{R}$.

(a) Demonstrate that if X is a nonempty proper subset of $\mathbf{R}$, then

$$\omega \leq card(\mathcal{T}_X).$$

(b) For any cardinal number κ satisfying the inequalities $\omega \leq \kappa \leq 2^{\omega}$, give an example of $X \subset \mathbf{R}$ such that $card(\mathcal{T}_X) = \kappa$.

7. Let G be a Lebesgue measurable subgroup of $\mathbf{R}$. Derive from the Steinhaus property that either $G = \mathbf{R}$ or G is of Lebesgue measure zero. Show that, in the latter case, $card(\mathcal{T}_G) = 2^{\omega}$.

Formulate and prove an analogous result for a group $G \subset \mathbf{R}$ possessing the Baire property.

8. Let $(E_1, \mathcal{S}_1, \mu_1)$ and $(E_2, \mathcal{S}_2, \mu_2)$ be any two spaces with σ-finite measures. Let X_1 be a μ_1-thick subset of E_1 and let X_2 be a μ_2-thick subset of E_2. Show that the set $X_1 \times X_2$ is $(\mu_1 \times \mu_2)$-thick in the product space $E_1 \times E_2$.

Formulate and prove an analogous result for the product of an arbitrary family of probability measures and for thick sets with respect to these measures.

9. Let $f : \mathbf{R} \to \mathbf{R}$ be a function. We say that f is convex in the sense of Jensen (or f satisfies the Jensen inequality) if

$$f((x + y)/2) \leq (1/2)(f(x) + f(y))$$

for all $x \in \mathbf{R}$ and $y \in \mathbf{R}$. Obviously, any solution of the Cauchy functional equation is convex in the sense of Jensen.

Suppose that a function $g : \mathbf{R} \to \mathbf{R}$ is Lebesgue measurable and satisfies the Jensen inequality. Demonstrate that g is continuous and, hence, is convex in the usual sense. In other words, show that

$$g(tx + (1 - t)y) \leq tg(x) + (1 - t)g(y)$$

for all $x \in \mathbf{R}$, $y \in \mathbf{R}$ and $t \in [0, 1]$.

This result was first obtained by Sierpiński.

10. Let $f : \mathbf{R} \to \mathbf{R}$ be a Lebesgue measurable function such that, for any $x \in \mathbf{R}$, $y \in \mathbf{R}$, $z \in \mathbf{R}$ satisfying the relation

$$x - 2y + z = 0,$$

we have

$$f(x) - 2f(y) + f(z) = 0.$$

Prove that f is linear; in other words, prove that f can be represented in the form

$$f(x) = ax + b \qquad (x \in \mathbf{R}),$$

where a and b are some fixed real coefficients.

11. Let E be a metric space, F be a complete metric space, X be a subset of E, and let $h : X \to F$ be a continuous function. Demonstrate that there exist a set $X' \subset E$ and a function $h' : X' \to F$, such that:

(a) $X \subset X'$;

(b) X' is a G_δ-subset of E;

(c) h' is a continuous extension of h.

This classical theorem on extensions of continuous functions is due to Lavrentiev and has important applications in descriptive set theory (see [125]).

12. We know that any nontrivial solution of the Cauchy functional equation is very bad from the point of view of continuity: such a solution turns out to be discontinuous at all points of $\mathbf{R}$. Here we wish to present another interesting and important example of a function (acting from $\mathbf{R}$ into $\mathbf{R}$) which is extremely discontinuous; namely, it is discontinuous on each subset of $\mathbf{R}$ of cardinality $\mathbf{c}$. This classical example is due to Sierpiński and Zygmund (see [203]). Actually, their result is essentially based on the Lavrentiev extension theorem formulated in Exercise 11.

Denote by α the first ordinal of cardinality $\mathbf{c}$. Let $\{f_\xi : \xi < \alpha\}$ be an enumeration of all those partial continuous functions which act from $\mathbf{R}$ into $\mathbf{R}$ and whose domains are uncountable G_δ-subsets of $\mathbf{R}$. Further, let $\{x_\xi : \xi < \alpha\}$ be an injective enumeration of all points of $\mathbf{R}$.

We are going to define a function

$$f : \mathbf{R} \to \mathbf{R}$$

which will be the required one; in other words, the restriction of f to any subset of $\mathbf{R}$ of cardinality $\mathbf{c}$ will be discontinuous. In order to do this, let us take a point x_ξ ($\xi < \alpha$) and consider the partial families

$$\{x_\zeta : \zeta < \xi\}, \quad \{f_\zeta : \zeta < \xi\}.$$

Note that

$$card(\{f_\zeta(x_\xi) \ : \ \zeta < \xi, \ x_\xi \in dom(f_\zeta)\}) < \mathbf{c}.$$

Therefore, we can choose a point

$$y_\xi \in \mathbf{R} \setminus \{f_\zeta(x_\xi) \ : \ \zeta < \xi, \ x_\xi \in dom(f_\zeta)\}$$

and put $f(x_\xi) = y_\xi$. In this way, f will be defined on the whole real line $\mathbf{R}$.

According to the construction, for each partial function f_ξ, we have

$$card(\{x \in \mathbf{R} : f(x) = f_\xi(x)\}) < \mathbf{c}.$$

Deduce from this fact that for any set $X \subset \mathbf{R}$ with $card(X) = \mathbf{c}$, the function $f|X : X \to \mathbf{R}$ is not continuous. It also turns out that f is not measurable in the Lebesgue sense and does not possess the Baire property. For more details, see Chapter 5 of this book.

Actually, the graph of f is totally imperfect in $\mathbf{R}^2$ and almost avoids all graphs of the functions from $\{f_\xi : \xi < \alpha\}$. Thus, we may say that the above-mentioned construction is aimed at determining a function which almost avoids all members from a given family of partial functions. Clearly, the same method works in a more general situation, for a family $\{g_\xi : \xi < \kappa\}$ of partial functions acting from an infinite set E into itself, where $card(E) = \kappa$.

Some other similar constructions will be discussed later, in Chapter 5. Namely, in that chapter we will be dealing with sets which almost avoid all members of a given σ-ideal of subsets of $\mathbf{R}$.

13. Let H be an arbitrary Hamel basis in $\mathbf{R}$. Prove that H is not an analytic subset of $\mathbf{R}$; in other words, H is not a continuous image of a Borel subset of $\mathbf{R}$. (Apply the Steinhaus property for λ-measurable sets and the fact that any analytic subset of $\mathbf{R}$ is λ-measurable.)

The result of Exercise 13 is due to Sierpiński.

It is useful to compare Exercise 13 with Exercise 5 of this chapter.

Chapter 4
The Fubini theorem and nonmeasurable sets

In this chapter we present several results which are closely connected with the classical Fubini theorem and the existence of nonmeasurable sets and functions.

Note that we have already given one application of this theorem in Chapter 2 for constructing a non Lebesgue-measurable function with the thick graph (see Theorem 4 therein).

First, let us recall the precise formulation of the Fubini theorem.

Suppose that $(E_1, \mathcal{S}_1, \mu_1)$ and $(E_2, \mathcal{S}_2, \mu_2)$ are two measure spaces such that both measures μ_1 and μ_2 are σ-finite, and let

$$f : E_1 \times E_2 \to \mathbf{R}$$

be a function integrable with respect to the completion μ of the product measure $\mu_1 \times \mu_2$. Then:

(a) for μ_1-almost all $x \in E_1$, the partial function

$$f_x : E_2 \to \mathbf{R}$$

defined by $f_x(y) = f(x, y)$ is integrable with respect to the completion of μ_2;

(b) for μ_2-almost all $y \in E_2$, the partial function

$$f_y : E_1 \to \mathbf{R}$$

defined by $f_y(x) = f(x, y)$ is integrable with respect to the completion of μ_1;

(c) the function

$$\phi_1 : x \to \int_{E_2} f_x(y) d\mu_2(y)$$

is integrable with respect to the completion of μ_1;

(d) the function

$$\phi_2 : y \to \int_{E_1} f_y(x)d\mu_1(x)$$

is integrable with respect to the completion of μ_2;

(e) the equalities

$$\int_{E_1 \times E_2} f(x,y)d\mu(x,y) = \int_{E_1} \phi_1(x)d\mu_1(x) = \int_{E_2} \phi_2(y)d\mu_2(y)$$

are valid; in other words, we may write in short

$$\int_{E_1 \times E_2} f(x,y)d\mu(x,y) = \int_{E_1}(\int_{E_2} f(x,y)d\mu_2(y))d\mu_1(x) =$$

$$\int_{E_2}(\int_{E_1} f(x,y)d\mu_1(x))d\mu_2(y).$$

Actually, in our further considerations we primarily need the following easy corollary of the Fubini theorem: if the given measures μ_1 and μ_2 are such that

$$\mu_1(E_1) < +\infty, \quad \mu_2(E_2) < +\infty$$

and a bounded function

$$f : E_1 \times E_2 \to \mathbf{R}$$

does not satisfy the equality

$$\int_{E_1}(\int_{E_2} f(x,y)d\mu_2(y))d\mu_1(x) = \int_{E_2}(\int_{E_1} f(x,y)d\mu_1(x))d\mu_2(y),$$

then f is not measurable with respect to μ (where μ stands again for the completion of the product measure $\mu_1 \times \mu_2$). This simple observation is the starting point for us in the present chapter.

Let us begin with the construction of a classical Sierpiński partition of the plane $\mathbf{R}^2 = \mathbf{R} \times \mathbf{R}$. The existence of such a partition is established under the Continuum Hypothesis and implies a number of interesting results and statements which can be successfully utilized in various areas of mathematics. Especially, they can be applied to some deep questions and problems from real analysis, measure theory and general topology (see [204], [102]).

Let ω denote, as usual, the least infinite ordinal number and let ω_1 denote the least uncountable ordinal number. It is a well-known fact that Sierpiński was the first mathematician who considered, in his classical paper [200], a partition $\{A, B\}$ of the product set $\omega_1 \times \omega_1$, defined as follows:

$$A = \{(\xi, \zeta) \ : \ \xi \leq \zeta < \omega_1\},$$

$$B = \{(\xi, \zeta) \ : \ \omega_1 > \xi > \zeta\}.$$

He observed that, for any $\xi < \omega_1$ and $\zeta < \omega_1$, the inequalities

$$card(A^{\zeta}) \leq \omega, \quad card(B_{\xi}) \leq \omega$$

are true, where

$$A^{\zeta} = \{\xi \ : \ (\xi, \zeta) \in A\},$$

$$B_{\xi} = \{\zeta \ : \ (\xi, \zeta) \in B\}.$$

In other words, each of the sets A and B can be represented as the union of a countable family of "curves" lying in the product set $\omega_1 \times \omega_1$. As mentioned above, this property of the partition $\{A, B\}$ leads to numerous interesting and unexpected consequences. For instance, it immediately follows from the existence of $\{A, B\}$ that if the Continuum Hypothesis

$$2^{\omega} = \omega_1$$

holds, then there exists a partition $\{A', B'\}$ of the euclidean plane $\mathbf{R}^2$, satisfying the relations:

1) for each straight line L in $\mathbf{R}^2$ parallel to the line $\mathbf{R} \times \{0\}$, the inequality

$$card(A' \cap L) \leq \omega$$

is valid;

2) for each straight line M in $\mathbf{R}^2$ parallel to the line $\{0\} \times \mathbf{R}$, the inequality

$$card(B' \cap M) \leq \omega$$

is valid.

Moreover, Sierpiński demonstrated that if a covering $\{A', B'\}$ of $\mathbf{R}^2$ with the above-mentioned properties 1) and 2) does exist, then the Continuum Hypothesis is true. Indeed, suppose that $\{A', B'\}$ is such a covering of $\mathbf{R}^2$. Choose an arbitrary subset X of $\mathbf{R}$ having cardinality ω_1 and put

$$Z = (X \times \mathbf{R}) \cap B'.$$

Then, according to relation 2), we must have

$$card(Z) \leq \omega \cdot \omega_1 = \omega_1.$$

On the other hand, let us show that

$$pr_2(Z) = \mathbf{R}.$$

In order to do this, take an arbitrary point $y \in \mathbf{R}$ and consider the straight line $\mathbf{R} \times \{y\}$. Relation 1) implies that

$$card(A' \cap (\mathbf{R} \times \{y\})) \leq \omega.$$

At the same time, we obviously have

$$card((X \times \mathbf{R}) \cap (\mathbf{R} \times \{y\})) = \omega_1.$$

Hence there exists a point $t \in \mathbf{R}$ such that

$$(t, y) \notin A', \quad (t, y) \in X \times \mathbf{R}.$$

Since $\{A', B'\}$ is a covering of $\mathbf{R}^2$, we infer that $(t, y) \in B'$ and, consequently,

$$(t, y) \in Z, \quad y \in pr_2(Z),$$

which yields the required equality $pr_2(Z) = \mathbf{R}$. We thus get

$$\mathbf{c} = card(\mathbf{R}) \leq card(Z) \leq \omega_1$$

and, finally, $2^\omega = \mathbf{c} = \omega_1$.

In other words, according to the Sierpiński result presented above, the Continuum Hypothesis is equivalent to the statement that there exists a partition $\{A', B'\}$ of the euclidean plane $\mathbf{R}^2$, satisfying relations 1) and 2).

Let us mention a straightforward and important consequence of the existence of a Sierpiński partition $\{A', B'\}$ of $\mathbf{R}^2$. For this purpose, introduce the sets

$$A'' = [0, 1]^2 \cap A', \quad B'' = [0, 1]^2 \cap B'.$$

Then we obtain the partition $\{A'', B''\}$ of $[0, 1]^2$ with the properties similar to the ones of $\{A', B'\}$. Let us consider the characteristic functions

$$f = \chi_{A''}, \quad g = \chi_{B''}.$$

It can easily be observed that there exist four iterated integrals

$$\int_0^1 dx\Big(\int_0^1 f(x,y)dy\Big), \qquad \int_0^1 dy\Big(\int_0^1 f(x,y)dx\Big),$$

$$\int_0^1 dx\Big(\int_0^1 g(x,y)dy\Big), \qquad \int_0^1 dy\Big(\int_0^1 g(x,y)dx\Big),$$

but we have

$$\int_0^1 dx\Big(\int_0^1 f(x,y)dy\Big) = \int_0^1 dy\Big(\int_0^1 g(x,y)dx\Big) = 1,$$

$$\int_0^1 dy\Big(\int_0^1 f(x,y)dx\Big) = \int_0^1 dx\Big(\int_0^1 g(x,y)dy\Big) = 0$$

and, therefore,

$$\int_0^1 \Big(\int_0^1 f(x,y)dy\Big)dx \neq \int_0^1 \Big(\int_0^1 f(x,y)dx\Big)dy,$$

$$\int_0^1 \Big(\int_0^1 g(x,y)dy\Big)dx \neq \int_0^1 \Big(\int_0^1 g(x,y)dx\Big)dy.$$

Thus, we infer that the statement of the classical Fubini theorem does not hold for each of the functions f and g. At the same time, it is obvious that the functions f and g are bounded on $[0,1]^2$. Taking these facts into account, we claim that both f and g are nonmeasurable in the Lebesgue sense; in other words, they are nonmeasurable with respect to the standard two-dimensional Lebesgue measure λ_2 on the plane $\mathbf{R}^2$.

Remark 1. We see that the Continuum Hypothesis implies the existence of a function f acting from $[0,1]^2$ into $[0,1]$ such that its iterated integrals differ from each other. It is not hard to verify that **CH** is not necessary for this conclusion. For instance, Martin's Axiom also implies the existence of such a function. And, moreover, we do not need here the full power of **MA**: it suffices to assume that each subset of $\mathbf{R}$, whose cardinality is strictly less than $\mathbf{c}$, is measurable in the Lebesgue sense. On the other hand, it was shown in [51] that there are models of set theory in which, for every function

$$g \ : \ [0,1]^2 \to [0,1],$$

the existence of the iterated integrals

$$\int_0^1 \Big(\int_0^1 g(x,y)dx\Big)dy, \qquad \int_0^1 \Big(\int_0^1 g(x,y)dy\Big)dx$$

implies their equality.

For some further results concerning iterated integrals and tightly connected with the Sierpiński partition, see [193].

Let us formulate several statements (also interesting from the point of view of measure theory) which follow from the corresponding properties of the Sierpiński partition $\{A, B\}$ of $\omega_1 \times \omega_1$.

(i) If $\mathcal{P}(\omega_1)$ denotes the σ-algebra of all subsets of ω_1, then the product σ-algebra $\mathcal{P}(\omega_1) \otimes \mathcal{P}(\omega_1)$ coincides with the σ-algebra $\mathcal{P}(\omega_1 \times \omega_1)$ of all subsets of $\omega_1 \times \omega_1$. In other words, we have the equality

$$\mathcal{P}(\omega_1) \otimes \mathcal{P}(\omega_1) = \mathcal{P}(\omega_1 \times \omega_1).$$

In order to establish this result, it is sufficient to consider an arbitrary embedding of ω_1 into the real line $\mathbf{R}$ and to apply the well-known fact that the graph Γ_f of any measurable function

$$f : (E, \mathcal{S}, \mu) \to \mathbf{R}$$

is a measurable subset of the product space $(E, \mathcal{S}) \times (\mathbf{R}, \mathcal{B}(\mathbf{R}))$, where $\mathcal{B}(\mathbf{R})$ denotes, as usual, the Borel σ-algebra of $\mathbf{R}$ (for more details, see Chapter 13).

From the equality $\mathcal{P}(\omega_1 \times \omega_1) = \mathcal{P}(\omega_1) \otimes \mathcal{P}(\omega_1)$ we can directly deduce the following important statement.

(ii) There does not exist a nonzero σ-finite diffused measure μ defined on the σ-algebra $\mathcal{P}(\omega_1)$.

Let us mention that this classical result is due to Ulam [222] who established the nonexistence of such a measure μ in another way, by applying a transfinite matrix of some special type (see, for instance, [222], [120], [155], [165] or Chapter 7 of this book).

In order to prove statement (ii) by using the properties of $\{A, B\}$, suppose for a moment that such a measure μ does exist and let us apply the Fubini theorem to the product measure $\mu \times \mu$ and to the sets A and B of the Sierpiński partition. Taking into account the countability of the corresponding sections of A and B, we immediately get the equalities

$$(\mu \times \mu)(A) = (\mu \times \mu)(B) = 0$$

and, consequently,

$$(\mu \times \mu)(A \cup B) = (\mu \times \mu)(\omega_1 \times \omega_1) = 0,$$

which yields a contradiction. Briefly, we have just demonstrated that ω_1 is not a real-valued measurable cardinal. See the precise definition of real-valued measurable cardinals below in this chapter; for more detailed information about measurability of cardinals, see also [50], [64], [127], [211] and [222]. This topic is extensively studied in modern combinatorial set theory.

Let us return to the Sierpiński partition and consider some other interesting facts closely related to it.

(iii) Assuming the Continuum Hypothesis, there exists a function

$$\phi \; : \; \mathbf{R} \to \mathbf{R}$$

such that

$$\mathbf{R}^2 = \cup\{g_n(\Gamma_\phi) \; : \; n < \omega\},$$

where Γ_ϕ denotes the graph of ϕ and g_n $(n < \omega)$ are some motions of the plane $\mathbf{R}^2$, each of which is either a translation or a rotation (about a point) whose corresponding angle is equal to $\pm\pi/2$.

The proof of this result is not difficult and we leave it to the reader as a useful exercise. (Hint: it suffices to apply a countable family of "curves" produced by a Sierpiński partition of $\mathbf{R}^2$.)

Let now X and Y be any two sets. We recall that a set-valued mapping is an arbitrary function of the type

$$F \; : \; X \to \mathcal{P}(Y),$$

where $\mathcal{P}(Y)$ denotes, as usual, the family of all subsets of Y. According to a well-known definition from general set theory, a subset Z of X is independent with respect to F if, for any two distinct elements $x \in Z$ and $y \in Z$, the relations

$$x \notin F(y), \quad y \notin F(x)$$

are valid.

We denote by the symbol $[\omega_1]^{\leq\omega}$ the family of all countable subsets of ω_1. Clearly,

$$[\omega_1]^{\leq\omega} \subset \mathcal{P}(\omega_1).$$

Now, we can formulate the following simple statement.

(iv) There exists a set-valued mapping

$$F \; : \; \omega_1 \to [\omega_1]^{\leq \omega}$$

such that no two-element subset of ω_1 is independent with respect to F.

In fact, the desired set-valued mapping F can be defined as follows:

$$F(\zeta) = A^\zeta \qquad (\zeta < \omega_1),$$

where A is the first component of the Sierpiński partition $\{A, B\}$ of $\omega_1 \times \omega_1$.

In connection with (iv), let us remark that if $[\omega_1]^{<\omega}$ denotes the family of all finite subsets of ω_1 and a set-valued mapping

$$F \; : \; \omega_1 \to [\omega_1]^{<\omega}$$

is given, then there always exists a subset Ξ of ω_1 satisfying the following two conditions:

1) $card(\Xi) = \omega_1$;

2) Ξ is independent with respect to F.

The reader can easily prove this result or derive it directly from the so-called $\triangle$-system lemma (see [120]). It is also reasonable to point out here that the $\triangle$-system lemma is a theorem of the theory **ZF & DC**.

Finally, let us mention that an analogous result (concerning the existence of large independent subsets) holds true for uncountable cardinal numbers, but the proof of this generalized result (due to Erdös and Hajnal) is more difficult and needs an additional argument.

There are many other interesting statements and facts which are related to the Sierpiński partition or can be obtained by using certain properties of this partition (see especially [28], [29], [39], [40], [89], [100], [153], [194], [195] and [204]).

Now, we want to consider one more statement related to the Sierpiński partition of $\omega_1 \times \omega_1$. This statement does not require additional set-theoretic hypotheses and establishes close connections between partitions of Sierpiński type and nonmeasurability in the Lebesgue sense.

Theorem 1. *In the theory* **ZF & DC**, *the assertion*

"there exists a well-ordering of **R**"*

implies the assertion

"there exists a subset of $\mathbf{R}$ *nonmeasurable in the Lebesgue sense"*.

Proof. Obviously, the existence of a well-ordering of $\mathbf{R}$ means that $\mathbf{R}$ can be represented as an injective family of all its points:

$$\mathbf{R} = \{x_\xi \ : \ \xi < \alpha\},$$

where α denotes some ordinal number of cardinality continuum. Also, it is clear that, in order to prove the existence of subsets of $\mathbf{R}$ nonmeasurable with respect to the standard one-dimensional Lebesgue measure $\lambda = \lambda_1$, it suffices to establish the existence of subsets of $\mathbf{R}^2$ nonmeasurable with respect to the standard two-dimensional Lebesgue measure λ_2 (since, as is well known, these two measures are isomorphic to each other).

Let $\beta \leq \alpha$ be the least ordinal number for which

$$\lambda^*(\{x_\xi \ : \ \xi < \beta\}) > 0,$$

where λ^* is the outer measure associated with λ. If the set $\{x_\xi \ : \ \xi < \beta\}$ is nonmeasurable with respect to λ, then we are done. Otherwise, we may write

$$\{x_\xi \ : \ \xi < \beta\} \in dom(\lambda),$$

$$\lambda(\{x_\xi \ : \ \xi < \beta\}) > 0$$

and, according to the definition of β, for each ordinal $\gamma < \beta$ we have

$$\lambda(\{x_\xi \ : \ \xi < \gamma\}) = 0.$$

Consider now a subset Z of $\mathbf{R}^2$ defined as follows:

$$Z = \{(x_\xi, x_\zeta) \ : \ \xi \leq \zeta < \beta\}.$$

We assert that Z is nonmeasurable with respect to λ_2. Indeed, suppose for a moment that $Z \in dom(\lambda_2)$. Then, considering the vertical and horizontal sections of Z and applying the Fubini theorem to Z, we get, on the one hand, the relation

$$\lambda_2(Z) > 0$$

and, on the other hand, the equality

$$\lambda_2(Z) = 0.$$

Since this is impossible, we conclude that Z is not λ_2-measurable, which also implies the existence of a non Lebesgue-measurable subset of the real line. The theorem has thus been proved.

In order to give other applications of the classical Fubini theorem to the existence of nonmeasurable sets, we need some auxiliary notions and statements.

First, let us recall the concept of a universal measure zero space, which will be rather useful in our further constructions.

Let E be a topological space all singletons of which are Borel subsets of E. We say that E is a universal measure zero space if there exists no nonzero σ-finite diffused Borel measure on E.

Let E be a topological space whose all singletons are Borel in E and let X be a subset of E. We say that X is a universal measure zero set in E if X is a universal measure zero space with respect to the induced topology.

The following auxiliary statement due to Szpilrajn (Marczewski) yields some characterization of universal measure zero subsets of $\mathbf{R}$ in terms of Lebesgue measure zero sets.

Lemma 1. *Let X be a subset of $\mathbf{R}$. Then these two assertions are equivalent:*

1) X is a universal measure zero set in $\mathbf{R}$;

2) each homeomorphic image of X lying in $\mathbf{R}$ is of Lebesgue measure zero.

Proof. Suppose first that X satisfies relation 1). Let Y be a subset of $\mathbf{R}$ homeomorphic to X. Fix any homeomorphism

$$f \, : \, X \to Y.$$

If $\lambda^*(Y) > 0$, then it is obvious that there exists a nonzero σ-finite diffused Borel measure μ on Y (produced by λ). Putting

$$\nu(Z) = \mu(f(Z)) \qquad (Z \in \mathcal{B}(X)),$$

we obtain a nonzero σ-finite diffused Borel measure ν on X. But this is impossible since X is a universal measure zero space. Therefore, the equality

$$\lambda(Y) = 0$$

must be valid, and the implication 1) $\Rightarrow$ 2) has been proved.

Let now X satisfy relation 2). We are going to demonstrate that relation 1) holds for X, too. Of course, without loss of generality, we may assume that our X is a subset of the unit segment $[0, 1]$. Suppose for a moment that X is not a universal measure zero space. Then there exists a Borel diffused probability measure μ on $[0, 1]$ such that

$$\mu^*(X) > 0.$$

We may also assume that μ does not vanish on any nonempty open subinterval of $[0, 1]$ (replacing, if necessary, μ by $(\mu + \lambda)/2$). Now, define a function

$$f \; : \; [0, 1] \to [0, 1]$$

by the formula

$$f(x) = \mu([0, x]) \qquad (x \in [0, 1]).$$

Evidently, f is an increasing homeomorphism from $[0, 1]$ onto itself. Let us put

$$\nu(Z) = \mu(f^{-1}(Z)) \qquad (Z \in \mathcal{B}([0, 1])).$$

In this way, we get a Borel probability measure ν on $[0, 1]$ such that

$$\nu^*(f(X)) = \mu^*(X) > 0.$$

Furthermore, it turns out that ν coincides with the standard Borel measure on $[0, 1]$. Indeed, for each interval $[a, b] \subset [0, 1]$, we may write

$$f^{-1}([a, b]) = \{t \in [0, 1] \; : \; \mu([0, t]) \in [a, b]\} = [c, d],$$

where

$$\mu([0, c]) = a, \qquad \mu([0, d]) = b.$$

Then we have

$$\nu([a, b]) = \mu(f^{-1}([a, b])) = \mu([c, d]) =$$

$$= \mu([0, d]) - \mu([0, c]) = b - a.$$

Consequently, the measures ν and λ are identical on the family of all subintervals of $[0, 1]$. Hence, these two measures coincide on the whole Borel σ-algebra of $[0, 1]$. Thus

$$\lambda^*(f(X)) = \nu^*(f(X)) > 0,$$

which contradicts relation 2). The contradiction obtained establishes the implication 2) $\Rightarrow$ 1) and ends the proof of Lemma 1.

Remark 2. Actually, Lemma 1 can be formulated in a much stronger form. Let E be a Polish space equipped with a Borel diffused probability measure ν and let X be a subset of E. Then the following two relations are equivalent:

1) X is a universal measure zero subset of E;

2) for any Borel isomorphism $f : E \to E$, we have

$$\nu^*(f(X)) = 0.$$

The proof of this equivalence is based on Lemma 1 and the well-known fact from topological measure theory, stating that all Borel diffused probability measures on Polish spaces (more generally, on Borel subsets of Polish spaces) are isomorphic to the restriction of λ to the Borel σ-algebra of $[0, 1]$.

Let E be an infinite set and let $\mathcal{S}$ be a σ-algebra of subsets of E. We shall say that $\mathcal{S}$ is admissible if it satisfies the following conditions:

(a) all singletons in E belong to $\mathcal{S}$;

(b) $\mathcal{S}$ is countably generated (that is $\mathcal{S}$ contains a countable subfamily which generates $\mathcal{S}$);

(c) there exists a nonzero σ-finite diffused measure whose domain coincides with $\mathcal{S}$.

Obviously, (c) can be replaced by the equivalent condition:

(c') there exists a diffused probability measure whose domain coincides with $\mathcal{S}$.

Also, it is not hard to see that the conjunction of (a) and (b) is equivalent to the following condition:

(ab) there exists a countable subfamily of $\mathcal{S}$ generating $\mathcal{S}$ and separating the points in E.

In this context, it is reasonable to recall that a family $\{X_i : i \in I\}$ of subsets of E separates the points in E if, for any two distinct points $y \in E$ and $z \in E$, there exists an index $i \in I$ such that

$$card(X_i \cap \{y, z\}) = 1.$$

Let us denote

$$\kappa = inf\{card(E) : E \text{ is a subset of } \mathbf{R} \text{ with } \lambda^*(E) > 0\}.$$

Clearly, we have

$$\kappa \geq \omega_1.$$

Further, suppose that a set $E \subset \mathbf{R}$ with $card(E) = \kappa$ and $\lambda^*(E) > 0$ is given. Equip this set with the topology induced by the standard topology of $\mathbf{R}$. Then the Borel σ-algebra of E may be regarded as an admissible σ-algebra of subsets of E. Indeed, conditions (a) and (b) are trivially fulfilled for $\mathcal{S} = \mathcal{B}(E)$. Condition (c) is valid for $\mathcal{S} = \mathcal{B}(E)$, too. Indeed, in view of the relation

$$\lambda^*(E) > 0,$$

we can easily define a nonzero σ-finite diffused Borel measure on E.

In the sequel, we also need the following lemma.

Lemma 2. *Let E be a set of cardinality κ, let $\mathcal{S}$ be an admissible σ-algebra of subsets of E and let ν be a nonzero σ-finite diffused measure on $\mathcal{S}$. Then, for every set $X \in \mathcal{S}$ with $card(X) < \kappa$, we have $\nu(X) = 0$.*

Proof. We may assume, without loss of generality, that

$$E \subset \mathbf{R}, \quad \lambda^*(E) > 0$$

and that ν is a nonzero diffused finite measure on $\mathcal{S}$. Suppose to the contrary that there exists a ν-measurable set $X \subset E$ with $card(X) < \kappa$, such that $\nu(X) > 0$. We can assume that $\nu(X) = 1$. Obviously, the family of sets

$$\mathcal{S}_X = \{X \cap Y \ : \ Y \in \mathcal{S}\}$$

is an admissible σ-algebra of subsets of X. Denote by $\{X_n \ : \ n < \omega\}$ a countable subfamily of $\mathcal{S}_X$ generating $\mathcal{S}_X$ and separating the points in X. Now, define a mapping

$$\phi \ : \ X \to 2^\omega$$

by the formula

$$\phi(x) = (i_n)_{n<\omega} \qquad (x \in X),$$

where $i_n = 1$ if $x \in X_n$, and $i_n = 0$ if $x \notin X_n$. Note that the mapping ϕ was introduced by Szpilrajn (Marczewski) in [215]. This mapping is usually called the characteristic function of the given family of sets $\{X_n \ : \ n < \omega\}$.

A straightforward verification shows that:

1) ϕ is an injection;

2) ϕ is a ν-measurable mapping from X into the Cantor space $C = 2^\omega$; moreover, we have the equality

$$\{\phi^{-1}(Z) : Z \in \mathcal{B}(2^\omega)\} = \mathcal{S}_X.$$

Therefore, we can define a Borel diffused probability measure μ on the Cantor space, by putting

$$\mu(Z) = \nu(\phi^{-1}(Z)) \qquad (Z \in \mathcal{B}(2^\omega)).$$

Accordingly, we get

$$\mu^*(\phi(X)) = 1 > 0.$$

Now, identifying 2^ω with $[0,1]$ by a Borel isomorphism, we obtain that $\phi(X)$ is not a universal measure zero subset of $[0,1]$. Hence, in view of Lemma 1, there exists a homeomorphic image of $\phi(X)$ lying on $[0,1]$ and having positive outer Lebesgue measure. But the latter fact yields a contradiction with the relation

$$card(\phi(X)) = card(X) < \kappa.$$

This contradiction completes the proof.

Let E be a set. We say that E admits a universal measure if there exists a nonzero σ-finite diffused measure defined on the family of all subsets of E. In this case, we also say that $card(E)$ is a real-valued measurable cardinal (in connection with these definitions, see [8], [50], [165], [211], [222]). We shall touch the problem of the existence of universal measures on uncountable sets in further sections of the book (see, for instance, Chapter 7). It is well known that the existence of real-valued measurable cardinals cannot be established within the theory **ZFC** (see [122], [127] or [222]). These cardinals belong to the class of so-called large cardinal numbers and are extensively studied in axiomatic set theory. Their influence on various properties of the classical Lebesgue measure is frequently unusual and unexpected.

Here we would like to present a remarkable result of Kunen [119] stating that if the cardinal **c** is real-valued measurable, then there exists a non Lebesgue-measurable subset of **R** whose cardinality is relatively small, namely, is strictly less than **c**. We shall give below a strengthened version of this result, due to Grzegorek and Prikry (compare [58] and [175]).

Theorem 2. *Let E be a set with $card(E) = \kappa$. Then there exists a countably generated σ-algebra of subsets of E, separating the points in E and not admitting a nonzero σ-finite diffused measure.*

Proof. We may assume, without loss of generality, that

$$E \subset \mathbf{R}, \quad 0 < \lambda^*(E) < +\infty.$$

As earlier, we equip E with the induced topology and denote by μ the Borel diffused measure on E produced by λ. Evidently,

$$0 < \mu(E) < +\infty.$$

We also fix a countable base $\{V_n : n < \omega\}$ of the topology of E. Let

$$E = \{e_\xi \ : \ \xi < \kappa\}$$

be an injective enumeration of all points of E (where, as usual, κ is identified with the least ordinal number of cardinality κ). In view of the definition of κ, for each ordinal $\alpha < \kappa$, we have

$$\mu^*(\{e_\xi : \xi < \alpha\}) = 0.$$

This implies that there exists an open set $U_\alpha \subset E$ satisfying the relations

$$\{e_\xi : \xi < \alpha\} \subset U_\alpha, \quad \mu(U_\alpha) \leq (1/2)\mu(E).$$

Consider now the product set $\kappa \times E$ and its subset

$$G = \cup\{\{\alpha\} \times U_\alpha : \alpha < \kappa\}.$$

In addition to this, for each $n < \omega$, introduce the set

$$A_n = \{\alpha < \kappa : V_n \subset U_\alpha\}.$$

Let us first verify that the equality

$$G = \cup\{A_n \times V_n : n < \omega\}$$

is true. Indeed, take any pair $(\alpha, e_\beta) \in G$. Then we have $e_\beta \in U_\alpha$. Since the family $\{V_n : n < \omega\}$ forms a base of the space E, there exists a natural number m such that $e_\beta \in V_m \subset U_\alpha$. This implies the relation

$$(\alpha, e_\beta) \in A_m \times V_m.$$

Therefore,

$$(\alpha, e_\beta) \in \cup\{A_n \times V_n : n < \omega\}.$$

We thus come to the inclusion

$$G \subset \cup\{A_n \times V_n : n < \omega\}.$$

In order to show the reverse inclusion, take any pair

$$(\alpha, e_\beta) \in \cup\{A_n \times V_n : n < \omega\}.$$

Then, for some natural number m, we have

$$(\alpha, e_\beta) \in A_m \times V_m,$$

from which it follows that

$$\alpha \in A_m, \quad e_\beta \in V_m, \quad V_m \subset U_\alpha, \quad (\alpha, e_\beta) \in \{\alpha\} \times U_\alpha.$$

Consequently,

$$\cup\{A_n \times V_n : n < \omega\} \subset G.$$

In this way, we get the required equality $G = \cup\{A_n \times V_n : n < \omega\}$. Now, since $\kappa = card(E)$ and E is a subset of $\mathbf{R}$, there exist countably generated σ-algebras of subsets of κ, separating the points of κ. Denote by $\mathcal{S}'$ one of such σ-algebras and let $\mathcal{S}$ be the σ-algebra of subsets of κ, generated by the family

$$\mathcal{S}' \cup \{A_n : n < \omega\}.$$

Then, obviously, $\mathcal{S}$ is also countably generated and separates the points of κ. Our goal is to demonstrate that $\mathcal{S}$ does not admit a nonzero σ-finite diffused measure. Suppose otherwise, and let ν be a probability diffused measure with $dom(\nu) = \mathcal{S}$. Then we may consider the product measure $\nu \times \mu$ on $\kappa \times E$. Evidently,

$$(\nu \times \mu)(\kappa \times E) = \nu(\kappa)\mu(E) = \mu(E).$$

As shown above, the set

$$G = \cup\{\{\alpha\} \times G_\alpha : \alpha < \kappa\}$$

belongs to the domain of $\nu \times \mu$. Now, let us try to apply the Fubini theorem to G. Fix any $e_\beta \in E$ and consider the horizontal section $G^{-1}(e_\beta)$ of G corresponding to e_β. Note that, for all $\alpha \in \,]\beta, \kappa[$, we have

$$e_\beta \in U_\alpha, \quad (\alpha, e_\beta) \in \{\alpha\} \times U_\alpha \subset G, \quad \alpha \in G^{-1}(e_\beta),$$

from which it follows (in view of Lemma 2) that

$$\nu(G^{-1}(e_\beta)) = \nu(\kappa) = 1.$$

In other words, all horizontal sections of G turn out to be of full measure (with respect to ν). Therefore, by the Fubini theorem, the equalities

$$(\nu \times \mu)(G) = (\nu \times \mu)(\kappa \times E) = \mu(E)$$

must be valid. On the other hand, if $\alpha < \kappa$, then the vertical section $G(\alpha)$ of G, corresponding to α, coincides with the set U_α, and, by the definition of U_α, we have

$$\mu(U_\alpha) \leq (1/2)\mu(E).$$

Taking this fact into account and utilizing the Fubini theorem once more, we get

$$(\nu \times \mu)(G) \leq (1/2)\mu(E)\nu(\kappa) = (1/2)\mu(E)$$

and, finally,

$$\mu(E) \leq (1/2)\mu(E)$$

which is impossible because of

$$0 < \mu(E) < +\infty.$$

The contradiction obtained ends the proof.

Let us formulate two immediate consequences of Theorem 2.

The first of them is due to Kunen (see [119]).

Theorem 3. *If all subsets of* **R** *of cardinality strictly less than* **c** *are Lebesgue measurable, then* **c** *is not a real-valued measurable cardinal.*

Proof. Indeed, suppose that all sets $X \subset \mathbf{R}$ with $card(X) < \mathbf{c}$ are Lebesgue measurable (or, equivalently, are of Lebesgue measure zero). Then we must have $\kappa = \mathbf{c}$. In accordance with the preceding theorem, there exists a countably generated σ-algebra of subsets of **R**, separating the points of **R** and not admitting a nonzero σ-finite diffused measure. The existence of such a σ-algebra obviously implies that **c** is not a real-valued measurable cardinal, and the proof is completed.

As known (see Appendix 1), Martin's Axiom implies that all subsets of **R** of cardinality strictly less than **c** are Lebesgue measurable (more precisely, are of Lebesgue measure zero). Taking into account Theorem 3, we readily infer that under this axiom, the cardinality of the continuum is not real-valued measurable.

Some stronger versions of Theorem 3 can be found in [50].

The following result is due to Grzegorek (see [58]).

Theorem 4. *There exists a universal measure zero subspace of* **R** *whose cardinality is equal to* κ.

Proof. The argument utilized in the proof of Theorem 2 yields the existence of a countably generated σ-algebra $\mathcal{S}$ of subsets of κ, which separates the points of κ and does not admit a nonzero σ-finite diffused measure.

Let $\{D_n : n < \omega\}$ be a countable family of subsets of κ, generating $\mathcal{S}$ and separating the points of κ. Further, let

$$\phi : \kappa \to 2^\omega$$

denote the Szpilrajn (Marczewski) characteristic function of the family $\{D_n : n < \omega\}$. As we know (see the proof of Lemma 2), the mapping ϕ is injective and

$$\{\phi^{-1}(Z) : Z \in \mathcal{B}(2^\omega)\} = \mathcal{S}.$$

Consider the set $\phi(\kappa)$. We assert that this set is a universal measure zero subspace of the Cantor space $C = 2^\omega$. Indeed, suppose to the contrary that there exists a Borel diffused probability measure on $\phi(\kappa)$. Then it is clear that there exists a Borel diffused probability measure μ on 2^ω such that

$$\mu^*(\phi(\kappa)) = \mu(2^\omega) = 1 > 0.$$

Now, we put

$$\nu(\phi^{-1}(Z)) = \mu(Z) \qquad (Z \in \mathcal{B}(2^\omega)).$$

It can easily be verified that the functional ν is well defined and turns out to be a diffused probability measure on the σ-algebra $\mathcal{S}$. Since such a measure cannot exist on $\mathcal{S}$, we claim that the set $\phi(\kappa)$ must be a universal measure zero subspace of 2^ω with cardinality κ. But this result immediately leads to the existence of a universal measure zero subspace of **R** with the same cardinality.

The theorem has thus been proved.

Remark 3. Of course, it is impossible to indicate the precise value of κ in the theory **ZFC**. We only can assert that $\kappa \geq \omega_1$, since all countable subsets of **R** are of Lebesgue measure zero. In any case, by applying Theorem

4 we obtain a universal measure zero subspace of $\mathbf{R}$ of cardinality ω_1. Some other classical constructions of such subspaces are presented in [125] and [147]. In a certain sense, this result cannot be strengthened. Indeed, there are models of set theory, in which any universal measure zero subspace of $\mathbf{R}$ has cardinality less than or equal to ω_1 (in this connection, see [147]).

Remark 4. The assumption that $\mathbf{c}$ is real-valued measurable implies a number of interesting consequences in classical Lebesgue measure theory. One of them was indicated in Theorem 3 of Kunen. Let us point out another one due to Jakab and Laczkovich [73]. Namely, they demonstrated in [73] that if the cardinality of the continuum is real-valued measurable, then there exists a countably additive functional μ defined on $dom(\lambda_2)$ such that:

(a) $\mu \neq \lambda_2$;

(b) $\mu(T) = 1$ for every square T in $\mathbf{R}^2$ congruent to $[0,1[^2$.

Indeed, the real-valued measurability of $\mathbf{c}$ implies that there exists a measure ν on $\mathbf{R}$ extending λ, with $dom(\nu) = \mathcal{P}(\mathbf{R})$ (see Exercise 9 of this chapter). By virtue of the Vitali theorem (see Chapter 1), ν cannot be invariant under all translations of $\mathbf{R}$. Hence, there are a set $X \in dom(\nu)$ and an element $h \in \mathbf{R}$ such that

$$\nu(X + h) \neq \nu(X).$$

Now, identifying the real line $\mathbf{R}$ with the subset $\mathbf{R} \times \{0\}$ of the euclidean plane $\mathbf{R}^2$, define a functional μ on $dom(\lambda_2)$ by the formula:

$$\mu(Z) = \lambda_2(Z) + \nu(Z \cap \mathbf{R}) - \nu((Z \cap \mathbf{R}) + h) \qquad (Z \in dom(\lambda_2)).$$

It can easily be verified that μ is countably additive,

$$\mu(X) = \nu(X) - \nu(X + h) \neq 0 = \lambda_2(X)$$

and $\mu(T) = 1$ for all those squares T in the plane $\mathbf{R}^2$ which are congruent to the unit square of this plane.

We thus see that μ is the required functional.

Conversely, the existence of a countably additive functional μ on $dom(\lambda_2)$ with properties (a) and (b) implies the real-valued measurability of $\mathbf{c}$ (for more details, see [73]). Therefore, one can claim that if $\mathbf{c}$ is not real-valued measurable, then every countably additive functional μ on $dom(\lambda_2)$ satisfying condition (b) necessarily coincides with λ_2. This is an interesting geometric version of the uniqueness property for λ_2. An easy induction on

n yields that the same result remains true for the euclidean space $\mathbf{R}^n$ where $n > 2$.

Remark 5. It is not difficult to show (within the theory **ZF** & **DC**) that there are many countably additive functionals μ on $dom(\lambda_1)$ satisfying the following relations:

(a) $\mu \neq \lambda_1$;

(b) $\mu(T) = 1$ for every interval $T \subset \mathbf{R}$ whose length is equal to 1.

Remark 6. Some nontrivial consequences of the real-valued measurability of $\mathbf{c}$ in classical descriptive set theory are discussed in [64] and [65].

EXERCISES

1. Let E_1 and E_2 be two sets such that:

$$card(E_1) \geq \omega_1, \quad card(E_2) \geq \omega_1,$$

$$card(E_1) + card(E_2) \geq \mathbf{c}.$$

Suppose also that there exists a partition $\{A, B\}$ of the product set $E_1 \times E_2$, satisfying the relations:

$$(\forall y \in E_2)(card(A \cap (E_1 \times \{y\})) \leq \omega),$$

$$(\forall x \in E_1)(card(B \cap (\{x\} \times E_2)) \leq \omega).$$

Under these assumptions, prove that:

(a) the Continuum Hypothesis is true;

(b) $card(E_1) = card(E_2) = \mathbf{c}$.

2. Define a function

$$f : [-1, 1]^2 \to \mathbf{R}$$

by putting:

$f(x, y) = xy/(x^2 + y^2)^2$ if $(x, y) \neq (0, 0)$;

$f(x, y) = 0$ if $(x, y) = (0, 0)$.

Demonstrate that:

(a) f is Lebesgue measurable but is not Lebesgue integrable;

(b) the iterated integrals for f do exist and

$$\int_{-1}^{1}(\int_{-1}^{1} f(x,y)dx)dy = \int_{-1}^{1}(\int_{-1}^{1} f(x,y)dy)dx = 0.$$

3. Give a detailed proof of the relation

$$\mathcal{P}(\omega_1 \times \omega_1) = \mathcal{P}(\omega_1) \otimes \mathcal{P}(\omega_1).$$

Note that an argument establishing this equality relies essentially on the Axiom of Choice, because the existence of an embedding of ω_1 into **R** cannot be proved in the theory **ZF** & **DC**. Moreover, as has been shown by Shelah and Raisonnier (see [176] and Chapter 15), the existence of such an embedding implies in **ZF** & **DC** the existence of a subset of **R** nonmeasurable in the Lebesgue sense.

4. Prove statement (iii) of this chapter.

Notice that a similar result within the theory **ZFC**, due to Davies, will be discussed in Chapter 6.

5. Recall that a certain topological analogue of the classical Fubini theorem (the so-called Kuratowski-Ulam theorem) is valid (see, for instance, [125] or [165]). By applying this analogue, prove in the theory **ZF** & **DC** that the assertion

"there exists a well-ordering of **R***"*

implies the assertion

"there exists a subset of **R** *without the Baire property"*.

In other words, Theorem 1 and the result formulated in this exercise show us that the existence of a well-ordering of the real line **R** immediately yields the existence of subsets of **R** having a very bad descriptive structure (from the point of view of Lebesgue measurability and the Baire property).

In this connection, let us indicate that the existence of a totally imperfect subset of **R** of cardinality **c** also implies, in the same theory **ZF** & **DC**, the existence of a non Lebesgue-measurable subset of **R** and the existence of a subset of **R** without the Baire property (compare Theorem 1 from Chapter 12).

6. Let E be a topological space whose all one-element subsets are Borel in E. Consider the family $\mathcal{I}_0(E)$ of all universal measure zero subspaces of E. Demonstrate that:

(a) $\mathcal{I}_0(E)$ has the hereditary property; in other words, if $X \in \mathcal{I}_0(E)$ and $Y \subset X$, then $Y \in \mathcal{I}_0(E)$;

(b) $\mathcal{I}_0(E)$ is countably additive; in other words, if $\{X_n : n < \omega\}$ is an arbitrary countable family of sets from $\mathcal{I}_0(E)$, then the set $\cup\{X_n : n < \omega\}$ also belongs to $\mathcal{I}_0(E)$.

In particular, if E is not universal measure zero, then $\mathcal{I}_0(E)$ forms a σ-ideal of subsets of E.

7. Let E_1 and E_2 be any two universal measure zero spaces. Show that the topological product $E_1 \times E_2$ is a universal measure zero space, too. Therefore, the class of all universal measure zero spaces is closed under the operation of finite products.

Give also an example of a countable family of universal measure zero spaces, whose topological product is not universal measure zero.

8. Let E be a set and let $\mathcal{S}$ be a σ-algebra of subsets of E, separating the points in E. Suppose also that a family of sets $\{X_i : i \in I\} \subset \mathcal{S}$ generates $\mathcal{S}$. Show that the same family $\{X_i : i \in I\}$ separates the points in E.

9. Prove that the following three assertions are equivalent:

(a) the cardinal $\mathbf{c}$ is real-valued measurable;

(b) there exists a measure ν on $\mathbf{R}$ extending λ such that

$$dom(\nu) = \mathcal{P}(\mathbf{R});$$

(c) there exists a measure ν on $[0, 1]$ such that

$$dom(\nu) = \mathcal{P}([0, 1])$$

and ν extends the restriction of λ to $dom(\lambda) \cap \mathcal{P}([0, 1])$.

10. Let λ_n denote the standard n-dimensional Lebesgue measure on the euclidean space $\mathbf{R}^n$. Let X and Y be some λ_n-measurable subsets of $\mathbf{R}^n$. For each vector $z \in \mathbf{R}^n$, put

$$f(z) = \lambda_n((X + z) \cap Y).$$

In this manner, the function

$$f : \mathbf{R}^n \to [0, +\infty]$$

is defined on $\mathbf{R}^n$. If at least one of the relations

$$\lambda_n(X) < +\infty, \qquad \lambda_n(Y) < +\infty$$

is true, then f takes only finite values. Show that, in such a situation, f is necessarily continuous (compare the Steinhaus property for λ_n). Furthermore, by applying the Fubini theorem, prove the equality

$$\int f(z)d\lambda_n(z) = \lambda_n(X)\lambda_n(Y).$$

In connection with this equality, let us remark that it can be generalized to the case where a Haar measure μ on a σ-compact locally compact topological group G is under consideration. Formulate and prove the corresponding generalized result for the completion of μ on G.

Chapter 5
Small nonmeasurable sets

Let $\mathcal{I}$ be a σ-ideal of subsets of the real line $\mathbf{R}$. In many situations, the following question arises and is of certain interest: does there exist a set $X \in \mathcal{I}$ nonmeasurable with respect to the Lebesgue measure λ on $\mathbf{R}$? In other words, we are interested whether there exist small sets (in the sense of $\mathcal{I}$) which are nonmeasurable in the Lebesgue sense. Of course, an analogous question can be posed for the Baire property: does there exist a set $X \in \mathcal{I}$ without the Baire property?

The above-mentioned questions (for various σ-ideals $\mathcal{I}$) will be central in this chapter. Namely, we will focus on those features of $\mathcal{I}$ which guarantee the existence of such subsets of $\mathbf{R}$.

In the previous chapter we were concerned with the following σ-ideal:

$$\mathcal{J} = [\mathbf{R}]^{<\mathbf{c}} = \{X \subset \mathbf{R} : card(X) < \mathbf{c}\}.$$

In particular, it was established that if we assume the real-valued measurability of the cardinal $\mathbf{c}$, then some subsets of $\mathbf{R}$ can be found belonging to $\mathcal{J}$ and nonmeasurable in the Lebesgue sense. Here we wish to discuss analogous situations for other natural σ-ideals of subsets of $\mathbf{R}$.

We begin our consideration with the so-called Sierpiński sets which were constructed by Sierpiński in 1924, under the assumption of the Continuum Hypothesis (see, for instance, [195], [125], [147], [165]).

Let us introduce the notion of a Sierpiński set and establish some interesting properties of them.

Let X be a subset of $\mathbf{R}$. We shall say that X is a Sierpiński set if the following relations are satisfied:

1) X is uncountable;

2) for each Lebesgue measure zero subset Y of $\mathbf{R}$, the intersection $X \cap Y$ is at most countable.

Many facts concerning Sierpiński sets are easily obtained from their definition. For example, we have:

(a) the family of all Sierpiński sets generates the σ–ideal $\mathcal{I}_S$ of subsets of the real line $\mathbf{R}$, invariant under the group of all those transformations of $\mathbf{R}$ which preserve the σ–ideal of all Lebesgue measure zero subsets of $\mathbf{R}$ (in particular, $\mathcal{I}_S$ is invariant with respect to the group of all translations of $\mathbf{R}$);

(b) the assumption

$$(Martin's\ Axiom)\ \&\ (2^\omega > \omega_1)$$

implies that there are no Sierpiński sets on the real line.

Indeed, (a) is almost trivial. To show the validity of (b), assume that Martin's Axiom with the negation of the Continuum Hypothesis hold and suppose that $X \subset \mathbf{R}$ is a Sierpiński set. Take a subset X' of X such that

$$card(X') = \omega_1 < \mathbf{c}.$$

Then, in virtue of Martin's Axiom (see Theorem 2 of Appendix 1), we must have

$$\lambda(X') = 0.$$

On the other hand, since X is a Sierpiński set, the relation

$$\omega_1 = card(X') = card(X \cap X') \leq \omega$$

must be fulfilled, which yields a contradiction.

We also have the following classical theorem due to Sierpiński.

Theorem 1. *Assume the Continuum Hypothesis. Then there exist Sierpiński subsets of the real line $\mathbf{R}$.*

Proof. According to our assumption $\mathbf{c} = \omega_1$, we can denote by $(X_\xi)_{\xi<\omega_1}$ the family of all λ–measure zero Borel subsets of $\mathbf{R}$. Applying the method of transfinite recursion, let us define an injective family of points

$$\{x_\xi : \xi < \omega_1\} \subset \mathbf{R}.$$

Suppose that, for an ordinal $\zeta < \omega_1$, the partial family of points $\{x_\xi : \xi < \zeta\}$ has already been defined. Consider the set

$$X'_\zeta = (\cup\{X_\xi : \xi < \zeta\}) \cup \{x_\xi : \xi < \zeta\}.$$

Obviously, this set is of Lebesgue measure zero. Consequently, there exists a point $x \in \mathbf{R} \setminus X'_\zeta$. We put $x_\zeta = x$. Proceeding in this manner, we are able to construct the family $\{x_\xi : \xi < \omega_1\}$. In view of our construction, the following relations are true:

1) $\{x_\xi : \xi < \omega_1\}$ is an injective family;

2) for any two ordinals $\alpha < \omega_1$ and $\beta < \omega_1$, such that $\alpha < \beta$, we have $x_\beta \notin X_\alpha$.

Let us define

$$X = \{x_\xi : \xi < \omega_1\}$$

and let us demonstrate that X is a Sierpiński subset of $\mathbf{R}$. Indeed, relation 1) implies

$$card(X) = \omega_1 > \omega.$$

Thus, X is an uncountable set. Relation 2) shows that, for each ordinal $\alpha < \omega_1$, we have

$$card(X \cap X_\alpha) \leq card(\{x_\xi : \xi \leq \alpha\}) \leq \omega,$$

from which it follows that, for any set $Y \subset \mathbf{R}$ of Lebesgue measure zero, the set $X \cap Y$ is at most countable (since Y is contained in some X_α). This completes the proof.

Remark 1. In connection with Theorem 1, it should be noted that the existence of Sierpiński subsets of $\mathbf{R}$ is possible in some models of set theory, where the Continuum Hypothesis does not hold. More precisely, there are models of **ZFC** in which the negation of the Continuum Hypothesis is valid and there exist Sierpiński sets of cardinality $\mathbf{c}$ (see [121]). Evidently, in such models we also have non Lebesgue-measurable subsets of $\mathbf{R}$ with cardinality $\omega_1 < \mathbf{c}$.

By applying a dual argument (to the σ-ideal of all first category subsets of $\mathbf{R}$), the following classical result of Luzin [134] can be established.

Theorem 2. *Assume that the Continuum Hypothesis holds. Then there exists a subset X of $\mathbf{R}$ such that:*

1) X is uncountable;

2) for any first category set $Y \subset \mathbf{R}$, we have

$$card(X \cap Y) \leq \omega.$$

We leave to the reader a detailed proof of Theorem 2. A set X mentioned in this theorem is usually called a Luzin set. It can easily be observed that the family of all Luzin sets generates the σ-ideal $\mathcal{I}_L$ of subsets of $\mathbf{R}$, which is invariant with respect to all transformations of $\mathbf{R}$ preserving the σ-ideal $\mathcal{K}(\mathbf{R})$ of all first category sets in $\mathbf{R}$. Therefore, Luzin sets are small in the sense of $\mathcal{I}_L$, and it can easily be shown that these sets are also of Lebesgue measure zero. A much stronger result is formulated in Exercise 20 of this chapter.

Remark 2. There is a general statement due to Sierpiński and Erdös which establishes that under certain additional set-theoretical hypotheses (in particular, under the Continuum Hypothesis or Martin's Axiom), the σ-ideal of all first category subsets of $\mathbf{R}$ is isomorphic to the σ-ideal of all Lebesgue measure zero subsets of $\mathbf{R}$. An isomorphism between these two classical σ-ideals is purely set-theoretical and does not have good properties. However, the existence of such an isomorphism enables us to obtain automatically many theorems for the Lebesgue measure by starting with the corresponding theorems for the Baire category (and conversely). In particular, applying this general statement, we can easily deduce Theorem 2 from Theorem 1 (and, conversely, Theorem 1 from Theorem 2).

A detailed proof of the Sierpiński-Erdös statement mentioned above (the so-called Sierpiński-Erdös Duality Principle) is given in the well-known textbook [165] where numerous applications of this principle are presented as well. Some more general version of the Duality Principle is formulated and proved in [19].

The next easy result describes several properties of Sierpiński sets.

Theorem 3. *Every Sierpiński set is of first category in* $\mathbf{R}$. *No uncountable subset of a Sierpiński set is Lebesgue measurable.*

Proof. Let X be a Sierpiński set. Let $\mathcal{I}(\lambda)$ denote, as usual, the σ-ideal of all Lebesgue measure zero subsets of $\mathbf{R}$. As known (see [165] or Exercise 7 of this chapter), the σ–ideals $\mathcal{K}(\mathbf{R})$ and $\mathcal{I}(\lambda)$ are orthogonal; in other words, there exists a partition $\{A, B\}$ of $\mathbf{R}$ such that

$$A \in \mathcal{K}(\mathbf{R}), \quad B \in \mathcal{I}(\lambda).$$

Simultaneously, we have the inequality

$$card(X \cap B) \leq \omega$$

and the inclusion

$$X \subset A \cup (X \cap B),$$

from which we immediately obtain that $X \in \mathcal{K}(\mathbf{R})$.

Suppose now that Y is an uncountable subset of a Sierpiński set X (hence Y is a Sierpiński set, as well). Since $X \cap Y$ is uncountable, we observe that $Y \notin \mathcal{I}(\lambda)$. Suppose to the contrary that Y is Lebesgue measurable. Then the inequality $\lambda(Y) > 0$ must be valid, and we can find an uncountable set $Z \subset Y$ of Lebesgue measure zero (cf. Exercise 1 of this chapter). Since the set

$$X \cap Z = Z$$

is uncountable, we get a contradiction with the definition of the Sierpiński set X. The contradiction obtained ends the proof of Theorem 3.

It is important to note that if we replace the Continuum Hypothesis by Martin's Axiom (which is a much weaker assertion than **CH**), then we can prove the existence (in $\mathbf{R}$) of some analogues of Sierpiński and Luzin sets. Namely, if Martin's Axiom holds, then there exists a set $X \subset \mathbf{R}$ such that:

1) $card(X) = \mathbf{c}$;
2) for each set $Y \in \mathcal{I}(\lambda)$, we have

$$card(Y \cap X) < \mathbf{c}.$$

A set X with the above properties is called a generalized Sierpiński subset of the real line.

Similarly, if Martin's Axiom holds, then there exists a set $X \subset \mathbf{R}$ such that:

(1) $card(X) = \mathbf{c}$;
(2) for each set $Z \in \mathcal{K}(\mathbf{R})$, we have

$$card(Z \cap X) < \mathbf{c}.$$

A set X with the above properties is usually called a generalized Luzin subset of the real line.

Let us remark that for the existence of generalized Sierpiński sets or generalized Luzin sets, we do not need the full power of Martin's Axiom. In fact, the existence of generalized Sierpiński and Luzin sets is implied by some additional set-theoretical assumptions which are essentially weaker than Martin's Axiom. In this connection, see Exercise 8 where the corresponding result is formulated even for abstract σ-ideals of sets.

Some interesting facts about generalized Sierpiński sets and generalized Luzin sets are presented in Exercises 9 - 13 of this chapter.

Let us return to nonmeasurable sets and functions and let us give one application of a generalized Luzin set to the construction of a function (acting from $\mathbf{R}$ into $\mathbf{R}$) which has extremely bad properties from the point of view of measure theory. First, we must formulate the corresponding definition.

Let E be a set and let f be a function acting from E into $\mathbf{R}$. We shall say that f is absolutely nonmeasurable if it is nonmeasurable with respect to every nonzero σ-finite diffused measure μ on E.

Let us underline that in this definition, the domain of μ is not a fixed σ-algebra of subsets of E (apriori, $dom(\mu)$ may be an arbitrary σ-algebra of subsets of E, containing all singletons in E).

Recall that in Chapter 2 a function $f : \mathbf{R} \to \mathbf{R}$ was constructed with the λ_2-thick graph in $\mathbf{R}^2$. Clearly, such an f is nonmeasurable in the Lebesgue sense, but we cannot assert that f is absolutely nonmeasurable. Moreover, it turns out that f is measurable with respect to an appropriate extension of the Lebesgue measure λ on $\mathbf{R}$ (see Exercise 22).

However, we have the following statement.

Theorem 4. *Suppose that Martin's Axiom holds. Then there exists an injective function*

$$f : \mathbf{R} \to \mathbf{R}$$

which simultaneously is absolutely nonmeasurable.

Proof. We know that Martin's Axiom implies the existence of a generalized Luzin subset of $\mathbf{R}$. Let X be such a subset. Since we have

$$card(X) = \mathbf{c} = card(\mathbf{R}),$$

there exists a bijection

$$f : \mathbf{R} \to X.$$

Obviously, we can consider f as an injection from $\mathbf{R}$ into itself. Let us verify that f is the required function. Suppose, for a moment, that our f is not absolutely nonmeasurable. Then there exists a nonzero σ-finite diffused measure μ on $\mathbf{R}$ such that f is μ-measurable; in other words, for any Borel subset B of $\mathbf{R}$, the relation

$$f^{-1}(B) \in dom(\mu)$$

is satisfied. Equivalently, for any Borel subset B' of X, we have

$$f^{-1}(B') \in dom(\mu).$$

Without loss of generality, we may assume that our μ is a probability measure. Now, for each Borel subset B' of X, we put

$$\nu(B') = \mu(f^{-1}(B')).$$

In this manner, we obtain a Borel diffused probability measure ν on X, which is impossible since X is a universal measure zero space (see Exercise 20 of this chapter). The contradiction ends the proof of Theorem 4.

A much stronger result is presented in Exercise 10.

Remark 3. The preceding theorem was formulated and proved under the assumption that Martin's Axiom is valid. In this connection, it is reasonable to mention here that the existence of an absolutely nonmeasurable function (acting from $\mathbf{R}$ into $\mathbf{R}$) cannot be established within the theory $\mathbf{ZFC}$. Indeed, if the cardinality of the continuum is real-valued measurable, then such functions do not exist.

At the same time, one can demonstrate (in $\mathbf{ZFC}$) that there exists an injective absolutely nonmeasurable function

$$f \; : \; \omega_1 \to \mathbf{R}.$$

In order to show this fact, it suffices to pick a universal measure zero subspace X of $\mathbf{R}$ with $card(X) = \omega_1$ and then to take as f any bijection acting from ω_1 onto X. Notice that the question of the existence of a universal measure zero space $X \subset \mathbf{R}$ of cardinality ω_1 was discussed in Chapter 4 where it was also established that the required X does always exist. This circumstance immediately follows from a much stronger result presented in the same chapter.

Let E be a topological space and let X be a subset of E. We recall (see Chapter 2) that X is totally imperfect in E if X contains no nonempty perfect subset of E. An interesting class of totally imperfect subsets of $\mathbf{R}$ was introduced and investigated by Szpilrajn (Marczewski) (see [213]). We are going to discuss here some properties of these sets. In our further considerations, we will call them Marczewski sets.

Let E be a Polish topological space and let $X \subset E$. We say that X is a Marczewski subset of E if for each nonempty perfect set $P \subset E$, there exists a nonempty perfect set $P' \subset E$ such that

$$P' \subset P, \qquad P' \cap X = \emptyset.$$

It immediately follows from this definition that every Marczewski set is totally imperfect in E and that any subset of a Marczewski set is a Marczewski set, too. Also, it can easily be observed that any set $Y \subset E$ with $card(Y) < \mathbf{c}$ is a Marczewski set. Indeed, let us take an arbitrary nonempty perfect set $P \subset E$. Then, as known (compare Exercise 1 of Chapter 2), there exists a disjoint family $\{P_i \; : \; i \in I\}$ consisting of nonempty perfect sets and satisfying the relations

$$card(I) = \mathbf{c}, \qquad (\forall i \in I)(P_i \subset P).$$

Now, since $card(Y) < card(I)$, it is clear that there exists at least one index $i_0 \in I$ such that $P_{i_0} \cap Y = \emptyset$, and thus Y is a Marczewski set.

Let us recall the classical result of Alexandrov and Hausdorff stating that every uncountable Borel set in a Polish topological space contains a subset homeomorphic to the Cantor discontinuum, hence it contains a nonempty perfect subset (see [125]). Taking this result into account, we can give another equivalent definition of Marczewski sets. Namely, we may say that a set X lying in a Polish space E is a Marczewski set if for each uncountable Borel subset B of E, there exists an uncountable Borel set $B' \subset E$ satisfying the relations

$$B' \subset B, \qquad B' \cap X = \emptyset.$$

In some situations, the second definition is more convenient. For instance, let E_1 and E_2 be two Polish spaces and let

$$f \; : \; E_1 \to E_2$$

be a Borel isomorphism between them. Then, for a set $X \subset E_1$, the following two assertions are equivalent:

1) X is a Marczewski set in E_1;

2) $f(X)$ is a Marczewski set in E_2.

In other words, the Borel isomorphism f yields a one-to-one correspondence between Marczewski sets in the given spaces E_1 and E_2. This fact is rather useful. For instance, suppose that we need to construct a Marczewski subset of E_1 having some additional properties which are invariant under Borel isomorphisms. Sometimes it turns out that such a set can easily be constructed in E_2. Let us denote it by X'. Then we apply the Borel isomorphism f^{-1} to X' and obtain the required Marczewski set $f^{-1}(X')$ in the space E_1.

Later, we shall demonstrate the usefulness of this method, showing that there exist Marczewski subsets of $\mathbf{R}$ nonmeasurable in the Lebesgue sense (respectively, lacking the Baire property).

One simple (but important) fact concerning Marczewski sets is presented in the following auxiliary statement.

Lemma 1. *Let $\{X_k : k < \omega\}$ be a countable family of Marczewski subsets of a Polish space E. Then $\cup\{X_k : k < \omega\}$ is a Marczewski set, too. In particular, if the space E is uncountable, then the family of all Marczewski subsets of E forms a σ-ideal in the Boolean algebra of all subsets of E.*

Proof. Fix a nonempty perfect set $P \subset E$. Since X_0 is a Marczewski set, there exists a nonempty perfect set $P_0 \subset E$ such that

$$P_0 \subset P, \quad P_0 \cap X_0 = \emptyset, \quad diam(P_0) < 1,$$

where $diam(P_0)$ stands for the diameter of P_0. Further, since X_1 is also a Marczewski set, there exist nonempty perfect sets $P_{00} \subset E$ and $P_{01} \subset E$ such that

$$P_{00} \subset P_0, \qquad P_{01} \subset P_0,$$

$$diam(P_{00}) < 1/2, \qquad diam(P_{01}) < 1/2,$$

$$P_{00} \cap X_1 = \emptyset, \quad P_{01} \cap X_1 = \emptyset, \quad P_{00} \cap P_{01} = \emptyset.$$

Continuing in this manner, we will be able to define a dyadic system

$$\{P_{j_1\ldots j_k} : j_1 = 0, \ j_2 \in \{0,1\}, \ldots, j_k \in \{0,1\}, \ 1 \leq k < \omega\}$$

of nonempty perfect sets in E whose diameters converge to zero, and

$$P_{j_1\ldots j_k} \cap X_{k-1} = \emptyset$$

for each natural number $k \geq 1$. Now, putting

$$D_k = \cup\{P_{j_1\ldots j_k} : j_1 = 0, j_2 \in \{0,1\}, \ldots, j_k \in \{0,1\}\},$$

$$D = \cap\{D_k : 1 \leq k < \omega\},$$

we obtain a nonempty perfect set $D \subset E$ satisfying the relation

$$D \cap (\cup\{X_k \; : \; k < \omega\}) = \emptyset.$$

This shows that $\cup\{X_k \; : \; k < \omega\}$ is a Marczewski subset of E, and the proof is completed.

We thus see that in an uncountable Polish topological space E, the family of all Marczewski subsets of E forms a σ-ideal. It is usually called the Marczewski σ-ideal in E and plays an essential role in classical point set theory (cf. [15], [155]). Therefore, Marczewski subsets of E can be regarded as a certain type of small sets in E.

In point set theory we frequently deal with different types of small sets which generate proper σ-ideals in an original space E. For instance, we have already mentioned the σ-ideal $\mathcal{I}_S$ generated by all Sierpiński subsets of $\mathbf{R}$ (respectively, the σ-ideal $\mathcal{I}_L$ generated by all Luzin subsets of $\mathbf{R}$). Besides, in Chapter 4 we have considered the σ-ideal of all universal measure zero subsets of $\mathbf{R}$. Various properties of these subsets and deep connections between them are also discussed in subsequent chapters of the book. Note that valuable information about different kinds of small sets can be found in [125], [147] and [172].

In general, if we deal with some class of subsets of $\mathbf{R}$ which are small in a certain sense, then it is not easy to establish the existence of a set belonging to this class and nonmeasurable in the Lebesgue sense (or without the Baire property). For establishing the existence of small nonmeasurable sets, an additional nontrivial argument is usually needed.

For example, let us return to the two classical σ-ideals:

$\mathcal{I}(\lambda)$ = the σ-ideal of all λ-measure zero subsets of $\mathbf{R}$;

$\mathcal{K}(\mathbf{R})$ = the σ-ideal of all first category subsets of $\mathbf{R}$.

As pointed out earlier, these two σ-ideals are orthogonal: there exists a partition $\{A, B\}$ of $\mathbf{R}$ such that

$$A \in \mathcal{I}(\lambda), \quad B \in \mathcal{K}(\mathbf{R}).$$

Having the partition $\{A, B\}$, we immediately obtain the existence of a non Lebesgue-measurable set belonging to $\mathcal{K}(\mathbf{R})$ and the existence of a Lebesgue measure zero set without the Baire property. Indeed, let X be a Bernstein subset of $\mathbf{R}$. We put

$$X_0 = A \cap X, \quad X_1 = B \cap X.$$

Then it is easy to check that:

1) $X_0 \in \mathcal{I}(\lambda)$ and X_0 does not possess the Baire property;

2) $X_1 \in \mathcal{K}(\mathbf{R})$ and X_1 is not measurable in the Lebesgue sense.

Let us return to the question formulated in the beginning of the chapter. We would like to consider this question more thoroughly for the Marczewski σ-ideal $\mathcal{I}_M$ on $\mathbf{R}$. In other words, it is natural to ask whether there exist Marczewski subsets of $\mathbf{R}$ nonmeasurable in the Lebesgue sense (or without the Baire property). This problem was originally raised by Marczewski many years ago. A solution of the problem was independently obtained by Corazza [26] and Walsh [227]. Here we present a detailed proof of their result.

First of all, we need one auxiliary proposition which is helpful in many analogous situations. The proof of this proposition is similar to the argument used in the proof of Theorem 4 from Chapter 2.

Lemma 2. *Let $\{Z_j \ : \ j \in J\}$ be a family of subsets of the plane $\mathbf{R}^2$, such that:*

1) $card(J) \leq \mathbf{c}$;

2) for each index $j \in J$, the set of all $x \in pr_1(Z_j)$ satisfying the relation

$$card(Z_j(x)) = \mathbf{c}$$

is of cardinality $\mathbf{c}$.

Then there exist a set-valued mapping

$$F \ : \ J \to \mathcal{P}(\mathbf{R})$$

and an injective family of points $\{x_j \ : \ j \in J\} \subset \mathbf{R}$ such that, for any index $j \in J$, we have the equalities

$$F(j) = Z_j(x_j), \qquad card(F(j)) = \mathbf{c}.$$

Proof. We may assume without loss of generality that $card(J)$ is equal to $\mathbf{c}$. Also, we can identify the set J with the least ordinal number α for which $card(\alpha) = \mathbf{c}$. Now, we are going to define a set-valued mapping F and a family $\{x_\xi \ : \ \xi < \alpha\}$ by using the method of transfinite recursion. Suppose that, for an ordinal $\beta < \alpha$, the partial families

$$\{F(\xi) \ : \ \xi < \beta\}, \qquad \{x_\xi \ : \ \xi < \beta\}$$

have already been constructed. Consider the set Z_β. According to our assumption, the set of all those points $x \in pr_1(Z_\beta)$ for which

$$card(Z_\beta(x)) = \mathbf{c}$$

is of cardinality continuum. Since

$$card(\{x_\xi \ : \ \xi < \beta\}) \leq card(\beta) < \mathbf{c},$$

there exists a point $x \in \mathbf{R}$ such that

$$(\forall \xi < \beta)(x \neq x_\xi), \qquad card(Z_\beta(x)) = \mathbf{c}.$$

Therefore, we can put

$$F(\beta) = Z_\beta(x), \qquad x_\beta = x.$$

In this way, we are able to define F and $\{x_\xi \ : \ \xi < \alpha\}$ with the required properties. The lemma has thus been proved.

By utilizing the previous lemma, it is not difficult to establish the following statement.

Theorem 5. *There exists a Marczewski subset of* $\mathbf{R}^2$ *nonmeasurable in the Lebesgue sense and lacking the Baire property.*

Proof. Let α be again the least ordinal number with $card(\alpha) = \mathbf{c}$ and let $\{Z_\xi \ : \ \xi < \alpha\}$ denote the family of all those Borel subsets of $\mathbf{R}^2$ which have strictly positive λ_2-measure (where λ_2 stands, as usual, for the classical two-dimensional Lebesgue measure on $\mathbf{R}^2$). Applying Lemma 2 to $\{Z_\xi \ : \ \xi < \alpha\}$, we can find a set-valued mapping F and an injective family $\{x_\xi \ : \ \xi < \alpha\}$ of points of $\mathbf{R}$ with the corresponding properties. Let now $\{P_\xi \ : \ \xi < \alpha\}$ be the family of all nonempty perfect subsets of $\mathbf{R}^2$. For each $\beta < \alpha$, we put

$$y_\beta \in F(\beta) \setminus \cup\{P_\xi(x_\beta) \ : \ \xi \leq \beta, \ card(P_\xi(x_\beta)) \leq \omega\}.$$

Note that y_β can be defined because of the relations

$$card(F(\beta)) = \mathbf{c},$$

$$card(\cup\{P_\xi(x_\beta) \ : \ \xi \leq \beta, card(P_\xi(x_\beta)) \leq \omega\}) \leq card(\beta) + \omega < \mathbf{c}.$$

Let us check that

$$D_0 = \{(x_\xi, y_\xi) \ : \ \xi < \alpha\}$$

is a Marczewski set nonmeasurable with respect to λ_2. Indeed, D_0 can be regarded as the graph of a partial function acting from $\mathbf{R}$ into $\mathbf{R}$ and the construction of D_0 shows that D_0 is a λ_2-thick subset of $\mathbf{R}^2$. Consequently (compare the proof of Theorem 4 from Chapter 2), we may assert that D_0 is nonmeasurable in the Lebesgue sense. It remains to demonstrate that D_0 is a Marczewski set. To do so, take an arbitrary nonempty perfect subset P of $\mathbf{R}^2$. We must verify that P contains a nonempty perfect set whose intersection with D_0 is empty. If there exists at least one point $x \in \mathbf{R}$ for which

$$card(P(x)) = \mathbf{c},$$

then there is nothing to prove. Suppose now that

$$(\forall x \in \mathbf{R})(card(P(x)) \leq \omega).$$

For some $\beta < \alpha$, we have $P_\beta = P$. Taking into account the definition of points y_ξ $(\xi < \alpha)$, we get

$$card(D_0 \cap P_\beta) \leq card(\beta) + \omega < \mathbf{c}.$$

This last relation readily implies that there exists a nonempty perfect subset of $P = P_\beta$ which does not have common points with D_0.

In a similar way, starting with the family of all those Borel subsets of the plane $\mathbf{R}^2$ which do not belong to the σ-ideal $\mathcal{K}(\mathbf{R}^2)$, we can construct a Marczewski set $D_1 \subset \mathbf{R}^2$ thick in the Baire category sense. Then it is not hard to see that

$$D = D_0 \cup D_1$$

is a Marczewski set in $\mathbf{R}^2$ nonmeasurable in the Lebesgue sense and without the Baire property. This finishes the proof of Theorem 5.

One can easily infer from Theorem 5 the existence of Marczewski subsets of $\mathbf{R}$ nonmeasurable in the Lebesgue sense (respectively, without the Baire property). Indeed, consider a Borel isomorphism

$$\phi \ : \ \mathbf{R} \to \mathbf{R}^2.$$

It is a well-known fact (see, for instance, [19]) that ϕ can be chosen in such a way that, simultaneously, ϕ will be an isomorphism between the measures λ

and λ_2. Now, we claim that if X is a λ_2-nonmeasurable Marczewski subset of $\mathbf{R}^2$, then $\phi^{-1}(X)$ is a λ-nonmeasurable Marczewski subset of $\mathbf{R}$.

Analogously, a Borel isomorphism ϕ can be chosen in such a way that the Baire category of sets (consequently, the Baire property of sets) will be preserved by ϕ. Now, we claim that if X is a Marczewski subset of $\mathbf{R}^2$ without the Baire property, then $\phi^{-1}(X)$ is a Marczewski subset of $\mathbf{R}$ without the Baire property.

A much stronger result is contained in Exercise 16.

One more example of a non Lebesgue-measurable function acting from $\mathbf{R}$ into $\mathbf{R}$ and lacking the Baire property, can be obtained by using the corresponding results of Chapter 3. Namely, let us recall the classical theorem of Sierpiński and Zygmund, stating that there exists a function

$$f : \mathbf{R} \to \mathbf{R}$$

satisfying the following condition: for each subset X of $\mathbf{R}$ with $card(X) = \mathbf{c}$, the restriction $f|X$ is not continuous on X (see Exercise 12 of Chapter 3).

By starting with this condition, it is not difficult to show that f is not Lebesgue measurable and does not possess the Baire property.

Indeed, suppose for a moment that f has the Baire property. Then, according to a well-known theorem of general topology (see [125]), we can find a dense G_δ-subset A of $\mathbf{R}$ such that $f|A$ is continuous. Obviously,

$$card(A) = \mathbf{c},$$

and we obtain a contradiction with the fact that f is a Sierpiński-Zygmund function. Thus, f does not possess the Baire property.

Now, let us demonstrate that f is not measurable in the Lebesgue sense. We will prove a more general result asserting that f is not measurable with respect to the completion of any nonzero σ-finite diffused Borel measure on $\mathbf{R}$. Let μ be such a measure and let μ' denote the completion of μ. Suppose for a while that f is μ'-measurable. Since an analogue of the classical Luzin theorem holds true for μ', we can find a closed subset B of $\mathbf{R}$ with $\mu'(B) > 0$ such that the restricted function $f|B$ is continuous. Taking into account the diffusedness of μ' and the inequality $\mu'(B) > 0$, we infer at once that B is uncountable and hence $card(B) = \mathbf{c}$. This yields again a contradiction with the fact that f is a Sierpiński-Zygmund function. Thus, f cannot be measurable with respect to μ'.

EXERCISES

1. Let X be an uncountable λ-measurable subset of $\mathbf{R}$. Demonstrate that there exists a set Y satisfying the following conditions:

(a) $Y \subset X$ and $card(Y) \geq \omega_1$;

(b) Y is of λ-measure zero.

Formulate and prove an analogous result for Polish topological spaces and the completions of σ-finite Borel measures on these spaces.

2. Let $\mathcal{T}_d$ denote the density topology on $\mathbf{R}$ (see [165], [219] or Exercise 12 of Chapter 1). Show that a set $Z \subset \mathbf{R}$ is a Sierpiński subset of $\mathbf{R}$ if and only if Z is a Luzin set in the space $(\mathbf{R}, \mathcal{T}_d)$. This means that Z is uncountable and for every first category set Y in $(\mathbf{R}, \mathcal{T}_d)$, the intersection $Z \cap Y$ is at most countable.

3. Let X be a Sierpiński subset of $\mathbf{R}$ considered as a topological space with respect to the induced topology. Show that any Borel subset of X is simultaneously an F_σ-set and a G_δ-set in X. In particular, each countable subset of X is a G_δ-set in X.

4. Let X be a Sierpiński subset of $\mathbf{R}$. As mentioned above, all countable subsets of X are G_δ-sets in X. Applying this fact, demonstrate that for any nonempty perfect set $P \subset \mathbf{R}$, the set $X \cap P$ is of first category in P. This result strengthenes the corresponding part of Theorem 3.

5. Let X be a Sierpiński set on $\mathbf{R}$. Equip X with the topology induced by the density topology $\mathcal{T}_d$ of $\mathbf{R}$. Prove that the topological space X is non-separable and hereditarily Lindelöf. The latter means that each subspace of X is Lindelöf; in other words, any open covering of a subspace contains a countable subcovering.

6. Assume that the Continuum Hypothesis holds. Let X be a Sierpiński set on $\mathbf{R}$. Equip X with the topology induced by the standard topology of $\mathbf{R}$. Prove that

$$\mathcal{A}(X) = \mathcal{B}(X),$$

where $\mathcal{A}(X)$ denotes the class of all analytic subsets of X and $\mathcal{B}(X)$ denotes, as usual, the class of all Borel subsets of X.

In connection with Exercise 6, let us remark that a more general result can be obtained for so-called Sierpiński topological spaces (compare Exercise 12 below).

7. Two ideals $\mathcal{I}_1$ and $\mathcal{I}_2$ of subsets of a set E are called orthogonal if there exist sets $A \in \mathcal{I}_1$ and $B \in \mathcal{I}_2$ such that $A \cup B = E$. Obviously, in this definition we may additionally assume that $A \cap B = \emptyset$. Check that the classical σ-ideals $\mathcal{I}(\lambda)$ and $\mathcal{K}(\mathbf{R})$ are orthogonal.

In this context, let us recall a more general fact stating that if E is a metric space of second category whose topological weight is not real-valued measurable and μ is a nonzero σ-finite diffused Borel measure on E, then the σ-ideals $\mathcal{I}(\mu)$ and $\mathcal{K}(E)$ are orthogonal (see, for instance, [165]).

Let $\mathcal{J}_1$ and $\mathcal{J}_2$ be two orthogonal σ-ideals of subsets of $\mathbf{R}$, each of which is invariant with respect to the group of all translations of $\mathbf{R}$. Let A_1 and A_2 be two subsets of $\mathbf{R}$ satisfying the relations

$$A_1 \notin \mathcal{J}_1, \quad A_2 \notin \mathcal{J}_2.$$

Demonstrate that:

(1) there exists a set $B_1 \in \mathcal{J}_1$ for which we have

$$B_1 - A_2 = \cup\{B_1 - a \; : \; a \in A_2\} = \mathbf{R};$$

(2) there exists a set $B_2 \in \mathcal{J}_2$ for which we have

$$B_2 - A_1 = \cup\{B_2 - a \; : \; a \in A_1\} = \mathbf{R}.$$

Further, put:

$\mathcal{J}_1 = $ the σ-ideal of all first category subsets of $\mathbf{R}$;

$\mathcal{J}_2 = $ the σ-ideal of all Lebesgue measure zero subsets of $\mathbf{R}$.

Deduce from (1) and (2) that if X is an arbitrary Luzin set on $\mathbf{R}$ and Y is an arbitrary Sierpiński set on $\mathbf{R}$, then the equalities

$$card(X) = card(Y) = \omega_1$$

are true. We thus conclude that the simultaneous existence in $\mathbf{R}$ of Luzin and Sierpiński sets immediately implies that the cardinality of these sets is as minimal as possible (in fact, it is equal to the least uncountable cardinal). This result was first obtained by Rothberger (see [183]).

8. Let E be a set and let $\mathcal{J}$ be a σ-ideal of subsets of E, containing all singletons in E. We denote:

$cov(\mathcal{J}) = $ the smallest cardinality of a covering of E by sets belonging to $\mathcal{J}$;

$cof(\mathcal{J}) = $ the smallest cardinality of a base of $\mathcal{J}$.

Prove that if the equalities

$$cov(\mathcal{J}) = cof(\mathcal{J}) = card(E)$$

are fulfilled, then there exists a subset D of E such that

$$card(D) = card(E)$$

and, for any set $Z \in \mathcal{J}$, we have

$$card(Z \cap D) < card(E).$$

In particular, if our original set E coincides with the real line $\mathbf{R}$ and $\mathcal{J}$ is the σ-ideal of all first category subsets of $\mathbf{R}$ (respectively, the σ-ideal of all Lebesgue measure zero subsets of $\mathbf{R}$), then we obtain, under Martin's Axiom, the existence of a generalized Luzin subset of $\mathbf{R}$ (respectively, the existence of a generalized Sierpiński subset of $\mathbf{R}$).

Some nontrivial facts about generalized Luzin and Sierpiński sets are presented in the next three exercises.

9. Assume that the Continuum Hypothesis holds. Prove that there exists a set $X \subset \mathbf{R}$ satisfying the following conditions:

(a) X is a vector space over the field $\mathbf{Q}$;

(b) X is an everywhere dense Luzin subset of $\mathbf{R}$.

Show also that there exists a set $Y \subset \mathbf{R}$ satisfying the following conditions:

(c) Y is a vector space over the field $\mathbf{Q}$;

(d) Y is an everywhere dense Sierpiński subset of $\mathbf{R}$.

Moreover, by assuming Martin's Axiom, formulate and prove analogous results for generalized Luzin sets and for generalized Sierpiński sets.

In addition, infer from these results (assuming Martin's Axiom again) that there exist an isomorphism f of the additive group $\mathbf{R}$ onto itself and a generalized Luzin set X in $\mathbf{R}$ such that $f(X)$ is a generalized Sierpiński set in $\mathbf{R}$.

10. Let E be a set and let $f : E \to \mathbf{R}$ be a function. Show that the following two assertions are equivalent:

(a) f is absolutely nonmeasurable;

(b) $ran(f)$ is a universal measure zero subset of $\mathbf{R}$ and $card(f^{-1}(t)) \leq \omega$ for all $t \in \mathbf{R}$.

In particular, if $card(E) > \mathbf{c}$, then no function acting from E into $\mathbf{R}$ is absolutely nonmeasurable.

Suppose that Martin's Axiom holds. Starting with a generalized Luzin set which is a vector space over $\mathbf{Q}$ (see the previous exercise), demonstrate that there exists an injective group homomorphism from $\mathbf{R}$ into $\mathbf{R}$ which is an absolutely nonmeasurable function. Conclude from this result that there are absolutely nonmeasurable solutions of the classical Cauchy functional equation (see Chapter 3).

11. Assume Martin's Axiom. By applying a generalized Luzin set which is a vector space over $\mathbf{Q}$ (see Exercise 9), prove that there exist two σ-algebras $\mathcal{S}_1$ and $\mathcal{S}_2$ of subsets of $\mathbf{R}$, satisfying the following conditions:

1) $\mathcal{B}(\mathbf{R}) \subset \mathcal{S}_1 \cap \mathcal{S}_2$;

2) both $\mathcal{S}_1$ and $\mathcal{S}_2$ are countably generated σ-algebras;

3) there exists a measure μ_1 on $\mathcal{S}_1$ extending the standard Borel measure on $\mathbf{R}$ and invariant under all isometric transformations of $\mathbf{R}$;

4) there exists a measure μ_2 on $\mathcal{S}_2$ extending the standard Borel measure on $\mathbf{R}$ and invariant under all isometric transformations of $\mathbf{R}$;

5) there is no nonzero σ-finite diffused measure defined on the σ-algebra generated by $\mathcal{S}_1 \cup \mathcal{S}_2$.

12. Let X be an uncountable topological space such that all one-element subsets of X are Borel in X. We shall say that X is a Sierpiński space if X contains no universal measure zero subspace with cardinality equal to $card(X)$. Show that:

(a) any generalized Sierpiński subset of $\mathbf{R}$ is a Sierpiński space;

(b) if X is a Sierpiński space of cardinality ω_1, then

$$\mathcal{A}(X) = \mathcal{B}(X),$$

where $\mathcal{A}(X)$ denotes the class of all analytic subsets of X (that is the class of all those sets which can be obtained by applying the A-operation to various A-systems consisting of Borel subsets of X) and $\mathcal{B}(X)$ denotes, as usual, the Borel σ-algebra of X;

(c) if X_1 and X_2 are two Sierpiński spaces and X is their topological sum, then X is a Sierpiński space, too;

(d) if X is a Sierpiński space, Y is a topological space such that $card(Y) = card(X)$, all one-element subsets of Y are Borel in Y and there exists a Borel surjection from X onto Y, then Y is a Sierpiński space, too. Consequently, if X is a Sierpiński subset of $\mathbf{R}$ and

$$f : X \to \mathbf{R}$$

is a Borel mapping such that

$$card(f(X)) = card(X),$$

then $f(X)$ turns out to be a Sierpiński subspace of $\mathbf{R}$.

13. By assuming Martin's Axiom and applying the method of transfinite recursion, construct a generalized Sierpiński subset X of $\mathbf{R}$ such that

$$X + X = \mathbf{R}.$$

Infer from this equality that there exists a continuous surjection from the product space $X \times X$ onto $\mathbf{R}$. Further, by starting with this property of X and taking into account assertion (d) of the preceding exercise, demonstrate that the topological product $X \times X$ is not a Sierpiński space. Conclude from this fact that the topological product of two Sierpiński spaces is not, in general, a Sierpiński space.

14. Let H be a Hilbert space (over the field $\mathbf{R}$) whose Hilbert dimension is equal to $\mathbf{c}$ (in particular, the cardinality of H equals $\mathbf{c}$, too). Assuming that $\mathbf{c}$ is not real-valued measurable, show that there exists a subset X of H satisfying the following conditions:
  (1) $card(X) = \mathbf{c}$;
  (2) X is everywhere dense in H (in particular, X is nonseparable);
  (3) X is a universal measure zero subspace of H.
  Suppose now that $\mathbf{c}$ is not cofinal with ω_1. In other words, suppose that $\mathbf{c}$ cannot be represented in the form

$$\mathbf{c} = \sum_{\xi < \omega_1} \kappa_\xi,$$

where all cardinal numbers κ_ξ $(\xi < \omega_1)$ are strictly less than $\mathbf{c}$.
  By starting with the fact that there exists an ω_1-sequence of nowhere dense subsets of the space H, covering this space, demonstrate that there is no generalized Luzin subset of H. In other words, demonstrate that there is no set $Y \subset H$ satisfying these two relations:
  (a) $card(Y) = \mathbf{c}$;

(b) for each first category set $Z \subset H$, the inequality

$$card(Y \cap Z) < \mathbf{c}$$

is fulfilled.

15. Show that there exists a Marczewski subset D of $\mathbf{R}^2$ such that:

(a) D does not possess the Baire property;

(b) D is not measurable with respect to the completion of the product measure $\mu \times \nu$, where μ and ν are any two nonzero σ-finite diffused Borel measures on $\mathbf{R}$.

16. Prove that there exists a Borel isomorphism

$$\psi \ : \ \mathbf{R} \to \mathbf{R}^2$$

such that:

(a) ψ preserves the Baire category of sets (consequently, ψ preserves the Baire property);

(b) ψ is an isomorphism between the measures λ and λ_2.

By starting with this fact and applying Theorem 5, show that there exists a Marczewski subset of the real line, nonmeasurable in the Lebesgue sense and without the Baire property.

17. Let K be a compact subset of $\mathbf{R}^2$. Demonstrate that there exists a Borel mapping

$$\phi \ : \ pr_1(K) \to \mathbf{R}$$

such that the graph of ϕ is contained in K.

This simple result is a very particular case of general statements concerning the existence of measurable selectors (see, for instance, [49], [128] and [144]).

In addition, give an example of a compact subset P of $\mathbf{R}^2$ for which there exists no continuous mapping

$$\psi \ : \ pr_1(P) \to \mathbf{R}$$

such that the graph of ψ is contained in P.

Rich information about continuous selectors and their applications can be found in [145], [146] and [181].

18. By using the result of the previous exercise, show that the graph of any Sierpiński-Zygmund function is a Marczewski subset of the euclidean plane $\mathbf{R}^2$.

19. Construct, by using the method of transfinite recursion, a Sierpiński-Zygmund function whose graph is a λ_2-thick set in the plane $\mathbf{R}^2$. Applying a similar method, construct a Sierpiński-Zygmund function whose graph is a thick subset of $\mathbf{R}^2$ in the category sense.

Deduce directly from these results that there are Marczewski subsets of $\mathbf{R}^2$ nonmeasurable in the Lebesgue sense (respectively, without the Baire property).

On the other hand, prove that there exists a Sierpiński-Zygmund function whose graph is a λ_2-measure zero subset of $\mathbf{R}^2$. Analogously, prove that there exists a Sierpiński-Zygmund function whose graph is a first category set in $\mathbf{R}^2$.

The next two exercises indicate some simple relationships between three classical σ-ideals on $\mathbf{R}$.

20. Verify that the σ-ideal $\mathcal{I}_L$ generated by all Luzin subsets of $\mathbf{R}$ is properly contained in the σ-ideal of all universal measure zero subsets of $\mathbf{R}$.

Assuming Martin's Axiom, prove an analogous statement for the σ-ideal generated by all generalized Luzin subsets of $\mathbf{R}$.

21. Verify that the σ-ideal of all universal measure zero subsets of $\mathbf{R}$ is properly contained in the σ-ideal $\mathcal{I}_M$ of all Marczewski subsets of $\mathbf{R}$.

22. Let $f : \mathbf{R} \to \mathbf{R}$ be a function whose graph is λ_2-thick in $\mathbf{R}^2$. Prove that there exists a measure μ on $\mathbf{R}$ extending λ and having the property that f is measurable with respect to μ.

Moreover, let g be a homomorphism of the additive group $\mathbf{R}$ into itself, such that the graph of g is λ_2-thick in $\mathbf{R}^2$. Prove that there exists a measure ν on $\mathbf{R}$ satisfying the following relations:

(a) ν is an extension of λ;

(b) ν is quasi-invariant under the group of all isometric transformations of $\mathbf{R}$;

(c) g is measurable with respect to ν.

23. Let $\Phi : \mathbf{R} \times \mathbf{R} \to \mathbf{R}$ be a function of two variables. We shall say that this function is sup-measurable if, for every Lebesgue measurable function $\phi : \mathbf{R} \to \mathbf{R}$, the superposition Φ_ϕ defined by

$$\Phi_\phi(x) = \Phi(x, \phi(x)) \qquad (x \in \mathbf{R})$$

is Lebesgue measurable, too. Verify that:

(a) any Borel function $\Phi : \mathbf{R} \times \mathbf{R} \to \mathbf{R}$ is sup-measurable;

(b) there exists a λ_2-measurable function $\Phi : \mathbf{R} \times \mathbf{R} \to \mathbf{R}$ which is not sup-measurable;

(c) a function $\Phi : \mathbf{R} \times \mathbf{R} \to \mathbf{R}$ is sup-measurable if and only if the functions Φ_ϕ are measurable in the Lebesgue sense for all continuous functions $\phi : \mathbf{R} \to \mathbf{R}$.

Suppose that Z is a Sierpiński subset of the euclidean plane $\mathbf{R}^2$ (in other words, Z is uncountable and $card(Z \cap T) \leq \omega$ for all λ_2-measure zero sets T in $\mathbf{R}^2$). Show that the characteristic function of Z is sup-measurable but is not λ_2-measurable. Conclude from this fact that, under the Continuum Hypothesis, there exist sup-measurable functions which are not Lebesgue measurable. Prove an analogous result under Martin's Axiom, starting with a generalized Sierpiński subset of $\mathbf{R}^2$.

We say that a function $\Phi : \mathbf{R} \times \mathbf{R} \to \mathbf{R}$ satisfies the Carathéodory conditions if:

(i) for each $x \in \mathbf{R}$, the partial function $\Phi(x, \cdot) : \mathbf{R} \to \mathbf{R}$ defined by

$$\Phi(x, \cdot)(y) = \Phi(x, y) \qquad (y \in \mathbf{R})$$

is continuous;

(ii) for each $y \in \mathbf{R}$, the partial function $\Phi(\cdot, y) : \mathbf{R} \to \mathbf{R}$ defined by

$$\Phi(\cdot, y)(x) = \Phi(x, y) \qquad (x \in \mathbf{R})$$

is Lebesgue measurable.

Demonstrate that any function

$$\Phi : \mathbf{R} \times \mathbf{R} \to \mathbf{R}$$

satisfying the Carathéodory conditions is sup-measurable and Lebesgue measurable.

Note that sup-measurable functions play a significant role in various topics of analysis (especially, in the theory of differential equations and optimization).

Recently, Shelah and Roslanowski have announced that the statement

"all sup-measurable functions are measurable in the Lebesgue sense"

is consistent with the theory **ZFC**.

24. Let X be a subset of $\mathbf{R}$ nonmeasurable in the Lebesgue sense. Define a function of two variables

$$\Phi_1 : \mathbf{R} \times \mathbf{R} \to \mathbf{R}$$

by putting

$$\Phi_1(x, x) = 0 \qquad (x \in X),$$

$$\Phi_1(x, x) = 1 \qquad (x \in \mathbf{R} \setminus X),$$

$$\Phi_1(x, y) = 1 \qquad (x \in \mathbf{R},\ y \in \mathbf{R},\ x \neq y).$$

Verify that Φ_1 is λ_2-measurable but is not sup-measurable.

Let $f : \mathbf{R} \to \mathbf{R}$ be an injective function whose graph Γ_f does not intersect the set

$$\triangle = \{(x, x) : x \in \mathbf{R}\}$$

and is a λ_2-thick subset of the plane $\mathbf{R}^2$ (show the existence of such a function), and let Ψ denote the characteristic function of Γ_f. Put

$$\Phi_2 = 1 - \Psi,$$

$$\Phi = \Phi_1 + \Phi_2.$$

Deduce from the properties of Φ_1 and Φ_2 that the function Φ satisfies the following relations:

(a) $ran(\Phi) = \{1, 2\}$;
(b) for any $y \in \mathbf{R}$, the partial function $\Phi(\cdot, y)$ is lower semi-continuous;
(c) for any $x \in \mathbf{R}$, the partial function $\Phi(x, \cdot)$ is lower semi-continuous;
(d) Φ is not λ_2-measurable;
(e) Φ is not sup-measurable.

Chapter 6
Strange subsets of the Euclidean plane

We begin this chapter with consideration of non Lebesgue-measurable subsets of the euclidean plane $\mathbf{R}^2$, which have small linear sections. We will investigate these sets from the point of view of measurability with respect to appropriate invariant extensions of the Lebesgue measure on $\mathbf{R}^2$. Our interest in such subsets of the plane is inspired by a classical Sierpiński partition of $\mathbf{R}^2$ whose properties were thoroughly discussed in Chapter 4.

In the literature there are many examples of paradoxical subsets of a finite-dimensional euclidean space, having small sections by certain affine hyperplanes of this space.

One of the earliest examples is due again to Sierpiński who constructed a function

$$f \; : \; \mathbf{R} \to \mathbf{R}$$

such that its graph is a λ_2-thick (or λ_2-massive) subset of the plane $\mathbf{R}^2$. This construction was presented in Chapter 2. Here λ_2 denotes, as usual, the standard two-dimensional Lebesgue measure on $\mathbf{R}^2$, and we recall that a subset X of $\mathbf{R}^2$ is λ_2-thick (or λ_2-massive) in $\mathbf{R}^2$ if the inner λ_2-measure of the set $\mathbf{R}^2 \setminus X$ is equal to zero (see the corresponding definition in [62]). In particular, the λ_2-thickness of the graph Γ_f of f implies that Γ_f is non-measurable with respect to λ_2 and, therefore, f is not measurable in the Lebesgue sense (see Theorem 4 from Chapter 2). At the same time, any straight line in $\mathbf{R}^2$ parallel to the line $\{0\} \times \mathbf{R}$ meets Γ_f at exactly one point.

Another interesting example was given by Mazurkiewicz who constructed a subset Y of $\mathbf{R}^2$ having the property that for each straight line l in $\mathbf{R}^2$, the set $l \cap Y$ consists of exactly two points.

In the sequel, any such subset of the plane will be called a Mazurkiewicz set.

The descriptive structure of a Mazurkiewicz set turned out to be rather complicated. In general, one cannot assert that a Mazurkiewicz set is necessarily nonmeasurable with respect to λ_2. Indeed, there are Mazurkiewicz subsets of the plane which have λ_2-measure zero (see Exercise 1 of this chapter). Moreover, by using the classical Fubini theorem, one can easily conclude that if a Mazurkiewicz set is λ_2-measurable, then it must be of λ_2-measure zero.

Slightly changing the argument of Mazurkiewicz, we shall demonstrate below the existence of a Mazurkiewicz set which is λ_2-thick (consequently, it is nonmeasurable with respect to λ_2). Actually, the same argument works not only for the plane $\mathbf{R}^2$ and the measure λ_2 on it, but also for the euclidean n-dimensional space $\mathbf{R}^n$, where $n \geq 2$, and for the standard n-dimensional Lebesgue measure λ_n on this space.

In fact, various analogous constructions of sets in $\mathbf{R}^n$ which have small sections but are large in some sense can be carried out by utilizing the method of Mazurkiewicz (compare Exercise 3 of this chapter).

Theorem 1. *Let $n \geq 2$. There exists a subset X of $\mathbf{R}^n$ such that:*

1) each affine hyperplane in $\mathbf{R}^n$ intersects X in exactly n points; consequently, X is a set of points in general position;

2) X is thick with respect to λ_n.

Proof. Let α be the least ordinal number of cardinality continuum (denoted by $\mathbf{c}$), let $\{L_\xi : \xi < \alpha\}$ be the family of all affine hyperplanes in $\mathbf{R}^n$ and let $\{B_\xi : \xi < \alpha\}$ be the family of all those Borel subsets of $\mathbf{R}^n$ which have strictly positive λ_n-measure. We are going to construct, using the transfinite recursion method, a family $\{X_\xi : \xi < \alpha\}$ of subsets of $\mathbf{R}^n$ satisfying the following conditions:

(1) if $\zeta < \xi < \alpha$, then $X_\zeta \subset X_\xi$ (in other words, this family is increasing by inclusion);

(2) $card(X_\xi) \leq card(\xi) + \omega$ for any ordinal $\xi < \alpha$;

(3) for each $\xi < \alpha$, the set X_ξ is a set of points in general position in the space $\mathbf{R}^n$;

(4) if $\xi < \alpha$, then $card(X_\xi \cap L_\xi) = n$;

(5) if $\xi < \alpha$, then $X_\xi \cap B_\xi \neq \emptyset$.

Suppose that, for an ordinal $\xi < \alpha$, the partial family of sets

$$\{X_\zeta : \zeta < \xi\}$$

has already been constructed. Consider the set B_ξ. Let B_ξ' be a subset of B_ξ of cardinality continuum, whose all points are in general position.

Note that the existence of such a subset of B_ξ can easily be established by utilizing Lemma 1 of this chapter, which is presented below.

Further, let us denote

$$Y_\xi = \cup\{X_\zeta : \zeta < \xi\}.$$

Evidently, Y_ξ is a set of points in general position and

$$card(Y_\xi) \leq card(\xi) + \omega.$$

Consequently, we may write

$$card(Y_\xi \cap L_\xi) \leq n.$$

Now, by using ordinary recursion, it is not hard to define a finite subset Z_ξ of L_ξ such that:

(i) $card((Y_\xi \cup Z_\xi) \cap L_\xi) = n$;

(ii) $Y_\xi \cup Z_\xi$ is a set of points in general position.

Note that in order to define Z_ξ, it suffices to apply a simple geometric fact stating that no affine linear manifold L in $\mathbf{R}^n$ can be covered by a family of affine linear manifolds whose cardinality is strictly less than $\mathbf{c}$ and whose members all have dimension strictly less than $dim(L)$ (compare Lemma 1 below in which a more general result is formulated).

Further, since

$$card(Y_\xi \cup Z_\xi) < \mathbf{c}, \quad card(B'_\xi) = \mathbf{c}$$

and B'_ξ is a set of points in general position, there exists a point $x_\xi \in B'_\xi$ for which the union $\{x_\xi\} \cup Y_\xi \cup Z_\xi$ is also a set of points in general position. So, we may put

$$X_\xi = \{x_\xi\} \cup Y_\xi \cup Z_\xi.$$

Proceeding in this way, we are able to construct the required family of sets $\{X_\xi : \xi < \alpha\}$.

Finally, we define

$$X = \cup\{X_\xi : \xi < \alpha\}.$$

It can easily be verified, in view of our construction, that the set X satisfies relations 1) and 2) of the theorem. This finishes the proof.

In particular, for $n = 2$, the theorem proved above yields a Mazurkiewicz subset of the euclidean plane $\mathbf{R}^2$, which is thick with respect to the Lebesgue measure λ_2. Hence, this subset turns out to be nonmeasurable with respect to λ_2. Now, according to the standard construction of extending the Lebesgue measure (see, for instance, Exercise 10 of Chapter 2), any Mazurkiewicz set can be included in the domain of a translation-invariant extension of λ_2, and we can assert that any Mazurkiewicz set necessarily becomes a set of measure zero with respect to such an extension (compare Exercise 4 of this chapter).

The same result remains valid for those extensions of λ_2 which are invariant under the group of all isometric transformations (motions) of $\mathbf{R}^2$.

Actually, we are interested in subsets of $\mathbf{R}^2$ having small linear sections in connection with the following natural question: how small are these subsets from the point of view of invariant extensions of the Lebesgue measure λ_2?

We are going to demonstrate in our further considerations that there are invariant extensions of λ_2, concentrated on sets with small linear sections in all directions. In fact, a more general result will be obtained stating that the corresponding sets can have small sections by all analytic curves lying in $\mathbf{R}^2$.

Note that our further argument is applicable to the space $\mathbf{R}^n$, where $n \geq 2$, and to the standard Lebesgue measure λ_n on this space.

In the sequel, we need the following simple auxiliary statement which is useful in many analogous situations.

Lemma 1. *Let Z be a λ_n-measurable subset of the Euclidean space $\mathbf{R}^n$ with $\lambda_n(Z) > 0$ and let $\{M_i \ : \ i \in I\}$ be a family of analytic manifolds in $\mathbf{R}^n$, such that:*

1) $card(I)$ is strictly less than $\mathbf{c}$;

2) for each index $i \in I$, the dimension of M_i is strictly less than n.

Then the relation

$$Z \setminus \cup \{M_i \ : \ i \in I\} \neq \emptyset$$

is satisfied.

In particular, Lemma 1 states that the space $\mathbf{R}^n$ cannot be covered by a small (in the sense of cardinality) family of analytic submanifolds of $\mathbf{R}^n$ whose dimensions are strictly less than n. The proof of this lemma is not difficult and can be carried out by induction on n. Here the classical Fubini theorem plays an essential role. We leave the corresponding purely technical details to the reader.

Some applications of Lemma 1 to certain questions of the geometry of euclidean spaces may be found in [87]. Note that the lemma does not hold true for topological manifolds in $\mathbf{R}^n$. For example, there are models of set theory in which the plane $\mathbf{R}^2$ can be covered by a family of Jordan curves, whose cardinality is strictly less than $\mathbf{c}$. (Here a Jordan curve is any homeomorphic image of the unit circumference $\mathbf{S}_1 \subset \mathbf{R}^2$.)

Starting with the above-mentioned lemma, we are able to establish the following result.

Theorem 2. *Let G be the group of all analytic diffeomorphisms of the space $\mathbf{R}^n$ ($n \geq 1$) onto itself. Then there exists a subset X of $\mathbf{R}^n$ such that:*
1) X is almost G-invariant; in other words, we have

$$(\forall g \in G)(card(g(X) \triangle X) < \mathbf{c}),$$

where the symbol $\triangle$ denotes the operation of symmetric difference of sets;
2) $card(X) = \mathbf{c}$ and X is λ_n-thick in $\mathbf{R}^n$;
3) for any analytic manifold M in $\mathbf{R}^n$ with $dim(M) < n$, we have

$$card(M \cap X) < \mathbf{c}.$$

Proof. We use the method developed in [75], [82] and [100]. Let α denote the least ordinal number of cardinality continuum. Since the equality $card(G) = \mathbf{c}$ holds, we may write

$$G = \cup\{G_\xi \; : \; \xi < \alpha\},$$

where $\{G_\xi \; : \; \xi < \alpha\}$ is some α-sequence of subgroups of G, satisfying these two conditions:
(a) for each ordinal $\xi < \alpha$, we have the inequality

$$card(G_\xi) \leq card(\xi) + \omega;$$

(b) the family $\{G_\xi \; : \; \xi < \alpha\}$ is increasing by inclusion.
Also, we introduce the family $\{Y_\xi \; : \; \xi < \alpha\}$ consisting of all those Borel subsets of $\mathbf{R}^n$ which are of strictly positive λ_n-measure. Finally, we denote by $\{M_\xi \; : \; \xi < \alpha\}$ the family of all analytic manifolds in $\mathbf{R}^n$ whose dimensions are strictly less than n.

Let us now define, by the method of transfinite recursion, an injective α-sequence $\{x_\xi \, : \, \xi < \alpha\}$ of points in $\mathbf{R}^n$, satisfying the following relations:

(i) for any $\xi < \alpha$, the point x_ξ belongs to Y_ξ;

(ii) for any $\xi < \alpha$, we have

$$x_\xi \notin G_\xi((\cup\{M_\zeta \, : \, \zeta \leq \xi\}) \cup \{x_\zeta \, : \, \zeta < \xi\}).$$

Note that Lemma 1 guarantees, at each ξ-step of our recursion, the existence of a point x_ξ. So the recursion can be continued up to α. Proceeding in this way, we are able to construct the required α-sequence of points $\{x_\xi \, : \, \xi < \alpha\}$. Now, putting

$$X = \{G_\xi(x_\xi) \, : \, \xi < \alpha\},$$

we can easily check that the set X is the desired one. (All details are left to the reader.) This finishes the proof of Theorem 2.

The next statement is a trivial consequence of the theorem just proved.

Theorem 3. *Let G denote the group of all isometric transformations (motions) of the space $\mathbf{R}^n$. Then there exists a subset X of $\mathbf{R}^n$ such that:*

1) X is almost G-invariant;

2) $card(X) = \mathbf{c}$ and X is λ_n-thick in $\mathbf{R}^n$;

3) for each analytic manifold M in $\mathbf{R}^n$ with $dim(M) < n$, we have

$$card(X \cap M) < \mathbf{c}.$$

In particular, for any affine hyperplane L in $\mathbf{R}^n$, the inequality

$$card(X \cap L) < \mathbf{c}$$

is true.

It immediately follows from Theorem 3 that if the Continuum Hypothesis holds, then, for each analytic manifold M in $\mathbf{R}^n$ with $dim(M) < n$, we have

$$card(X \cap M) \leq \omega,$$

where X is the set from Theorem 3.

Let us point out one straightforward application of Theorem 3.

Various statements related to the classical Fubini theorem lead to the following intriguing question: does there exist a measure μ_n on $\mathbf{R}^n$ extending the Lebesgue measure λ_n, invariant under the group of all isometric transformations of $\mathbf{R}^n$ and concentrated on some subset X of $\mathbf{R}^n$ with small sections by all hyperplanes?

Here the smallness of sections of X means that, for any hyperplane L in $\mathbf{R}^n$, the cardinality of $L \cap X$ is strictly less than the cardinality of the continuum.

Theorem 3 immediately yields a positive answer to this question. Indeed, let G be the group of all isometric transformations of $\mathbf{R}^n$ and consider the G-invariant σ-ideal of subsets of $\mathbf{R}^n$, generated by the set $\mathbf{R}^n \setminus X$, where X is the set from Theorem 3. We denote this σ-ideal by $\mathcal{J}$. Then, for any set $Z \in \mathcal{J}$, the inner λ_n-measure of Z is equal to zero (since X is almost G-invariant and λ_n-thick in $\mathbf{R}^n$). Taking this fact into account and applying the standard methods of extending invariant measures (see, for instance, [82], [100], [212] and [214]), we infer that there exists a measure μ_n on $\mathbf{R}^n$ satisfying the relations:

1) μ_n is complete and extends λ_n;

2) μ_n is invariant under the group G;

3) $\mathcal{J} \subset dom(\mu_n)$;

4) for each set $Z \in \mathcal{J}$, we have $\mu_n(Z) = 0$.

Relation 4) implies at once that

$$\mu_n(\mathbf{R}^n \setminus X) = 0.$$

In other words, our measure μ_n is concentrated on the set X. At the same time, we know that X has small sections by all hyperplanes in $\mathbf{R}^n$ and, moreover, by all analytic manifolds in $\mathbf{R}^n$ whose dimensions are strictly less than n. In addition to this, the measure μ_n being complete and metrically transitive has the so-called uniqueness property (compare Exercise 9 from Chapter 9).

In particular, for $n = 2$, we obtain that there exists a complete measure μ_2 on the euclidean plane $\mathbf{R}^2$, such that:

(1) μ_2 is an extension of the Lebesgue measure λ_2;

(2) μ_2 is invariant under the group of all isometric transformations of $\mathbf{R}^2$;

(3) μ_2 is concentrated on some subset X of $\mathbf{R}^2$ having the property that all linear sections of X are small; more precisely, the cardinality of each linear section of X is strictly less than $\mathbf{c}$.

Note that the question concerning the existence of a measure μ_2 on the plane, satisfying conditions (1) - (3), was formulated by R.D.Mabry (personal communication). Moreover, we see that the above-mentioned support X of μ_2 has a much stronger property: for every analytic curve l in $\mathbf{R}^2$, the cardinality of $l \cap X$ is strictly less than $\mathbf{c}$. Let us point out that in [96] an analogous question was considered for extensions of λ_2 which are invariant under the group of all translations of $\mathbf{R}^2$. Namely, it was demonstrated in [96] that there exists a translation-invariant extension of λ_2 concentrated on a subset of $\mathbf{R}^2$ whose all linear sections are at most countable.

Remark 1. If the Continuum Hypothesis holds, then we can also conclude that the above-mentioned subset X of $\mathbf{R}^2$ has the property that all its linear sections are at most countable. Obviously, the same is true for the sections of X by analytic curves lying in $\mathbf{R}^2$. Thus, in this case, the measure μ_2 is concentrated on a set with countable linear sections. In this context, the following question seems to be natural: does there exist a measure ν_2 on the plane, extending λ_2, invariant under the group of all isometric transformations of $\mathbf{R}^2$ and concentrated on a set with finite linear sections? It turns out that the answer to this question is negative. The proof of the corresponding result (and a more general statement) can be found in [94] (see also Exercise 10 of this chapter).

Remark 2. By assuming some additional set-theoretical axioms, it is possible to obtain a much stronger result than Theorem 2. Namely, let us suppose that, for any cardinal $\kappa < \mathbf{c}$, the space $\mathbf{R}^n$ ($n \geq 1$) cannot be covered by a κ-sequence of λ_n-measure zero sets. In fact, it suffices to suppose the validity of this assumption only for $n = 1$, which means that the real line $\mathbf{R}$ cannot be covered by a family of Lebesgue measure zero sets, whose cardinality is strictly less than $\mathbf{c}$. For instance, this hypothesis follows directly from Martin's Axiom (see [64] or Appendix 1). We say that a group G of transformations of $\mathbf{R}^n$ is admissible if each element g from G preserves the σ-ideal of all λ_n-measure zero subsets of $\mathbf{R}^n$. There are many natural examples of admissible groups. For instance, if G coincides with the group of all affine transformations of $\mathbf{R}^n$, then G is admissible. If G coincides with the group of all diffeomorphisms of $\mathbf{R}^n$, then G is admissible, too. Evidently, the cardinality of these two groups is equal to $\mathbf{c}$. It is not hard to check that there are also admissible groups G with $card(G) = 2^{\mathbf{c}}$.

Let us fix an admissible group G with $card(G) = \mathbf{c}$. Assuming the above-mentioned hypothesis and applying an argument similar to the proof of Theorem 2, we easily obtain the following statement: if $n \geq 1$, then **there** exists a subset X of $\mathbf{R}^n$ with $card(X) = \mathbf{c}$, such that:

1) X is almost G-invariant;
2) X is λ_n-thick in $\mathbf{R}^n$;
3) for any λ_n-measure zero subset Z of $\mathbf{R}^n$, we have

$$card(Z \cap X) < \mathbf{c}.$$

Condition 3) shows, in particular, that the above-mentioned set X is a generalized Sierpiński subset of $\mathbf{R}^n$. (For the definition and various properties of Sierpiński sets, see [155] or [165] where the dual objects to the Sierpiński sets - the so-called Luzin sets - are discussed as well; the corresponding material about these sets is also presented in Chapter 5 of this book.)

Now, we wish to return to a problem of Luzin posed by him many years ago (see [195]). We have already been slightly in touch with this problem in Chapter 4 where its close connection with a Sierpiński partition of the plane $\mathbf{R}^2$ was briefly discussed. Here we are going to consider this question in more details. First of all, let us recall the formulation of this beautiful problem. Namely, Luzin asked whether there exists a function

$$\phi : \mathbf{R} \to \mathbf{R}$$

such that the whole plane $\mathbf{R}^2$ can be covered by countably many isometric copies of the graph of ϕ. We have already mentioned that, under **CH**, this problem admits a positive solution even in a much stronger form (compare statement (iii) and Exercise 4 of Chapter 4). Actually, a positive answer to the Luzin question follows from the existence of a Sierpiński type partition for the euclidean plane. Here we are going to present the full solution of this problem, within the theory **ZFC**. Recall that the final result was obtained by Davies. (See his two works [31] and [32] devoted to the Luzin problem.)

In the sequel, we need some auxiliary notions and facts about special subsets of $\mathbf{R}^2$.

Let X be a subset of $\mathbf{R}^2$ and let p be a straight line in $\mathbf{R}^2$. We shall say that X is uniform with respect to p if, for each line p' parallel to p, we have

$$card(p' \cap X) \leq 1.$$

Obviously, for a set $X \subset \mathbf{R}^2$ and for any two parallel straight lines p and q in $\mathbf{R}^2$, the following assertions are equivalent:

1) X is uniform with respect to p;

2) X is uniform with respect to q.

Further, we shall say that a set $X \subset \mathbf{R}^2$ is uniform if there exists at least one straight line p in $\mathbf{R}^2$ such that X is uniform with respect to p.

It is not hard to observe that the next two assertions are also equivalent:

(1) there exists a function $\phi : \mathbf{R} \to \mathbf{R}$ such that the plane $\mathbf{R}^2$ can be covered by countably many isometric copies of the graph Γ_ϕ of ϕ;

(2) the plane $\mathbf{R}^2$ can be covered by countably many uniform subsets of $\mathbf{R}^2$.

Therefore, the Luzin problem is reduced to the problem of the existence of a countable family of uniform sets, whose union is identical with $\mathbf{R}^2$. In our further considerations, we will deal with the latter problem.

Let p be an arbitrary straight line in $\mathbf{R}^2$ and let z be a point of $\mathbf{R}^2$. We denote by $p(z)$ the unique line in $\mathbf{R}^2$ which is parallel to p and contains z.

A countable family $P = (p_k)_{k<\omega}$ of straight lines in $\mathbf{R}^2$ will be called admissible if all these lines are pairwise distinct and all of them contain the origin of $\mathbf{R}^2$.

Fix an admissible family $P = (p_k)_{k<\omega}$ of straight lines in $\mathbf{R}^2$.

Let Z be a subset of $\mathbf{R}^2$. We shall say that Z is P-closed if, for any two points $z_1 \in Z$, $z_2 \in Z$ and for any two distinct lines $p_{k_1} \in P$, $p_{k_2} \subset P$, the unique common point of the lines $p_{k_1}(z_1)$ and $p_{k_2}(z_2)$ also belongs to Z.

The following auxiliary proposition is almost evident.

Lemma 2. *Let P be an admissible family of straight lines in $\mathbf{R}^2$ and let Z be a subset of $\mathbf{R}^2$. Then there exists a smallest (with respect to inclusion) P-closed subset $[Z]$ of $\mathbf{R}^2$ containing Z. Moreover, we have the inequality*

$$card([Z]) \leq card(Z) + \omega.$$

In particular, if Z is infinite, then

$$card([Z]) = card(Z).$$

An easy proof of this statement is left to the reader.

Lemma 3. *Let $P = (p_k)_{k<\omega}$ be an admissible family of straight lines in $\mathbf{R}^2$ and let Z be an arbitrary infinite P-closed subset of $\mathbf{R}^2$. Then this* *set can be well-ordered by some relation $\preceq$ in such a way that, for any point $z \in Z$, the set of natural numbers*

$$\{k < \omega \ : \ \text{the line } p_k(z) \text{ contains at least one point } z' \prec z\}$$

is always finite.

Proof. We use the method of transfinite induction (on $card(Z)$). The case when $card(Z) = \omega$ is trivial: here it suffices to equip Z with any ordering isomorphic to the canonical well-ordering of ω. Suppose now that the assertion of the lemma has already been established for all infinite P-closed sets Z with $card(Z) < \kappa$, where κ is some uncountable cardinal.

Take any P-closed set Z such that $card(Z) = \kappa$. Of course, we may identify κ with the least ordinal of the same cardinality. Let

$$Z = \{z_\xi : \xi < \kappa\}$$

be an injective enumeration of all elements of Z and, for each ordinal $\xi < \kappa$, let

$$Z_\xi = \{z_\zeta : \zeta \leq \xi\}.$$

Further, for any point $z \in Z$, we put

$$\xi(z) = inf\{\xi \ : \ \xi \geq \omega, \ z \in [Z_\xi]\}.$$

We shall say that $\xi(z)$ is the index of z (in Z). Let us mention that this notion is well defined because the equality

$$Z = \cup\{[Z_\xi] \ : \ \omega \leq \xi < \kappa\}$$

holds true in view of the P-closedness of Z. By Lemma 2, if $\omega \leq \xi < \kappa$, then

$$card([Z_\xi]) = card(\xi) < \kappa$$

and, taking into account the inductive assumption, we can equip the set $[Z_\xi]$ with a well-ordering $\preceq_\xi$ in such a way that for every point $z \in [Z_\xi]$, the set

$$\{k < \omega \ : \ p_k(z) \text{ contains at least one point } z' \prec_\xi z\}$$

is necessarily finite. Now, we are able to introduce a suitable well-ordering $\preceq$ for the set Z. Namely, if $z \in Z$ and $z' \in Z$ are two distinct points, then we put $z' \prec z$ if and only if

$$(\xi(z') \prec \xi(z)) \ \vee \ (\exists \xi < \kappa)(\xi = \xi(z) = \xi(z') \ \& \ z' \prec_\xi z).$$

Let us check that our relation $\prec$ possesses the required properties.

First of all, by the definition of $\prec$, it is not hard to see that the corresponding relation $\preceq$ is a well-ordering of Z.

Let now $z \in Z$ and $\xi = \xi(z)$. We must verify that the set of natural numbers

$$\{k < \omega \ : \ p_k(z) \ contains \ at \ least \ one \ point \ z' \prec z\}$$

is finite. Suppose to the contrary that the above-mentioned set is infinite. Then there exist pairwise distinct lines

$$q_1 \in P, \ q_2 \in P, \ ..., \ q_m \in P, \ ...$$

such that, for some points

$$z_1 \in Z, \ z_2 \in Z, \ ..., \ z_m \in Z, \ ...,$$

the relations

$$z_1 \in q_1(z), \ z_2 \in q_2(z), \ ..., \ z_m \in q_m(z), \ ...,$$

$$z_1 \prec z, \ z_2 \prec z, \ ..., \ z_m \prec z, \ ...$$

are fulfilled. Observe now that, according to the inductive assumption for the well-ordering $\preceq_\xi$, the equalities

$$\xi(z_m) = \xi(z) = \xi$$

can be satisfied only for finitely many natural numbers m. This immediately implies that there are at least two distinct points from $\{z_1, z_2, ..., z_m, ...\}$ whose indices are strictly less than ξ. Without loss of generality, we may assume that

$$\xi(z_1) < \xi, \quad \xi(z_2) < \xi, \quad z_1 \neq z_2.$$

Let us denote

$$\eta = max(\xi(z_1), \xi(z_2)).$$

Then, obviously, we get

$$\{z_1, z_2\} \subset [Z_\eta].$$

Furthermore, it is evident that the straight lines $q_1(z_1)$ and $q_2(z_2)$ meet each other at the point z. Therefore, we must have $z \in [Z_\eta]$, from which it follows

$$\xi(z) \leq \eta = max(\xi(z_1), \xi(z_2)) < \xi = \xi(z),$$

and the last relation yields a contradiction.

The contradiction obtained completes the proof of Lemma 3.

Lemma 4. *Let $P = (p_k)_{k<\omega}$ be an admissible family of straight lines in $\mathbf{R}^2$. Then there exists a family $(Z_k)_{k<\omega}$ of subsets of $\mathbf{R}^2$, satisfying the following relations:*

1) each set Z_k $(k < \omega)$ is uniform with respect to the line p_k;

2) $\cup\{Z_k : k < \omega\} = \mathbf{R}^2$.

Proof. Clearly, the plane $\mathbf{R}^2$ is a P-closed set. Therefore, by virtue of the preceding lemma, $\mathbf{R}^2$ can be endowed with a well-ordering $\preceq$ such that, for each point $z \in \mathbf{R}^2$, the set

$$\{k < \omega \ : \ p_k(z) \text{ contains at least one point } z' \prec z\}$$

is finite. Keeping in mind this circumstance, we may denote by $k(z)$ the smallest natural number k for which the straight line $p_k(z)$ contains no point $z' \prec z$.

Now, for any natural number k, we put

$$Z_k = \{z \in \mathbf{R}^2 : k(z) = k\}.$$

Let us check that the sets Z_k $(k < \omega)$ are the required ones. First, it is evident that

$$\cup\{Z_k : k < \omega\} = \mathbf{R}^2.$$

Therefore, it remains to demonstrate that every set Z_k is uniform with respect to the line p_k. To see this, suppose otherwise and pick any two distinct points $z \in Z_k$ and $z' \in Z_k$ for which $p_k(z) = p_k(z')$. Since the disjunction

$$z \prec z' \ \vee \ z' \prec z$$

trivially holds, we may assume without loss of generality that $z' \prec z$. Under this assumption, we get

$$z' \in p_k(z') = p_k(z) = p_{k(z)}(z), \quad z' \prec z,$$

which contradicts the definition of $k(z)$. The contradiction obtained ends the proof of the lemma.

As mentioned earlier, Lemma 4 yields a positive solution of the Luzin problem. So, we can formulate the following statement first obtained by Davies [31].

Theorem 4. *There exist a function*

$$\phi \; : \; \mathbf{R} \to \mathbf{R}$$

and a countable family $(g_n)_{n<\omega}$ *of motions of the euclidean plane* $\mathbf{R}^2$, *such that*

$$\cup\{g_n(\Gamma_\phi) \; : \; n < \omega\} = \mathbf{R}^2,$$

where Γ_ϕ *denotes the graph of* ϕ.

Thus, we conclude that the problem of Luzin is solvable within the theory **ZFC**. Some interesting results closely related to this problem are indicated in exercises below (see also Chapter 14).

A generalization of Theorem 4 for euclidean spaces of higher dimension can be found in [32].

EXERCISES

1. Show that there is a nowhere dense set $Z \subset \mathbf{R}^2$ of λ_2-measure zero, satisfying the relation

$$card(Z \cap l) = \mathbf{c}$$

for all straight lines $l \subset \mathbf{R}^2$.

Starting with this fact and applying the method of transfinite induction, demonstrate that there exists a nowhere dense Mazurkiewicz set of λ_2-measure zero.

2. For any straight line l in the plane $\mathbf{R}^2$, let $\kappa(l)$ be a cardinal number such that

$$2 \leq \kappa(l) < \omega.$$

Prove that there is a subset X of $\mathbf{R}^2$ satisfying the relation

$$card(l \cap X) = \kappa(l)$$

for each straight line l in $\mathbf{R}^2$.

This beautiful result is due to Sierpiński and it generalizes some theorems of Mazurkiewicz and Bagemihl.

Prove, in the theory **ZF**, that there exists a set $Y \subset \mathbf{R}^2$ satisfying the relations:

(a) $card(Y) = \mathbf{c}$;

(b) $card(Y \cap l) = \omega$ for each straight line l in $\mathbf{R}^2$.

3. By applying the method analogous to the Mazurkiewicz argument, show that there exists a subset X of $\mathbf{R}^2$ satisfying the following condition: every circumference in $\mathbf{R}^2$ intersects X in exactly three points. Show additionally that X can be assumed to be a λ_2-thick set in $\mathbf{R}^2$ (consequently, X can be assumed to be a nonmeasurable set with respect to λ_2).

Generalize this result to the case of $\mathbf{R}^n$ where $n \geq 2$.

4. Let μ be a σ-finite translation-invariant measure on $\mathbf{R}^2$ and let X be a Mazurkiewicz subset of $\mathbf{R}^2$. Demonstrate that the relation $X \in dom(\mu)$ implies the relation $\mu(X) = 0$ (compare Exercise 10 below).

5. By applying the method of transfinite recursion, construct two sets A and B on the plane $\mathbf{R}^2$, such that:

(i) $card(A) = card(B) = \mathbf{c}$;

(ii) $card(g(A) \cap h(B)) = 1$ for any two motions g and h of $\mathbf{R}^2$.

Furthermore, utilizing the techniques of Hamel bases for $\mathbf{R}$ (see Chapter 3), construct two subsets A' and B' of $\mathbf{R}$ having the properties analogous to (i) and (ii).

6. Give a detailed proof of Lemma 1.

7. Show that the following two assertions are equivalent:

(1) there exists a function $\phi : \mathbf{R} \to \mathbf{R}$ such that the plane $\mathbf{R}^2$ can be covered by countably many isometric copies of the graph Γ_ϕ of ϕ;

(2) the plane $\mathbf{R}^2$ can be covered by countably many uniform subsets of $\mathbf{R}^2$.

8. Let $\phi : \mathbf{R} \to \mathbf{R}$ be a solution of the Luzin problem considered in this chapter and let M_2 denote the group of all motions of $\mathbf{R}^2$. Demonstrate that there exists no nonzero σ-finite M_2-invariant (more generally, M_2-quasi-invariant) measure μ on $\mathbf{R}^2$ for which $\Gamma_\phi \in dom(\mu)$.

In other words, the graph of ϕ turns out to be an absolutely nonmeasurable set with respect to the class of all nonzero σ-finite M_2-quasi-invariant measures on $\mathbf{R}^2$. (Absolutely nonmeasurable sets will be discussed later, in Chapter 11.)

9. Show that the plane $\mathbf{R}^2$ cannot be covered by a finite family of uniform subsets of $\mathbf{R}^2$. Give a direct proof of this fact. On the other hand, deduce this result from the classical Banach theorem on the existence of a universal finitely additive M_2-invariant measure on $\mathbf{R}^2$ extending the

Lebesgue measure λ_2. (As in the previous exercise, the symbol M_2 denotes the group of all motions of $\mathbf{R}^2$.)

The Banach theorem mentioned above (with its proof) is presented in many text-books and monographs (see, for example, [226] or [82]).

10. Let E be a set and let G be a group of transformations of E. We shall say that a set $X \subset E$ is negligible with respect to the class of all σ-finite G-invariant (G-quasi-invariant) measures on E if the following two conditions are satisfied:

(a) there exists a nonzero σ-finite G-invariant (G-quasi-invariant) measure μ on E such that $X \in dom(\mu)$;

(b) for any σ-finite G-invariant (G-quasi-invariant) measure ν on E, we have

$$X \in dom(\nu) \Rightarrow \nu(X) = 0.$$

Sometimes, we shall say in short that $X \subset E$ is G-negligible if X satisfies conditions (a) and (b).

Show that, in general, the union of two G-negligible subsets of E is not G-negligible in E.

Denote by T_2 the group of all translations of the euclidean plane $\mathbf{R}^2$. Let Y be a subset of $\mathbf{R}^2$ such that every straight line parallel to the line $\{0\} \times \mathbf{R}$ intersects Y in finitely many points. Demonstrate that Y is negligible with respect to the class of all σ-finite T_2-invariant (T_2-quasi-invariant) measures on $\mathbf{R}^2$. In particular, infer that:

(1) each uniform subset of $\mathbf{R}^2$ is negligible with respect to the same class of measures;

(2) there exists no nonzero σ-finite T_2-quasi-invariant measure on $\mathbf{R}^2$ concentrated on a subset of $\mathbf{R}^2$ whose all linear sections are finite.

Undoubtedly, negligible sets are of interest for the general theory of invariant (quasi-invariant) measures. However, in Exercise 4 of Chapter 10 the notion of an absolutely negligible set will be introduced which is more delicate and useful than the above-mentioned notion of negligible set.

11. Let E be a vector space over $\mathbf{R}$. As usual, we denote by $conv(\cdot)$ the operation of taking the convex hull of a subset of E. In other words, for any set $X \subset E$, the symbol $conv(X)$ denotes the smallest convex set in E containing X.

Let Z be a subset of E. We shall say that Z is convexly independent (in E) if, for each point $z \in Z$, we have

$$z \notin conv(Z \setminus \{z\}).$$

For example, the set of all vertices of a convex polygon in $\mathbf{R}^2$ is convexly independent.

By using the Ramsey combinatorial theorem (see [179]), prove that for each natural number $n \geq 2$, there exists a natural number $N(n)$ having the following property:

Any set $X \subset \mathbf{R}^2$ of points in general position, with $card(X) \geq N(n)$, contains a convexly independent subset whose cardinality is equal to n.

This beautiful result is due to Erdös and Szekeres [44]. They also conjectured that the smallest possible value of $N(n)$ is $2^{n-2} + 1$. So far, there is no significant progress in proving (or disproving) their conjecture.

By using an infinite version of the Ramsey theorem [179] (more precisely, by using its countable version), demonstrate that every infinite set $X \subset \mathbf{R}^2$ of points in general position contains an infinite convexly independent subset.

12. By applying the method of transfinite induction, show that there exists a set $X \subset \mathbf{R}^2$ having the following properties:

(a) $card(X) = \mathbf{c}$;

(b) X is a set of points in general position;

(c) no subset of X of cardinality $\mathbf{c}$ is convexly independent.

For this purpose, denoting by α the least ordinal of cardinality $\mathbf{c}$, take an injective family $\{L_\xi : \xi < \alpha\}$ consisting of all convex curves in $\mathbf{R}^2$ and construct by transfinite recursion a subset of $\mathbf{R}^2$ which almost avoids the members L_ξ of this family (compare the classical constructions of Sierpiński and Luzin sets presented in Chapter 5).

Finally, conclude that if the Continuum Hypothesis holds, then there exists a set in $\mathbf{R}^2$ of cardinality $\mathbf{c}$, whose points are in general position and which contains no uncountable convexly independent subset.

13. Suppose that the cardinal $\mathbf{c}$ is real-valued measurable. Starting with the result of Kunen (see Chapter 4), demonstrate that there exists a family $\{L_i : i \in I\}$ of Jordan curves in $\mathbf{R}^2$, satisfying the following relations:

(a) $card(I) < \mathbf{c}$;

(b) $\cup\{L_i : i \in I\} = \mathbf{R}^2$.

14. Let E be a set and let $\{X_i : i \in I\}$ be an injective family of nonempty subsets of E. We shall say that this family is a homogeneous covering of E if there exists a cardinal number κ such that

$$card(\{i \in I : x \in X_i\}) = \kappa$$

for all elements $x \in E$. In this case, we also say that κ is the index of homogeneity of $\{X_i : i \in I\}$ and that $\{X_i : i \in I\}$ is a κ-homogeneous covering of E.

For example, if $\kappa = 1$, then the κ-homogeneity of $\{X_i : i \in I\}$ simply means that $\{X_i : i \in I\}$ is a partition of E.

Show that there exists no partition of the plane $\mathbf{R}^2$ consisting of Jordan curves.

On the other hand, show within the theory $\mathbf{ZF}$ that, for every natural number $k \geq 1$, there exists a $2k$-homogeneous covering of $\mathbf{R}^2$ all whose elements are circumferences congruent to the unit circumference $\mathbf{S}_1 \subset \mathbf{R}^2$.

Moreover, for every natural number $k \geq 2$, construct a k-homogeneous covering of $\mathbf{R}^2$ consisting of pairwise congruent circumferences. (Use the method of transfinite recursion.)

Generalize these results to the case of the euclidean space $\mathbf{R}^n$ where $n \geq 2$. Namely, prove that:

(a) there exists no partition of $\mathbf{R}^n$ whose all elements are homeomorphic images of the unit sphere $\mathbf{S}_{n-1} \subset \mathbf{R}^n$;

(b) for any natural number $k \geq n$, there exists a k-homogeneous covering of $\mathbf{R}^n$ whose all elements are spheres congruent to $\mathbf{S}_{n-1}$.

Also, prove that for $n \geq 1$ and for any natural number k, there is no k-homogeneous covering of $\mathbf{R}^n$ whose all elements are balls in $\mathbf{R}^n$.

15. Let $\mathcal{P}$ be a family of straight lines in the plane $\mathbf{R}^2$, such that

$$card(\{p \in \mathcal{P} : z \in p\}) = \mathbf{c}$$

for every point $z \in \mathbf{R}^2$. Show that, for each natural number $k \geq 2$, there exists a subfamily of $\mathcal{P}$ which is a k-homogeneous covering of $\mathbf{R}^2$.

Formulate and prove a suitable analogue of this result for the euclidean space $\mathbf{R}^n$ where $n \geq 3$.

16. Let us identify the plane $\mathbf{R}^2$ with the field $\mathbf{C}$ of all complex numbers and consider two mappings

$$\phi : \mathbf{C} \to \mathbf{C}, \quad \psi : \mathbf{C} \to \mathbf{C}$$

defined as follows:

$$\phi(z) = z + 1 \quad (z \in \mathbf{C}), \quad \psi(z) = e^i z \quad (z \in \mathbf{C}).$$

Check that ϕ is a translation of $\mathbf{R}^2$ and ψ is a rotation of $\mathbf{R}^2$ about its origin 0. Denote by G the semigroup of motions of $\mathbf{R}^2$, generated by $\{\phi, \psi\}$, and put $X = G(0)$. Taking into account the fact that e^i is a transcendental number, demonstrate that

$$\phi(X) \cap \psi(X) = \emptyset, \quad \phi(X) \cup \psi(X) = X.$$

In other words, the countable unbounded set $X \subset \mathbf{R}^2$ admits a decomposition into two sets which are G-congruent with X. (Such a decomposition is impossible for a nonempty bounded subset X of $\mathbf{R}^2$.)

This result is due to Mazurkiewicz and Sierpiński. It does not rely on the Axiom of Choice (compare Chapter 14).

Chapter 7
Some special constructions of nonmeasurable sets

In this chapter, we deal with some other constructions of nonmeasurable sets and functions. The approach developed here uses spaces which significantly differ from the real line $\mathbf{R}$ (and from the euclidean space $\mathbf{R}^n$ where $n \geq 2$). Specific properties of new spaces enable us to obtain the existence of nonmeasurable sets by methods essentially distinct from the classical ones which were discussed earlier. (We mean the Vitali construction, the Bernstein construction and nonmeasurable sets associated with Hamel bases of $\mathbf{R}$.)

We begin with one remarkable result which is due to Sierpiński, but formulate and prove it in a slightly different manner.

As usual, we identify the set $\mathbf{N}$ of all natural numbers with the first infinite ordinal ω.

For any natural number n, let us consider a function

$$f_n : \mathbf{R} \to \{0, 1\}$$

defined by the formula

$$f_n(x) = [2^n x] - 2[2^{n-1} x] \qquad (x \in \mathbf{R}),$$

where $[t]$ stands for the largest natural number which does not exceed $t \in \mathbf{R}$. Note that if, for some natural number l and for some odd integer k, we have $x = k/2^l$, then

$$f_l(x) = 1, \qquad (\forall p \in \,]l, \omega[)(f_p(x) = 0).$$

In particular, we see that

$$ran(f_n) = \{0, 1\} \qquad (n \in \mathbf{N}).$$

121

Thus, we have the sequence $\{f_n : n \in \mathbf{N}\}$ of functions acting from $\mathbf{R}$ into $\{0, 1\}$. If we equip the set $\{0, 1\}$ with the discrete topology, then the product space $\{0, 1\}^{\mathbf{R}}$ may be regarded as a compact topological space by virtue of the classical Tychonoff theorem. Therefore, for our sequence $\{f_n : n \in \mathbf{N}\}$, there exists at least one accumulation point in $\{0, 1\}^{\mathbf{R}}$. Take any such point and denote it by f. Sierpiński showed that f cannot be Lebesgue measurable.

Theorem 1. *The function f is not measurable in the Lebesgue sense.*

Proof. From the properties of functions of our sequence, we easily infer that f satisfies the following conditions:

1) $f(x + r) = f(x)$ for any $x \in \mathbf{R}$ and for any rational number r of the form $r = k/2^l$, where k is an odd integer and l is a natural number;

2) $f(-x) = 1 - f(x)$ for any nonzero $x \in \mathbf{R}$ which cannot be represented in the form $k/2^l$ where k is an odd integer and l is a natural number.

These two properties of the function f are completely sufficient to demonstrate that f is necessarily nonmeasurable in the Lebesgue sense (compare the proof of Theorem 6 from Chapter 1). Indeed, suppose to the contrary that f is Lebesgue measurable. Then condition 1) and the metrical transitivity of the Lebesgue measure λ, with respect to any dense subgroup of $\mathbf{R}$, imply that f must be equivalent to a constant function. Since

$$ran(f) \subset \{0, 1\},$$

we have either $f(x) = 0$ for almost all $x \in \mathbf{R}$ or $f(x) = 1$ for almost all $x \in \mathbf{R}$. On the other hand, taking into account the relation

$$f(-x) = 1 - f(x)$$

for almost all $x \in \mathbf{R}$, we claim that if f is equivalent to 0, then it must be equivalent to 1 as well, and conversely. This is impossible, of course. The contradiction obtained shows that f is nonmeasurable in the Lebesgue sense.

Remark 1. In fact, the preceding argument establishes a much stronger result. Indeed, return to the sequence of functions $\{f_n : n \in \mathbf{N}\}$. Since, for each natural number n, the set of discontinuity points of the function f_n is locally finite, we claim that f_n belongs to the first Baire class and, in particular, is a Borel function. Hence, f_n is Lebesgue measurable as well. The proof of Theorem 1 shows that no subsequence of this sequence is point-wise convergent because the pointwise limit of any sequence of Lebesgue measurable functions must be Lebesgue measurable, too.

In this context, let us recall that the problem of the existence of a sequence of Lebesgue measurable functions, whose all accumulation points (in the Tychonoff topology) are nonmeasurable in the Lebesgue sense, was first raised by Banach and was positively solved by Sierpiński who presented the construction described above.

Remark 2. It is well known that Sierpiński devoted his numerous works to those point-sets on **R** which are nonmeasurable in the Lebesgue sense (see, for instance, [196], [197] and [201]). He also gave various constructions of such sets and investigated purely logical and set-theoretical aspects of the existence of sets with bad descriptive properties.

One of his ingenious constructions starts with the existence of a nontrivial ultrafilter in the Boolean algebra of all subsets of **N**. The corresponding argument is outlined in Exercise 2 of the present chapter. Note that this construction turned out to be fruitful for further deep investigations of measurability properties of filters in the Boolean algebra $\mathcal{P}(\mathbf{N})$ (see, for instance, [151], [217], [218]). We will be concerned with such properties in Chapter 15 of the book where a famous result of Raisonnier and Shelah will be proved, stating that in the theory **ZF** & **DC** the inequality $\omega_1 \leq \mathbf{c}$ implies the existence of a non Lebesgue-measurable subset of **R**.

Some relatively simple examples of nonmeasurable sets and functions can be constructed by using infinite-dimensional Banach spaces, in particular, infinite-dimensional Hilbert spaces.

The following example illustrates the said above.

Example 1. Let E be a Hilbert space (over **R**) whose Hilbert dimension is equal to the cardinality of the continuum, that is all orthonormal bases of E are of cardinality $\mathbf{c}$. So, we may denote by $(e_t)_{t \in [0,1]}$ a fixed orthonormal basis in E. Let us consider a function

$$f : [0,1] \to E$$

defined by the formula

$$f(t) = e_t \qquad (t \in [0,1]).$$

This function possesses rather interesting properties. First of all, let us check that for any open (closed) ball $B \subset E$, the preimage $f^{-1}(B)$ is a

Lebesgue measurable subset of $[0, 1]$. Indeed, if e is the centre of an open ball B, then

$$e = \sum_{n \in \mathbf{N}} a_n e_{t_n}$$

where $(a_n)_{n \in \mathbf{N}}$ is some sequence of nonzero real numbers, for which

$$\sum_{n \in \mathbf{N}} a_n^2 < +\infty,$$

and $(t_n)_{n \in \mathbf{N}}$ is an injective sequence of points of $[0, 1]$. Obviously, we may write

$$f^{-1}(B) = \{t \in [0, 1] : ||e_t - e|| < r\}$$

where $|| \cdot ||$ is the norm in E and r stands for the radius of B. It is easy to see that $f^{-1}(B)$ is Lebesgue measurable if and only if the set

$$T = \{t \in [0, 1] \setminus \{t_n : n \in \mathbf{N}\} : ||e_t - e|| < r\}$$

is Lebesgue measurable. But, for any

$$t \in [0, 1] \setminus \{t_n : n \in \mathbf{N}\},$$

we have the equality

$$||e_t - e|| = (1 + \sum_{n \in \mathbf{N}} a_n^2)^{1/2}.$$

Denote

$$a = (1 + \sum_{n \in \mathbf{N}} a_n^2)^{1/2}.$$

If $a < r$, then

$$T = [0, 1] \setminus \{t_n : n \in \mathbf{N}\}.$$

If $a \geq r$, then $T = \emptyset$.

In both these cases T turns out to be Lebesgue measurable. Consequently, $f^{-1}(B)$ is Lebesgue measurable, too.

We can also assert that f is weakly (or scalarly) measurable; in other words, for every continuous linear functional

$$h : E \to \mathbf{R},$$

the real-valued function $h \circ f$ is Lebesgue measurable. Indeed, it suffices to observe that the set of all those $t \in [0,1]$ for which

$$(h \circ f)(t) \neq 0$$

is at most countable, hence $h \circ f$ is equivalent to zero.

At the same time, our f is not Lebesgue measurable. In other words, there exists an open set $U \subset E$ for which $f^{-1}(U)$ is not Lebesgue measurable. To see this, let us take any subset T' of $[0,1]$ nonmeasurable in the Lebesgue sense. Then the set $\{e_t : t \in T'\}$ is closed in E and, obviously,

$$f^{-1}(\{e_t : t \in T'\}) = T'.$$

Now, putting

$$U = E \setminus \{e_t : t \in T'\},$$

we conclude that U is open in E and the set

$$f^{-1}(U) = [0,1] \setminus T'$$

is not measurable in the Lebesgue sense.

Example 2. Consider an infinite-dimensional separable Hilbert space E (over $\mathbf{R}$). The norm in this space will be denoted by $\| \cdot \|$. Let B be the closed unit ball in E:

$$B = \{e \in E : \|e\| \leq 1\}.$$

It will be shown in Exercise 7 of Chapter 11 that this ball is an absolutely nonmeasurable set with respect to the class of all nonzero σ-finite translation-invariant measures on E. In other words, it can be demonstrated that there does not exist a nonzero σ-finite translation-invariant measure μ on E for which $B \in dom(\mu)$. Starting with this fact, it is not difficult to construct an absolutely nonmeasurable subset of $\mathbf{R}$ which, in particular, is nonmeasurable in the Lebesgue sense. Indeed, let us consider our E as a vector space over the field $\mathbf{Q}$ of all rational numbers. Clearly, we can represent E as a direct sum of vector spaces each of which is isomorphic to $\mathbf{Q}$; in other words, we can write

$$E = \sum_{i \in I} Q_i,$$

where all Q_i ($i \in I$) are isomorphic to $\mathbf{Q}$. Since $card(E)$ is equal to the cardinality of the continuum, we have $card(I) = \mathbf{c}$. But we also know that the real line $\mathbf{R}$ (considered as a vector space over $\mathbf{Q}$) admits an analogous representation

$$\mathbf{R} = \sum_{j \in J} Q'_j,$$

where all Q'_j $(j \in J)$ are isomorphic to $\mathbf{Q}$ and $card(J) = \mathbf{c}$. Thus, we conclude that E and $\mathbf{R}$ are isomorphic as vector spaces over the field $\mathbf{Q}$. Let

$$\phi : E \to \mathbf{R}$$

be an isomorphism between these two vector spaces and let $X = \phi(B)$. Then it is not hard to demonstrate that there exists no nonzero σ-finite translation-invariant measure μ on $\mathbf{R}$ for which $X \in dom(\mu)$. Indeed, suppose otherwise and let μ be such a measure. Put

$$S' = \{\phi^{-1}(Y) : Y \in dom(\mu)\}$$

and define a functional

$$\mu' : S' \to \mathbf{R} \cup \{+\infty\}$$

by the formula

$$\mu'(Y') = \mu(\phi(Y')) \qquad (Y' \in S').$$

An easy verification shows that μ' is a nonzero σ-finite measure on E invariant under all translations of E and such that $B \in dom(\mu')$. But the existence of μ' is impossible, which yields the required result.

Let us consider one more construction of nonmeasurable sets.

We now start with the unit circumference $\mathbf{S}_1 \subset \mathbf{R}^2$ regarded as a commutative group and equipped with the standard Lebesgue probability measure λ_1. For the sake of simplicity, we denote

$$\Gamma = \mathbf{S}_1, \qquad \mu = \lambda_1.$$

We also denote by $+$ the group operation in Γ. Actually, the construction presented below is applicable to a more general situation, but here we restrict our considerations to the measure space

$$(\Gamma, dom(\mu), \mu).$$

We shall say that a subset A of the commutative group Γ is stable if it is closed under the operation $+$; in other words, for any two elements $a \in A$ and $b \in A$, we have $a + b \in A$.

Notice that there exist infinite stable subsets of Γ which are not subgroups of Γ. Indeed, take an element d from Γ of infinite order (that is $nd \neq 0$ for all natural numbers $n > 0$). Obviously, there are uncountably many $d \in \Gamma$ having this property. Then the set $D = \{nd : n \in \mathbf{N}\}$ is infinite and stable, but is not a subgroup of Γ (because $d \in D$ and $-d \notin D$).

By uzing the Kuratowski-Zorn lemma, it can easily be shown that there exist maximal (with respect to the inclusion relation) stable subsets of Γ containing the set D introduced above and not containing the element $-d$. Let A be such a maximal subset of Γ. Hence, there are elements in A of infinite order. In particular, A is everywhere dense in Γ regarded as a commutative topological group.

Theorem 2. *For the set A, the equality*

$$A \cup (-A) = \Gamma$$

is valid. Moreover, A is nonmeasurable in the Lebesgue sense.

Proof. First, let us establish the above-mentioned equality. Suppose to the contrary that

$$A \cup (-A) \neq \Gamma$$

and choose an element

$$b \in \Gamma \setminus (A \cup (-A)).$$

It is not difficult to check that the set

$$A' = \cup\{nb + A : n \in \mathbf{N}\}$$

is stable and properly contains the set A. This immediately implies the relation $-d \in A'$ or, equivalently,

$$-d = a_1 + nb$$

for some $a_1 \in A$ and $n \in \mathbf{N} \setminus \{0\}$.

Analogously, it is not hard to verify that the set

$$A'' = \cup\{m(-b) + A : m \in \mathbf{N}\}$$

is stable and properly contains A. This implies the relation $-d \in A''$ or, equivalently,

$$-d = a_2 + m(-b)$$

for some $a_2 \in A$ and $m \in \mathbf{N} \setminus \{0\}$. Consequently, we get

$$-(m+n)d = ma_1 + mnb + na_2 - nmb = ma_1 + na_2 \in A.$$

Taking into account the inclusion $D \subset A$, we have $(m+n-1)d \in A$, from which it follows that

$$-d = (m+n-1)d + (-(m+n)d) \in A + A \subset A,$$

and this is a contradiction since $-d \notin A$ according to the definition of A.

The contradiction obtained shows that the equality

$$A \cup (-A) = \Gamma$$

must be valid.

It remains to demonstrate that the set A is not measurable in the Lebesgue sense. Suppose otherwise: $A \in dom(\mu)$. Then the equality above implies at once that $\mu(A) > 0$ (actually, it implies that $\mu(A) \geq 1/2$ but, for our further purposes, the relation $\mu(A) > 0$ is sufficient). Now, utilizing the Steinhaus property for μ (see Exercise 8 of Chapter 1), we deduce that

$$int(A + A) \neq \emptyset,$$

where $int(A+A)$ denotes the interior of $A + A$. In view of the inclusion

$$A + A \subset A,$$

we then claim that $int(A) \neq \emptyset$. Remembering also that A is everywhere dense in Γ, we finally obtain

$$\Gamma = \cup \{a_k + A : k \in \mathbf{N}\},$$

where $\{a_k : k \in \mathbf{N}\}$ denotes some everywhere dense (in Γ) sequence of elements of A. Since

$$a_k + A \subset A \qquad (k \in \mathbf{N}),$$

we readily come to the equality $\Gamma = A$ which again contradicts the definition of A.

The theorem has thus been proved.

We now want to present one construction of a small (in the sense of cardinality) nonmeasurable set in a certain infinite-dimensional commutative compact topological group. We mean here the product group

$$G = \mathbf{S}_1^{\mathbf{c}},$$

where $\mathbf{S}_1$ is the unit circumference in $\mathbf{R}^2$. Since G is a compact commutative group, we have the canonical invariant measure on G, namely, the Haar probability measure which actually coincides with the product measure λ_1^c. Let us denote by μ the completion of the Haar measure on G and let $\mathcal{B}(G)$ be the Borel σ-algebra of G. We need some nontrivial properties of G and μ.

1. μ is a Radon measure; in other words, for any Borel set $X \subset G$, we have

$$\mu(X) = \sup\{\mu(K) : K \text{ is a compact subset of } X\}.$$

This fact is a particular case of the general statement saying that every Haar measure is Radon (see, for instance, [62] or [68]).

2. The cardinality of the family of all continuous real-valued functions on G is equal to $\mathbf{c}$.

This assertion follows from the Stone-Weierstrass theorem (see [77]). Indeed, it is not difficult to find a family F of real-valued continuous functions on G, having the following properties:

a) $card(F) = \mathbf{c}$;

b) F separates the points in G;

c) F is an algebra with respect to the addition and multiplication of functions.

Now, the above-mentioned Stone-Weierstrass theorem implies that each real-valued continuous function f on G belongs to the closure of F in the topology of uniform convergence. In other words, there exists a sequence $\{f_n : n < \omega\}$ of functions from F, such that

$$f = \lim_{n \to +\infty} f_n,$$

where the corresponding convergence is uniform. This circumstance implies at once that the cardinality of the family of all continuous real-valued functions on G does not exceed

$$(card(F))^\omega = \mathbf{c}^\omega = \mathbf{c}$$

and, consequently, is equal to $\mathbf{c}$.

3. Let $\mathcal{B}_0(G)$ stand for the smallest σ-algebra of subsets of G with respect to which all continuous real-valued functions on G are measurable.

This σ-algebra is usually called the Baire σ-algebra of G. The preceding property obviously yields that

$$card(\mathcal{B}_0(G)) = \mathbf{c}.$$

Note that the Baire σ-algebra of a topological space is always contained in the Borel σ-algebra of the same space and, in general, does not coincide with it.

4. For any set $X \in \mathcal{B}(G)$, the equality

$$\mu(X) = sup\{\mu(Y) \ : \ Y \subset X, \ Y \in \mathcal{B}_0(G)\}$$

holds. This property of μ can be established by using the standard argument from Haar measure theory (for details, see [62] and [68]).

In particular, taking into account the fact that $\mathcal{B}_0(G)$ is a σ-algebra of sets, we conclude that for any set $X \in \mathcal{B}(G)$, there exists a set $Y \in \mathcal{B}_0(G)$ such that

$$Y \subset X, \quad \mu(Y) = \mu(X).$$

Now, we are able to construct a small subset of G nonmeasurable with respect to μ. More precisely, we are going to define a set Z in G for which

$$card(Z) \leq \mathbf{c} < card(G)$$

and, in addition to this, Z is μ-thick in G.

In order to carry out our construction, fix an injective family $\{Y_i : i \in I\}$ consisting of all sets from $\mathcal{B}_0(G)$ with strictly positive μ-measure. Applying the Axiom of Choice, take an arbitrary selector of $\{Y_i : i \in I\}$ and denote it by Z. Since $card(I) = \mathbf{c}$, we have the relation

$$card(Z) \leq \mathbf{c} < 2^{\mathbf{c}} = card(G).$$

Keeping in mind property 4 and the definition of Z, we easily obtain that Z is μ-thick in G; in other words,

$$\mu_*(G \setminus Z) = 0$$

or, equivalently,

$$\mu^*(Z) = \mu(G) = 1.$$

Suppose for a moment that Z is measurable with respect to μ. Then the previous relation leads to the equalities

$$\mu(Z) = \mu(G) = 1.$$

On the other hand, since $card(Z) < card(G)$, there are uncountably many pairwise disjoint translates of Z in G. Therefore, by virtue of the invariance of μ, we must have $\mu(Z) = 0$ which is impossible. The contradiction obtained shows that our set Z cannot be measurable with respect to μ.

Taking the group $[Z]$ generated by Z, we come to a subgroup of G of cardinality $\mathbf{c}$, which is μ-thick in G and hence is also μ-nonmeasurable.

Remark 3. The result just obtained is of some interest from the set-theoretical point of view. We see that, in certain infinite-dimensional compact topological groups, the situation can be found where a set of cardinality strictly less than the cardinality of the original group, turns out to be nonmeasurable with respect to the completion of the Haar measure. Let us stress that such a situation is realizable in **ZFC** theory. On the other hand, we know that for the real line **R** and for the classical Lebesgue measure λ on **R** an analogous situation is impossible (within **ZFC**). Indeed, under Martin's Axiom, all subsets of **R** whose cardinalities are strictly less than $\mathbf{c}$ turn out to be of Lebesgue measure zero. At the same time, in Cohen type models of set theory there are non Lebesgue-measurable subsets of **R** of cardinality strictly less than $\mathbf{c}$ (see, for instance, [121]). In this connection, let us also recall the result of Kunen stating that if the cardinal $\mathbf{c}$ is real-valued measurable, then there exists a subset Z of **R** such that

$$card(Z) < \mathbf{c}, \quad Z \notin dom(\lambda).$$

The detailed proof of this result was presented in Chapter 4.

Now, we are going to discuss several purely set-theoretical (combinatorial) constructions leading to the existence of nonmeasurable sets. Our starting point here is the classical result of Ulam [222] which has already been mentioned in previous chapters. Let us recall the precise formulation of this result:

There does not exist a nonzero σ-finite diffused measure defined on the σ-algebra $\mathcal{P}(\omega_1)$. In other words, ω_1 is not a real-valued measurable cardinal.

Actually, Ulam established the nonexistence of such a measure by applying a certain transfinite matrix of subsets of ω_1. We shall consider an Ulam matrix below. In this context, let us recall that the nonexistence of a nontrivial σ-finite diffused measure on $\mathcal{P}(\omega_1)$ can directly be derived from the equality

$$\mathcal{P}(\omega_1) \otimes \mathcal{P}(\omega_1) = \mathcal{P}(\omega_1 \times \omega_1)$$

which was stated by using the Sierpiński partition of the product set $\omega_1 \times \omega_1$ (see Chapter 4).

Remark 4. We have one more approach which leads us to the real-valued nonmeasurability of ω_1. Namely, we know from the results presented in Chapter 4 that there are uncountable universal measure zero subspaces of $\mathbf{R}$. Consequently, there exists a universal measure zero set $X \subset \mathbf{R}$ with $card(X) = \omega_1$. This fact easily implies that ω_1 cannot be real-valued measurable.

All remarks made above are useful for better understanding various aspects of the non-real-valued-measurability of the first uncountable cardinal. In the sequel, we shall show that this classical result has important consequences (see Chapters 9 and 13). One of immediate consequences can be formulated as follows: if the Continuum Hypothesis holds, then the cardinal $\mathbf{c}$ is not real-valued measurable and, furthermore, there exists a countably generated σ-algebra $\mathcal{S}$ of subsets of $\mathbf{R}$ such that all one-element sets in $\mathbf{R}$ belong to $\mathcal{S}$ and there is no nonzero σ-finite diffused measure on $\mathcal{S}$.

In connection with the latter statement, we recall that it was originally established by Banach and Kuratowski in their well-known work [8]. The method of [8] essentially differs from the ones mentioned above, and it is reasonable to underline here that in some sense the method of Banach and Kuratowski yields a stronger result (applicable to functionals more general than ordinary measures). Let us briefly describe their construction.

Consider the family $F = \omega^\omega$ of all functions acting from ω into ω. Let f and g be any two functions from F. We put $f \preceq g$ if and only if there exists a natural number $n = n(f,g)$ such that $f(m) \le g(m)$ for all natural numbers $m \ge n$. Evidently, the relation $\preceq$ is a pre-ordering of F. Now, if the Continuum Hypothesis holds, then it is not hard to define a subset $E = \{f_\xi \ : \ \xi < \omega_1\}$ of F satisfying the following two conditions:

(a) if f is an arbitrary function from F, then there exists an ordinal $\xi < \omega_1$ such that $f \preceq f_\xi$ (in other words, E is cofinal with F);

(b) for any ordinals ξ and ζ such that $\xi < \zeta < \omega_1$, the relation $f_\zeta \preceq f_\xi$ is not true.

Note that each of these two conditions implies the equality

$$card(E) = \omega_1.$$

Further, for any two natural numbers m and n, we put

$$E_{m,n} = \{f_\xi \in E \ : \ f_\xi(m) \leq n\}.$$

In this way, we get a double countable family of sets

$$(E_{m,n})_{m<\omega,\ n<\omega}.$$

It is easy to check that, for each $m < \omega$, we have the inclusions

$$E_{m,0} \subset E_{m,1} \subset \ ... \ \subset E_{m,n} \subset \ ...$$

and the equality
$$E = \cup\{E_{m,n} \ : \ n < \omega\}.$$

Also, conditions (a) and (b) immediately imply that, for an arbitrary function f from F, the set

$$E_{0,f(0)} \cap E_{1,f(1)} \cap \ ... \ \cap E_{m,f(m)} \cap \ ...$$

is at most countable.

Any family $(E_{m,n})_{m<\omega,n<\omega}$ of subsets of E, having these properties is usually called a Banach-Kuratowski matrix over E.

Starting with the above-mentioned properties of a Banach-Kuratowski matrix, one can easily infer that there does not exist a nonzero σ-finite diffused measure on E defined simultaneously for all sets $E_{m,n}$ where $m < \omega$ and $n < \omega$. In addition, it is not difficult to see that an analogous result remains true for many other functionals on E which are essentially more general than ordinary measures. For instance, let ν be a real-valued, positive (that is nonnegative) function defined on some class of subsets of E, closed under finite intersections. We say that ν is an admissible functional (on E) if the following conditions are fulfilled:

1) the family of all countable subsets of E is contained in $dom(\nu)$ and, for any countable set $X \subset E$, we have $\nu(X) = 0$;

2) if $\{Z_n \ : \ n < \omega\}$ is an increasing (with respect to inclusion) family of sets belonging to $dom(\nu)$, then the set $\cup\{Z_n \ : \ n < \omega\}$ also belongs to $dom(\nu)$, and

$$\nu(\cup\{Z_n \ : \ n < \omega\}) \leq sup\{\nu(Z_n) \ : \ n < \omega\};$$

3) if $\{Z_n \ : \ n < \omega\}$ is a decreasing (with respect to inclusion) family of sets belonging to $dom(\nu)$, then the set $\cap\{Z_n \ : \ n < \omega\}$ also belongs to $dom(\nu)$, and

$$\nu(\cap\{Z_n \ : \ n < \omega\}) \geq inf\{\nu(Z_n) \ : \ n < \omega\}.$$

Evidently, if ν is a finite diffused measure on E, then ν satisfies conditions 1), 2) and 3).

In general, admissible functionals need not have additivity properties similar to the corresponding properties of ordinary measures. However, the Banach-Kuratowski method works for such functionals and we can conclude that there does not exist a nonzero admissible functional on E defined simultaneously for all members of a Banach-Kuratowski matrix. Indeed, suppose for a moment that ν is an admissible functional on E not equal to zero and satisfying the inclusion

$$\{E_{m,n} : m < \omega, \ n < \omega\} \subset dom(\nu).$$

Without loss of generality, we may assume that $\nu(E) \neq 0$. Then, in view of condition 2), there exists a set $E_{0,n(0)}$ of a Banach-Kuratowski matrix, such that

$$\nu(E_{0,n(0)}) > \varepsilon$$

for some strictly positive real ε. Consider the family of sets

$$\{E_{0,n(0)} \cap E_{1,n} : n < \omega\}.$$

Utilizing the same condition 2), we will be able to find a set $E_{1,n(1)}$ of our matrix, such that

$$\nu(E_{0,n(0)} \cap E_{1,n(1)}) > \varepsilon.$$

Proceeding in this manner, we will define a sequence

$$E_{0,n(0)}, \ E_{1,n(1)}, \ \ldots, \ E_{m,n(m)}, \ \ldots$$

of members of our matrix, satisfying the relations

$$\nu(E_{0,n(0)} \cap E_{1,n(1)} \cap ... \cap E_{m,n(m)}) > \varepsilon$$

for all $m < \omega$. Now, applying condition 3), we must have

$$\nu(E_{0,n(0)} \cap E_{1,n(1)} \cap ... \cap E_{m,n(m)} \cap ...) \geq \varepsilon,$$

which is impossible by virtue of condition 1). The contradiction obtained shows us that at least one set $E_{m,n}$ of a Banach-Kuratowski matrix does not belong to $dom(\nu)$.

Now, let us introduce an Ulam transfinite matrix which is also very helpful in various questions concerning the existence of nonmeasurable sets.

Let E be a set and let

$$(E_{m,\xi})_{m<\omega,\xi<\omega_1}$$

be a double family of subsets of E. We say that this family is an Ulam matrix over E (more exactly, an Ulam $(\omega \times \omega_1)$-matrix over E) if the following relations are satisfied:

(1) for each natural number m, the partial family $(E_{m,\xi})_{\xi<\omega_1}$ is disjoint;
(2) for each ordinal $\xi < \omega_1$, we have the inequality

$$card(E \setminus \cup\{E_{m,\xi} \; : \; m < \omega\}) \leq \omega.$$

The sets $E_{m,\xi}$ are usually called terms (or members) of a given Ulam matrix.

The following auxiliary proposition (due to Ulam) turns out to be crucial in solving the problem of the existence of universal diffused probability measures on a set of cardinality ω_1.

Lemma 1. *If $card(E) = \omega_1$, then there exists an Ulam matrix over E.*

Proof. Of course, we may identify E with the first uncountable ordinal ω_1; in other words, we may take $E = \omega_1$. Since for any ordinal $\xi < \omega_1$, we have

$$card([0,\xi]) \leq \omega,$$

there exists a surjection

$$f_\xi \; : \; \omega \rightarrow [0,\xi].$$

We put

$$E_{m,\xi} = \{\zeta \in \,]\xi,\omega_1[\ :\ f_\zeta(m) = \xi\}$$

for all $m < \omega$ and for all $\xi < \omega_1$. Let us check that the obtained family of sets

$$(E_{m,\xi})_{m<\omega,\xi<\omega_1}$$

is an Ulam matrix over $E = \omega_1$. To see this, first fix a natural number m and show that

$$E_{m,\xi} \cap E_{m,\xi'} = \emptyset$$

for any two distinct ordinals $\xi < \omega_1$ and $\xi' < \omega_1$. Indeed, supposing that

$$\zeta \in E_{m,\xi} \cap E_{m,\xi'},$$

we must have

$$\xi < \zeta, \quad \xi' < \zeta, \quad f_\zeta(m) = \xi, \quad f_\zeta(m) = \xi',$$

which is impossible since f_ζ is a function and $\xi \neq \xi'$. Thus, the sets $E_{m,\xi}$ and $E_{m,\xi'}$ are disjoint. Fix now an ordinal $\xi < \omega_1$ and consider the set

$$E_\xi = \cup\{E_{m,\xi} : m < \omega\}.$$

We assert that

$$E \setminus E_\xi \subset [0,\xi].$$

Indeed, taking an arbitrary countable ordinal ζ strictly greater than ξ, we may write

$$\xi \in [0,\zeta] = f_\zeta(\omega)$$

and, hence, there exists a natural number m for which $f_\zeta(m) = \xi$ or, equivalently, $\zeta \in E_{m,\xi}$. This establishes the required relation $E \setminus E_\xi \subset [0,\xi]$. We thus conclude that

$$card(E \setminus E_\xi) \leq card([0,\xi]) \leq \omega.$$

This completes the proof of Lemma 1.

By starting with the preceding lemma, it is not hard to deduce the following classical result of Ulam [222].

Theorem 3. *The first uncountable cardinal ω_1 is not real-valued measurable.*

Proof. We have to show that for every set E with $card(E) = \omega_1$, there is no universal diffused probability measure on E. Suppose otherwise and denote by μ such a measure. Let $(E_{m,\xi})_{m<\omega,\xi<\omega_1}$ be an Ulam matrix over E. We are going to demonstrate that at least one of the members of this matrix must be nonmeasurable with respect to μ. Indeed, taking into account the definition of an Ulam matrix, we can easily derive from relation (2) that, for each ordinal $\xi < \omega_1$, there exists a natural number $m = m(\xi)$ such that $\mu(E_{m,\xi}) > 0$. Now, since ω is a countable set and ω_1 is an uncountable set, we conclude that there exist a natural number k and a subset Ξ of ω_1, satisfying the relations

$$card(\Xi) = \omega_1, \qquad (\forall \xi \in \Xi)(m(\xi) = k).$$

Consequently, applying property (1) of the Ulam matrix, we have

$$E_{k,\xi} \cap E_{k,\zeta} = \emptyset \quad (\xi \in \Xi, \ \zeta \in \Xi, \ \xi \neq \zeta),$$

$$\mu(E_{k,\xi}) > 0 \quad (\xi \in \Xi),$$

which contradicts the countable chain condition for μ (see Exercise 10 from Chapter 1). The contradiction obtained finishes the proof.

Remark 5. Later, we shall give several applications of Ulam matrices to other measurability problems (see, for instance, Chapter 13 where measurability properties of subgroups of an uncountable group are discussed).

An important application of an Ulam matrix to the concept of smallness of sets was given by Pfeffer and Prikry in their extensive article [172].

Note that some additional information on Ulam matrices is contained in Exercise 4 of this chapter. Namely, Exercise 4 presents a generalization of Ulam $(\omega \times \omega_1)$-matrices for sets of higher cardinalities. Let us also mention Exercise 6 in which deep connections between real-valued measurable cardinals and combinatorial properties of infinite trees are indicated.

EXERCISES

1. Let E be a Hilbert space (over $\mathbf{R}$) whose Hilbert dimension is equal to ω_1. Equip ω_1 with the Dieudonné measure ν (see Exercise 7 of Chapter 2) and consider a mapping

$$f : \omega_1 \to E$$

defined by

$$f(\xi) = e_\xi \qquad (\xi < \omega_1),$$

where $(e_\xi)_{\xi<\omega_1}$ is a fixed orthonormal basis in E. Show that:

(a) the set $f^{-1}(B)$ is ν-measurable for each open (closed) ball B in E;

(b) the mapping f is weakly measurable with respect to ν;

(c) the mapping f is not measurable with respect to ν.

2. Let $\mathbf{N}$ denote, as usual, the set of all natural numbers (identified with ω) and let $\mathcal{P}(\mathbf{N})$ stand for the family of all subsets of $\mathbf{N}$. This family can be regarded as a complete Boolean algebra with respect to the standard set-theoretical operations. An easy application of the Kuratowski-Zorn lemma yields the existence of a nontrivial ultrafilter U in $\mathcal{P}(\mathbf{N})$ (in other words, U is a maximal filter containing no singleton in $\mathbf{N}$).

Consider simultaneously the Cantor discontinuum $C = \{0,1\}^\omega$ regarded as a compact topological group with respect to the addition operation (modulo 2) and with respect to the product topology (starting with the discrete topology on $\{0,1\}$). Let μ be the Haar probability measure on C and let μ' denote the completion of μ. As well known (see [62] or [19]), μ' is isomorphic to the standard Lebesgue measure on the closed unit interval $[0,1]$.

For each set $X \subset \mathbf{N}$, we may define its characteristic function f_X. Obviously, we have

$$f_X \in \{0,1\}^\omega.$$

Let us put

$$U' = \{f_X : X \in U\}.$$

By applying the metrical transitivity of μ' and its invariance with respect to the symmetry in C, show that the set U' is not measurable with respect to μ'. Conclude from this fact that, in the theory $\mathbf{ZF}$ & $\mathbf{DC}$, the existence of a nontrivial ultrafilter in $\mathbf{N}$ implies the existence of a non Lebesgue-measurable subset of $\mathbf{R}$.

Formulate and prove an analogous result for the Baire property.

3. Starting with the existence of a nontrivial ultrafilter in $\mathcal{P}(\omega)$, show that the countable version of the Ramsey combinatorial theorem [179] implies its finite version (compare Exercise 11 from Chapter 6).

4. Let us identify every cardinal number κ with the smallest ordinal whose cardinality is equal to κ. Let us also denote by κ^+ the smallest cardinal number strictly greater than κ.

Take an arbitrary infinite set E such that $\kappa^+ = card(E)$. We say that a double family

$$(E_{\xi,\zeta})_{\xi<\kappa,\zeta<\kappa^+}$$

of subsets of E is an Ulam (κ, κ^+)-matrix over E if the following conditions are satisfied:

(a) $card(E \setminus \cup\{E_{\xi,\zeta} : \xi < \kappa\}) \leq \kappa$ for each ordinal $\zeta < \kappa^+$;

(b) $E_{\xi,\zeta} \cap E_{\xi,\zeta'} = \emptyset$ for each $\xi < \kappa$ and for any two distinct ordinals $\zeta < \kappa^+$ and $\zeta' < \kappa^+$.

Prove the existence of an Ulam (κ, κ^+)-matrix over E. Deduce from this fact that if κ is not a real-valued measurable cardinal, then κ^+ is not real-valued measurable either.

Let $\{\kappa_i : i \in I\}$ be a family of cardinal numbers. Show that if $card(I)$ and all κ_i $(i \in I)$ are not real-valued measurable, then the cardinal sum $\sum_{i \in I} \kappa_i$ is not real-valued measurable either.

Recall that an infinite cardinal number κ is regular if it cannot be represented as the sum of cardinals $\sum_{i \in I} \kappa_i$, where

$$card(I) < \kappa,$$

$$(\forall i \in I)(\kappa_i < \kappa).$$

For example, the cardinal κ^+ is regular whenever κ is infinite.

Recall also that a cardinal $\kappa = \omega_\alpha$ is inaccessible if it is regular and α is a nonzero limit ordinal.

Starting with the above-mentioned auxiliary results and applying the method of transfinite induction, demonstrate that all cardinal numbers strictly less than the first inaccessible cardinal are not real-valued measurable [222].

It is well known that the existence of inaccessible cardinals cannot be proved within the theory **ZFC** (see, for instance, [122] or [127]). Therefore, it is consistent with **ZFC** that all cardinal numbers are not real-valued measurable.

5. Let $(E, \preceq)$ be a partially ordered set. We say that it is a tree if E has the smallest element and, for each $e \in E$, the set

$$E^e = \{x \in E : x \prec e\}$$

is well ordered with respect to the induced ordering. The ordinal type of the above-mentioned set E^e is called the height of e and is denoted by $o(e)$.

For any ordinal α, the set

$$E_\alpha = \{e \in E : o(e) = \alpha\}$$

is called the α-th level of a given tree. In particular, the 0-th level is a singleton consisting of the smallest element of E (which is usually called the root of E). We also denote

$$o(E) = sup_{e \in E}(o(e) + 1)$$

and say that $o(E)$ is the height of E. Actually, the height of E coincides with the least element of the class of all those α for which the α-th level of E is empty.

We shall say that $(E, \preceq)$ is an α-tree if $o(E) = \alpha$.

A subset of E is called a branch in E if it is a maximal linearly ordered (hence, maximal well-ordered) subset of E.

Finally, we say that the tree property holds for $(E, \preceq)$ if there exists a branch in E intersecting all nonempty levels of E. Obviously, such a branch is necessarily maximal.

Demonstrate that if $(E, \preceq)$ is an ω-tree whose all levels are finite, then E has the tree property.

This assertion is due to König and is known as the König lemma. It yields useful consequences in graph theory, combinatorics and other areas of mathematics (for example, by using this lemma, one can prove that if all finite subgraphs of a given countable graph are planar, then the graph is planar, too). Notice also that the proof of the König lemma needs some weak form of the Axiom of Choice (see Appendix 1).

6. Let $(E, \preceq)$ be a tree satisfying the following conditions:
(a) $o(E)$ is the first real-valued measurable cardinal;
(b) all levels of E are of cardinality strictly less than $o(E)$.
Show that $(E, \preceq)$ has the tree property.

Clearly, this result is an analogue of the König lemma for those $o(E)$-trees whose all levels are of cardinality strictly less than $o(E)$. (In this connection, see Exercise 7 below.)

In fact, a more general result can be established. Namely, if κ is a cardinal number with a κ-additive diffused probability measure on $\mathcal{P}(\kappa)$ and $(T, \preceq)$ is a κ-tree whose all levels are of cardinality $< \kappa$, then $(T, \preceq)$ has the tree property. (Here κ is again identified with the least ordinal of cardinality κ.)

To establish the above-mentioned generalized result, let us define

$$T^*(x) = \{y \in T : x \preceq y\}$$

for any $x \in T$, and let T_α be the α-th level of T for any ordinal $\alpha < \kappa$. Denote by μ a κ-additive diffused probability measure on $\mathcal{P}(T)$ and put

$$t_0(x) = \mu(T^*(x)),$$

$$t_1(\alpha) = sup\{t_0(x) : x \in T_\alpha\}.$$

Since the equality

$$\mu(\cup\{T^*(x) : x \in T_\alpha\}) = 1$$

holds for each $\alpha < \kappa$, we infer, in view of the κ-additivity of μ, that

$$(\forall \alpha < \kappa)(t_1(\alpha) > 0).$$

Furthermore, if $\alpha \leq \beta < \kappa$, then

$$t_1(\alpha) \geq t_1(\beta).$$

Therefore, there exist an ordinal $\gamma < \kappa$ and a real $r > 0$ such that

$$(\forall \alpha \in [\gamma, \kappa[)(t_1(\alpha) = r).$$

Now, we define

$$S = \{x : (\exists \alpha \geq \gamma)(x \in T_\alpha \ \& \ t_0(x) > (1/2)r)\}.$$

Obviously, we get
$$S \cap T_\alpha \neq \emptyset$$

for all those $\alpha \geq \gamma$ which are strictly smaller than κ. In particular, we have $card(S) = \kappa$. The set S possesses also the following property:

$$(\forall y \in S)(\exists x \in S)(x \in T_\gamma \ \& \ x \preceq y).$$

Indeed, to verify this, take any $y \in S$. Then $y \in T_\alpha$ for some $\alpha \geq \gamma$ and $t_0(y) > (1/2)r$. Pick an element $x \preceq y$ from the level T_γ. Clearly,

$$T^*(y) \subset T^*(x),$$

from which it follows that $t_0(x) > (1/2)r$ and hence $x \in S$, as well. Taking into account the circumstance that $card(T_\gamma) < \kappa$, we deduce that

$$card(S \cap T^*(z)) = \kappa$$

for some $z \in T_\gamma$. Finally, let us demonstrate that any two distinct elements u and v from $S \cap T^*(z)$ are comparable with respect to $\preceq$. Supposing to the contrary that $u \npreceq v$ and $v \npreceq u$, we get

$$T^*(u) \cap T^*(v) = \emptyset,$$

$$T^*(u) \cup T^*(v) \subset T^*(z),$$

$$t_0(u) > (1/2)r, \qquad t_0(v) > (1/2)r$$

and, consequently,

$$t_0(z) \geq t_0(u) + t_0(v) > r$$

which is impossible. The obtained contradiction shows that $S \cap T^*(z)$ is a linearly ordered subset of T whose cardinality equals κ. Evidently, this subset can be included in some branch intersecting all nonempty levels of T.

7. Recall that the symbol $\omega^{<\omega_1}$ denotes the family of all functions f such that $dom(f)$ is a proper initial subinterval of ω_1 and $ran(f)$ is a subset of ω. Utilizing this notation, let us put

$$F = \{f \in \omega^{<\omega_1} \ : \ f \text{ is an injection}\}.$$

For any two functions $f \in F$ and $g \in F$, define $f \sim g$ if and only if

$$dom(f) = dom(g), \qquad card(\{\xi \in dom(f) : f(\xi) \neq g(\xi)\}) < \omega.$$

By using the method of transfinite recursion, construct an ω_1-sequence

$$\{f_\xi : \xi < \omega_1\} \subset F$$

satisfying the following relations:

(a) $dom(f_\xi) = [0, \xi[$ for each ordinal $\xi < \omega_1$;
(b) $f_\alpha|[0, \beta[\ \sim f_\beta$ for any two ordinals $\alpha < \omega_1$ and $\beta < \alpha$;
(c) $card(\omega \setminus ran(f_\xi)) = \omega$ for each ordinal $\xi < \omega_1$.

Note that if for some $\xi < \omega_1$, the function f_ξ is already defined, then the role of $f_{\xi+1}$ can be played by any function from F which extends f_ξ and whose domain coincides with $[0, \xi]$.

The case of limit countable ordinals needs a more delicate argument.

Let $\xi < \omega_1$ be a limit ordinal and suppose that the partial family of functions $\{f_\zeta : \zeta < \xi\}$ satisfying relations (a), (b), (c) has already been defined. Take an arbitrary strictly increasing sequence of ordinal numbers $\{\xi_n : n < \omega\}$ such that

$$lim_{n \to +\infty} \xi_n = \xi,$$

and construct (by ordinary recursion) a sequence $\{f'_n : n < \omega\} \subset F$ with the following properties:

(i) $f'_n \sim f_{\xi_n}$ for each $n < \omega$;

(ii) $f'_{n+1}\|[0, \xi_n[\ = f'_n$ for each $n < \omega$.

Now, let f'_ξ be the unique function defined on $[0, \xi[$ and extending all functions f'_n $(n < \omega)$. Note that the existence of f'_ξ follows directly from property (ii).

Verify that $f'_\xi \in F$ and $f'_\xi\|[0, \zeta[\ \sim f_\zeta$ for all ordinals $\zeta < \xi$.

Further, show that an injective mapping

$$f_\xi \ : \ [0, \xi[\ \to \omega$$

can be defined satisfying the next two conditions:

(iii) $f_\xi(\zeta) = f'_\xi(\zeta)$ for any ordinal $\zeta \in [0, \xi[\ \setminus \{\xi_n : n < \omega\}$;

(iv) the set $\omega \setminus ran(f_\xi)$ is infinite.

Check that f_ξ is the required one (in other words, relations (a), (b), (c) are valid for the partial family $\{f_\zeta : \zeta \leq \xi\}$).

Finally, having the ω_1-sequence of functions

$$\{f_\xi : \xi < \omega_1\} \subset F$$

with properties (a), (b), (c), let us put

$$T = \{f \in F \ : \ (\exists \xi < \omega_1)(f \sim f_\xi)\}$$

and equip T with its standard ordering $\preceq$ (in other words, $f \preceq g$ if and only if g extends f).

Demonstrate that the partially ordered set $(T, \preceq)$ is an ω_1-tree, all levels of $(T, \preceq)$ are countable and $(T, \preceq)$ does not possess the tree property.

$(T, \preceq)$ is usually called an Aronszajn tree. We thus see that the König lemma does not admit a natural generalization to all uncountable cardinals.

8. Starting with the results of the two preceding exercises, give one more proof of the fact that ω_1 is not a real-valued measurable cardinal.

9. Let X be a stationary subset of ω_1 (see Exercise 7 for Chapter 2). Show that there exists a partition $\{X_\xi : \xi < \omega_1\}$ of X such that every X_ξ is also a stationary subset of ω_1.

Chapter 8
The generalized Vitali construction

Here we wish to return to the classical result of Vitali [224] stating that there are subsets of the real line $\mathbf{R}$, nonmeasurable in the Lebesgue sense (compare Theorem 1 from Chapter 1). Moreover, as pointed out in Chapter 1, the argument of Vitali yields a much stronger statement according to which for an arbitrary Lebesgue measurable set $X \subset \mathbf{R}$ with strictly positive measure, there exists a subset of X nonmeasurable in the Lebesgue sense. (See also Exercise 4 of Chapter 1 where a more general assertion is formulated.) We have already mentioned that this classical result was generalized in many other directions. A natural way to extend it is to examine its possible analogues for a finite-dimensional euclidean space and for a given group of isometries (motions) of this space.

In the present chapter, we consider questions closely related to the above-mentioned result of Vitali, for various groups of motions acting on finite-dimensional euclidean spaces. Our main goal is to describe all those groups of motions of an euclidean space, for which an analogue of Vitali's result remains true. (We recall that, in the formulation of Vitali's theorem, the group of all translations of the real line is a basic group of transformations of this line.)

Let E denote a finite-dimensional euclidean space and let G be a subgroup of the group of all isometric transformations (motions) of E. Then the pair (E, G) can be regarded as a space equipped with a transformation group.

Throughout this chapter, we denote by K the open unit cube in E; in other words, K is a fixed open cube in E whose all edges have length 1 and whose one vertex coincides with the origin of E.

Let μ be a measure given on E and let $dom(\mu)$ denote the domain of μ. We shall say that μ is a G-measure on E if the following two conditions are satisfied:

(1) $K \in dom(\mu)$ and $\mu(K) = 1$;

(2) μ is invariant with respect to G; in other words, $dom(\mu)$ is a G-invariant σ-ring of subsets of E, and for each set $X \in dom(\mu)$ and for each transformation $g \in G$, we have $\mu(g(X)) = \mu(X)$.

For example, the standard Lebesgue measure on E (considered as a G-invariant measure) is a G-measure on E.

We shall say that a group G has the Vitali property if for every G-measure μ on E and for each μ-measurable set $X \subset E$ with $\mu(X) > 0$, there exists at least one μ-nonmeasurable subset of X.

We shall say that a group G has the weak Vitali property if for every G-measure μ on E, there exists at least one μ-nonmeasurable subset of E.

We are going to establish some necessary and sufficient conditions (formulated in terms of the pair (E, G)) under which a given group G has the Vitali property (the weak Vitali property, respectively). The material presented in this chapter is primarily based on the results obtained in [103].

First, we wish to recall some auxiliary notions and facts.

Let F be a nonempty set and let Γ be a group of transformations of F. We say that Γ acts freely on (in) F if for each point $x \in F$ and for any two distinct transformations $g \in \Gamma$ and $h \in \Gamma$, the relation $g(x) \neq h(x)$ is valid.

In particular, the group of all translations of an euclidean space E acts freely on E. More generally, if Γ is an arbitrary group, then the group of all left (right) translations of Γ acts freely on Γ.

Id_E denotes the identity transformation of an euclidean space E.

If L is an affine linear manifold in a space E, then $dim(L)$ denotes the dimension of L.

In particular, if

$$dim(E) = n,$$

then the original euclidean space E can be identified with the canonical product space $\mathbf{R}^n$. The latter space is equipped with the standard Lebesgue measure λ_n whose dimension also coincides with the dimension of E. Thus, we may suppose in the sequel that λ_n is given on E. Moreover, in those cases where the dimension of λ_n does not play an essential role, we write λ instead of λ_n (in order to shorten the notation). However, if we deal with the canonical space $\mathbf{R}^n$, then we necessarily utilize the precise notation λ_n for the standard n-dimensional Lebesgue measure on this space.

Recall that if μ is an arbitrary measure on a space E, then μ^* and μ_* stand, respectively, for the outer measure and for the inner measure associated with μ.

Let μ be a measure on E and let $\{X_i : i \in I\}$ be a family of subsets of E. We shall say that this family is almost disjoint with respect to μ if

$$\mu^*(X_i \cap X_j) = 0$$

for all $i \in I$, $j \in I$, $i \neq j$.

As usual, the symbol $\mathbf{c}$ denotes the cardinality of the continuum. Thus, $\mathbf{c} = card(E)$ in the case $dim(E) > 0$.

Now, let us formulate several auxiliary statements which we need in our further considerations.

Lemma 1. *Let E be an n-dimensional euclidean space, G be a group of affine transformations of E with $card(G) < \mathbf{c}$ and let Y be a subset of E such that:*

1) $card(Y) = \mathbf{c}$;

2) no $n + 1$ pairwise distinct points of Y belong to an affine hyperplane of E; in other words, Y is a set of points in general position.

Then there exists a point $y \in Y$ for which the group G acts freely on the G-orbit $G(y)$.

Proof. The argument presented below is not difficult. Actually, it suffices to take into account the elementary fact from the geometry of euclidean spaces, stating that any affine transformation of E is completely determined by its restriction to a subset of Y with cardinality $n + 1$.

Indeed, suppose to the contrary that for each point $y \in Y$, the given group G does not act freely on the orbit $G(y)$. Then, for any $y \in Y$, there exist two distinct transformations $g_y \in G$ and $h_y \in G$, such that

$$g_y(y) = h_y(y).$$

According to the assumption of our lemma,

$$card(G \times G) \leq card(G) + \omega < card(Y) = \mathbf{c}.$$

In view of the above-mentioned geometric fact, if $g \in G$, $h \in G$ and $g \neq h$, then

$$card(\{y \in Y : g(y) = h(y)\}) \leq dim(E) < \omega.$$

Observe now that there exist a pair $(g, h) \in G \times G$ of two distinct transformations and a set $Y_0 \subset Y$ with $card(Y_0) \geq \omega$, satisfying the relation

$$(\forall y \in Y_0)(g = g_y \ \& \ h = h_y).$$

Therefore, applying the same geometric fact, we must have $g = h$, which leads to a contradiction. This contradiction finishes the proof of the lemma.

Lemma 1 yields a number of useful consequences. For example, one such consequence is the next well-known statement.

Lemma 2. *Let G be a group of isometric transformations of a finite-dimensional euclidean space E. Then the following two conditions are equivalent:*

1) G is discrete;

2) for each point $e \in K$, the set $G(e)$ is locally finite in E.

Proof. Assume that relation 1) holds, and let us show that for any point $e \in E$, the orbit $G(e)$ is a locally finite subset of E. Suppose otherwise: there exist a point $e \in E$ and a countable family $\{g_n : n < \omega\}$ of transformations from G, such that the family of points $\{g_n(e) : n < \omega\}$ is injective and bounded in E. In particular, we obtain that all motions g_n $(n < \omega)$ are pairwise distinct. If 0 denotes the origin of E, then we have

$$||g_n(e) - g_n(0)|| = ||e - 0|| = ||e|| \qquad (n < \omega),$$

from which it follows that the family of points $\{g_n(0) : n < \omega\}$ is bounded in E, too. From this fact we easily infer that the family $\{g_n : n < \omega\}$ is relatively compact in the topological group of all motions of E. Therefore, there exists an isometric transformation g of E belonging to the closure of $\{g_n : n < \omega\}$. We may assume, without loss of generality, that

$$g = lim_{n \to +\infty} g_n$$

or, equivalently,

$$lim_{n \to +\infty} g_n \circ g^{-1} = Id_E.$$

Now, taking into account the last equality, it is not difficult to construct (by ordinary recursion) an injective sequence of motions

$$\{h_0, h_1, ..., h_m, ...\} \subset G,$$

such that

$$lim_{m \to +\infty} h_m = Id_E,$$

which contradicts condition 1). The contradiction obtained establises the implication 1) $\Rightarrow$ 2).

Suppose now that condition 2) holds. Then, obviously, we must have

$$card(G) \leq \omega.$$

Let K_0 be any subset of K satisfying the relations:

(a) $card(K_0) = \mathbf{c}$;

(b) K_0 is a set of points in general position.

The existence of the above-mentioned set K_0 is evident. In accordance with Lemma 1, there exists at least one point $e \in K_0$ such that the group G acts freely on the orbit $G(e)$. Suppose for a moment that condition 1) does not hold, that is our group G is not discrete. Then we can find an injective sequence $\{g_n : n < \omega\}$ of transformations from G, convergent to the identity transformation of E. In other words, we have

$$lim_{n \to +\infty} g_n = Id_E.$$

Now, it is clear that

$$lim_{n \to +\infty} g_n(e) = e$$

and, moreover, all points $g_n(e)$ $(n < \omega)$ are pairwise distinct. This immediately implies that the set $\{g_n(e) : n < \omega\}$ (and, consequently, the orbit $G(e)$) cannot be locally finite in E which contradicts condition 2). The contradiction obtained yields the required implication 2) $\Rightarrow$ 1).

Lemma 2 has thus been proved.

Lemma 3. *Let E be a finite-dimensional euclidean space, G be a group of motions of E and let K denote the open unit cube in E. Then the following three assertions are equivalent:*

1) $\cup\{g(K) : g \in G\} = E$;

2) for each G-measure μ on E and for an arbitrary bounded μ-measurable set $X \subset E$, we have $\mu(X) < +\infty$;

3) for every G-measure μ on E and for each μ-measurable set $X \subset E$ with $\mu(X) > 0$, there exists a μ-measurable set $Y \subset X$ such that

$$0 < \mu(Y) < +\infty;$$

in other words, μ is a semifinite measure.

Proof. Suppose that 1) is true. Let μ be any G-measure on E and let X be an arbitrary bounded μ-measurable subset of E. Let $cl(X)$ denote the closure of X. Obviously, the set $cl(X)$ is compact and the family

$\{g(K) \;:\; g \in G\}$ forms an open covering of $cl(X)$. Consequently, there exists a finite subset H of G such that

$$X \subset cl(X) \subset \cup\{h(K) \;:\; h \in H\}.$$

Therefore, we have

$$\mu(X) \le \mu(\cup\{h(K) \;:\; h \in H\}) \le card(H) < +\infty,$$

and assertion 2) is thus valid.

Suppose again that 1) is true and let X be an arbitrary μ-measurable subset of E with $\mu(X) > 0$. Clearly, we can write

$$E = \cup\{h(K) \;:\; h \in H\}$$

for some countable subset H of G. This equality implies

$$0 < \mu(X) = \mu(\cup\{h(K) \cap X \;:\; h \in H\}) \le \sum_{h \in H} \mu(h(K) \cap X).$$

Consequently, there exists an element $h \in H$ such that

$$\mu(h(K) \cap X) > 0.$$

Let us put
$$Y = h(K) \cap X.$$

Then we may write

$$0 < \mu(Y) \le \mu(h(K)) = \mu(K) = 1 < +\infty$$

and we see that 3) is fulfilled.

In this way, we have obtained the implications 1) $\Rightarrow$ 2) and 1) $\Rightarrow$ 3).

Finally, suppose that assertion 1) is not true. Then there exists a point $z \in E$ satisfying the relation

$$G(z) \cap (\cup\{g(K) \;:\; g \in G\}) = \emptyset.$$

Let us put
$$Z = \cup\{g(K) \;:\; g \in G\}$$

and consider the family of sets

$$\mathcal{S} = \{Z' \cup Z'' \;:\; Z' \subset Z, \;\; Z' \in dom(\lambda), \;\; Z'' \subset G(z)\}.$$

It can easily be checked that $\mathcal{S}$ is a G-invariant σ-ring of subsets of E and $K \in \mathcal{S}$. Now, we define a functional ν on $\mathcal{S}$ in the following manner:

$\nu(Z' \cup Z'') = \lambda(Z')$ if $Z'' = \emptyset$;

$\nu(Z' \cup Z'') = +\infty$ if $Z'' \neq \emptyset$.

Then it is not difficult to see that this functional is a G-measure on E. Moreover, we have the one-element ν-measurable set $\{z\}$ with

$$\nu(\{z\}) = +\infty.$$

The existence of such a set shows directly that the assertions 2) and 3) are false for ν. Thus, we have established the implications 2) $\Rightarrow$ 1) and 3) $\Rightarrow$ 1). This completes the proof of Lemma 3.

Lemma 4. *Let G be a group of motions of an euclidean space E. Then the following two relations are equivalent:*

1) for any point $e \in K$, the G-orbit $G(e)$ is not a locally finite subset of E;

2) for every G-measure μ on E and for each point $e \in K$, we have

$$\{e\} \in dom(\mu) \Rightarrow \mu(\{e\}) = 0.$$

Proof. 1) $\Rightarrow$ 2). Suppose that 1) is true. Let e be an arbitrary point of K and let μ be a G-measure on E such that $\{e\} \in dom(\mu)$. It is not hard to deduce from 1) that

$$card(G(e) \cap K) \geq \omega.$$

If $\mu(\{e\}) > 0$, then (taking into account the G-invariance of μ) we immediately obtain that $\mu(K) = +\infty$, which is impossible. Consequently,

$$\mu(\{e\}) = 0$$

and the implication 1) $\Rightarrow$ 2) has been established.

2) $\Rightarrow$ 1). Suppose that 2) is true. Let $e \in K$. We must show that the orbit $G(e)$ is not locally finite in E. Assume to the contrary that $G(e)$ is a locally finite subset of E. Then it is easy to define a G-measure ν on E such that

$$\{e\} \in dom(\nu), \qquad \nu(\{e\}) > 0.$$

Indeed, in order to construct a suitable ν, we start with the inequalities

$$0 < card(G(e) \cap K) < \omega$$

and denote

$$p = card(G(e) \cap K).$$

Further, for each subset X of E, we put:

$\nu(X) = (1/p)card(X \cap G(e))$ if $card(X \cap G(e)) < \omega$;

$\nu(X) = +\infty$ if $card(X \cap G(e)) = \omega$.

Then ν is the required G-measure (defined on the family of all subsets of E). Thus, we have established the implication 2) $\Rightarrow$ 1) and the proof of Lemma 4 is completed.

Lemma 5. *Let E be a finite-dimensional euclidean space, e be a point of E, and let L be an affine linear manifold in E containing e. Let $\{h_m : m < \omega\}$ be a countable family of motions of E, satisfying the following conditions:*

1) the family of points $\{h_m(e) : m < \omega\}$ is injective and bounded in E;

2) the family of manifolds $\{h_m(L) : m < \omega\}$ is finite.

Denote by H the group of motions, generated by $\{h_m : m < \omega\}$. Then there exists a countable family $\{g_n : n < \omega\} \subset H$ such that:

(a) $g_n(L) = L$ for all $n < \omega$;

(b) $\{g_n : n < \omega\}$ converges to some motion of E;

(c) the family $\{g_n|L : n < \omega\}$ is injective, where the symbol $g_n|L$ denotes the restriction of g_n to L.

Proof. Note at once that condition 1) implies the relative compactness of the set $\{h_m : m < \omega\}$ in the topological group of all motions of E (compare the proof of Lemma 2). Further, since the family of sets $\{h_m(L) : m < \omega\}$ is finite (in view of condition 2)), there are infinitely many indices m for which we have

$$h_m(L) = h(L),$$

where h is some fixed element of $\{h_m : m < \omega\}$. Without loss of generality, we may assume that the last equality holds for all natural numbers m. Let us put

$$g_n = h^{-1} \circ h_n \qquad (n < \omega).$$

Clearly, $g_n \in H$ for each natural number n and the family of motions $\{g_n : n < \omega\}$ is relatively compact, too. Hence, we may suppose that this family is converging to some motion of E. In other words, relation (b) is valid. Now, according to our definition,

$$g_n(L) = h^{-1}(h_n(L)) = L \qquad (n < \omega),$$

from which it follows that relation (a) holds true, too. Finally, for any two distinct natural numbers n and k, we can write

$$g_n(e) = h^{-1}(h_n(e)),$$

$$g_k(e) = h^{-1}(h_k(e))$$

and, in virtue of condition 1), we get

$$h^{-1}(h_n(e)) \neq h^{-1}(h_k(e))$$

which implies that $g_n(e) \neq g_k(e)$. Therefore,

$$g_n|L \neq g_k|L,$$

which yields the validity of relation (c).

This finishes the proof of Lemma 5.

Now, we can formulate and prove the following result.

Theorem 1. *Let G be a group of motions of an euclidean space E. Then G has the Vitali property if and only if these two conditions are satisfied:*

1) for any point $e \in K$, the G-orbit $G(e)$ is not locally finite in E;

2) $\cup\{g(K) \ : \ g \in G\} = E$.

Proof. Suppose that 1) and 2) are valid, and let us show that our G has the Vitali property. Let μ be an arbitrary G-measure on E and let X be an arbitrary μ-measurable subset of E with $\mu(X) > 0$. We must find a μ-nonmeasurable subset of X. Taking into account the argument used in the proof of Lemma 3, we may assume that X is a subset of K. Denote by $r = r(X)$ the least natural number such that there exists an affine linear manifold $L \subset E$ for which

$$dim(L) = r \ \& \ \mu_*(L \cap X) > 0.$$

Obviously, r is well defined. According to Lemma 4, we can assert that

$$r = dim(L) > 0.$$

Consider the set

$$T = L \cap X.$$

If this set is nonmeasurable with respect to μ, then there is nothing to prove. Therefore, we may assume that T belongs to $dom(\mu)$. This assumption implies at once that $\mu(T) > 0$. Fix a point $t \in T$. Applying condition 1) to t, we see that the orbit $G(t)$ is not locally finite in E. It can easily be deduced from this fact that there exists a countable family of motions

$$\{h_m \ : \ m < \omega\} \subset G$$

such that the corresponding family of points $\{h_m(t) \ : \ m < \omega\}$ is injective and

$$lim_{m \to +\infty} \ h_m(t) = t.$$

We may also suppose, without loss of generality, that $\{h_m \ : \ m < \omega\}$ converges to some motion of E.

Now, consider in E the family of affine linear manifolds

$$\{h_m(L) \ : \ m < \omega\}.$$

Only two cases are possible.

1. This family is infinite. In such a case we can suppose (without loss of generality) that

$$(\forall m < \omega)(\forall n < \omega)(m \neq n \Rightarrow h_m(L) \neq h_n(L)).$$

In particular, if $m \neq n$, then we have

$$dim(h_m(L) \cap h_n(L)) < dim(L).$$

Remembering the definition of the natural number r, we easily deduce that

$$\mu(h_m(T) \cap h_n(T)) = 0 \quad (m < \omega, \ n < \omega, \ m \neq n).$$

It immediately follows from the last relation that

$$\mu(\cup\{h_m(T) \ : \ m < \omega\}) = +\infty,$$

because $\mu(T) > 0$ and the countable family $\{h_m(T) \ : \ m < \omega\}$ is almost disjoint with respect to μ. On the other hand, since the sequence $\{h_m \ : \ m < \omega\}$ converges to some motion of E and the set T is bounded in E, the μ-measurable set

$$\cup\{h_m(T) \ : \ m < \omega\}$$

is bounded in E, too, and according to Lemma 3, the inequality

$$\mu(\cup\{h_m(T) \ : \ m < \omega\}) < +\infty$$

must be true. Thus, we come to a contradiction which shows us that case 1 is impossible.

2. The family $\{h_m(L) \: : \: m < \omega\}$ is finite. In this case, applying Lemma 5, we see that there exists a countable family of motions

$$\{g_n \: : \: n < \omega\} \subset G,$$

satisfying the following relations:

(a) $g_n(L) = L$ for all $n < \omega$;

(b) $\{g_n \: : \: n < \omega\}$ converges to some motion of E;

(c) the family $\{g_n | L \: : \: n < \omega\}$ is injective.

Denote by G' the subgroup of G generated by $\{g_n \: : \: n < \omega\}$. Obviously, we have

$$(\forall g \in G')(g(L) = L).$$

Let $\{Z_i \: : \: i \in I\}$ stand for the family of all those G'-orbits in L which have nonempty intersection with the set T. Let Y be a selector of the family of nonempty sets

$$\{Z_i \cap T \: : \: i \in I\}.$$

We are going to show that Y is nonmeasurable with respect to μ. Suppose otherwise: $Y \in dom(\mu)$. Then, taking into account the relations

$$\mu(T) > 0, \quad T \subset \cup\{g(Y) \: : \: g \in G'\}$$

and the countability of the group G', we obtain that $\mu(Y) > 0$. It is not difficult to check that for any two distinct natural numbers n and m, the set $g_n(Y) \cap g_m(Y)$ lies in an affine linear manifold whose dimension is strictly less than $r = dim(L)$. According to the definition of r, we can write

$$\mu(g_n(Y) \cap g_m(Y)) = 0 \quad (n < \omega, \; m < \omega, \; n \neq m).$$

Finally, let us consider the set

$$\cup\{g_n(Y) \: : \: n < \omega\}.$$

This set is μ-measurable and bounded in E. In view of Lemma 3, we get

$$\mu(\cup\{g_n(Y) \: : \: n < \omega\}) < +\infty.$$

On the other hand, as mentioned above, the family $\{g_n(Y) \: : \: n < \omega\}$ is almost disjoint with respect to μ and $\mu(Y) > 0$. Consequently, we must have

$$\mu(\cup\{g_n(Y) \: : \: n < \omega\}) = +\infty.$$

Thus, we obtained a contradiction which shows that the Vitali property follows from the conjunction of conditions 1) and 2).

It remains to establish that if at least one of the conditions 1) and 2) is not fulfilled, then our group G does not have the Vitali property.

Suppose first that condition 1) does not hold for G. Then there exists a point e of K for which the orbit $G(e)$ is a locally finite subset of E. Therefore,

$$0 < card(K \cap G(e)) < \omega.$$

Let us denote

$$p = card(K \cap G(e)).$$

For each set $X \subset E$, we put:

$\nu(X) = (1/p)card(X \cap G(e))$ if $card(X \cap G(e)) < \omega$;

$\nu(X) = +\infty$ if $card(X \cap G(e)) = \omega$.

Then it is easy to see that the functional ν is a G-measure defined on the family of all subsets of E. Consequently, G does not possess the Vitali property.

Suppose now that condition 2) does not hold, that is

$$\cup\{g(K) \ : \ g \in G\} \neq E.$$

Let z be an arbitrary point from the set $E \setminus \cup\{g(K) \ : \ g \in G\}$. Then

$$G(z) \cap (\cup\{g(K) \ : \ g \in G\}) = \emptyset.$$

As in the proof of Lemma 3, we denote

$$Z = \cup\{g(K) \ : \ g \in G\},$$

$$S = \{Z' \cup Z'' \ : \ Z' \subset Z, \ Z' \in dom(\lambda), \ Z'' \subset G(z)\}.$$

Further, we define a functional ν on S in the same manner as in the proof of Lemma 3. Namely, we put:

$\nu(Z' \cup Z'') = \lambda(Z')$ if $Z'' = \emptyset$;

$\nu(Z' \cup Z'') = +\infty$ if $Z'' \neq \emptyset$.

Then ν is a G-measure on E such that the set $\{z\}$ is ν-measurable, $\nu(\{z\}) > 0$ and $\{z\}$ does not contain ν-nonmeasurable subsets. Consequently, the group G does not possess the Vitali property.

Thus, the proof of Theorem 1 is completed.

Slightly changing an argument presented above, we can also obtain a characterization of all those groups of motions of an euclidean space, which have the weak Vitali property.

Theorem 2. *Let G be a subgroup of the group of all isometric transformations of an euclidean space E. Then the following two conditions are equivalent:*

1) for each point $e \in K$, the orbit $G(e)$ is not locally finite in E;

2) G possesses the weak Vitali property.

Proof. It suffices to establish only the implication 1) $\Rightarrow$ 2). Suppose that condition 1) is fulfilled, and let μ be an arbitrary G-measure on E. We are going to show that the cube K contains a subset nonmeasurable with respect to μ. Denote by $r = r(K)$ the least natural number such that there exists an affine linear manifold L in E for which we have

$$dim(L) = r \ \& \ \mu_*(L \cap K) > 0.$$

Let us put

$$T = L \cap K.$$

If T is nonmeasurable with respect to μ, then there is nothing to prove. So we can suppose that $T \in dom(\mu)$ and, consequently, $\mu(T) > 0$. Fix a point $t \in T$. According to 1), the orbit $G(t)$ is not locally finite in E. It follows from this fact that there exists a countable family $\{h_m \ : \ m < \omega\}$ of elements of G, satisfying the relations:

(a) the family of points $\{h_m(t) \ : \ m < \omega\}$ is injective;

(b) $lim_{m \to +\infty} h_m(t) = t$;

(c) the sequence $\{h_m \ : \ m < \omega\}$ converges to some motion of E.

We may assume, without loss of generality, that $\{h_m \ : \ m < \omega\}$ converges to Id_E (see Exercise 1). Then the family $\{h_m^{-1} \ : \ m < \omega\}$ converges to Id_E, too. Since K is an open subset of E, we can write

$$K \subset \cup\{\cap\{h_m^{-1}(K) \ : \ n < m < \omega\} \ : \ n < \omega\}.$$

Consequently, there exists a natural number m_0 such that

$$\mu(T \cap (\cap\{h_m^{-1}(K) \ : \ m_0 < m < \omega\})) > 0.$$

Let us denote

$$Z = T \cap (\cap\{h_m^{-1}(K) \ : \ m_0 < m < \omega\}).$$

Then, for each integer $m > m_0$, we have the inclusion

$$h_m(Z) \subset K$$

and, therefore,

$$\cup \{h_m(Z) \ : \ m_0 < m < \omega\} \subset K,$$

$$\mu(\cup \{h_m(Z) \ : \ m_0 < m < \omega\}) \leq 1.$$

Consider now the family of affine linear manifolds $\{h_m(L) \ : \ m_0 < m < \omega\}$ and suppose, for a while, that this family is infinite. Then we may assume (without loss of generality) that

$$h_m(L) \neq h_n(L) \qquad (m_0 < m < \omega, \ \ m_0 < n < \omega, \ \ m \neq n).$$

Remembering the definition of $r = dim(L)$, we obtain

$$\mu(h_m(Z) \cap h_n(Z)) = 0$$

for any two distinct integers $m > m_0$ and $n > m_0$. In fact, the last relation is true because the intersection of $h_m(Z)$ and $h_n(Z)$ lies in an affine linear manifold whose dimension is strictly less than r. Finally, we get the equality

$$\mu(\cup \{h_m(Z) \ : \ m_0 < m < \omega\}) = +\infty,$$

which yields a contradiction. Thus, the family $\{h_m(L) \ : \ m_0 < m < \omega\}$ must be finite.

Now, utilizing Lemma 5, we can find a countable family $\{g_n : n < \omega\}$ of transformations from G, satisfying the relations:

(d) $g_n(L) = L$ for all $n < \omega$;

(e) the sequence $\{g_n \ : \ n < \omega\}$ converges to some motion of E;

(f) the family $\{g_n|L \ : \ n < \omega\}$ is injective.

We may also assume, without loss of generality, that $\{g_n \ : \ n < \omega\}$ converges to Id_E (in view of Exercise 1). Moreover, since T is an open subset of L and the sequence $\{g_n^{-1} \ : \ n < \omega\}$ converges to Id_E, we can write

$$T \subset \cup \{\cap \{g_n^{-1}(T) \ : \ m < n < \omega\} \ : \ m < \omega\}.$$

Therefore, there exists a natural number n_0 such that

$$\mu((\cap \{g_n^{-1}(T) \ : \ n_0 < n < \omega\}) \cap T) > 0.$$

Let us put

$$T' = (\cap\{g_n^{-1}(T) \ : \ n_0 < n < \omega\}) \cap T$$

and let G' denote the subgroup of G generated by $\{g_n \ : \ n_0 < n < \omega\}$. Obviously, we have

$$(\forall g \in G')(g(L) = L).$$

Further, let $\{Z_i \ : \ i \in I\}$ be the family of all those G'-orbits in L which have nonempty intersection with T'. Let Y be a selector of the family of sets $\{Z_i \cap T' \ : \ i \in I\}$. We assert that Y does not belong to $dom(\mu)$.

Suppose otherwise: $Y \in dom(\mu)$. Then, taking into account the inclusion

$$T' \subset \cup\{g(Y) \ : \ g \in G'\}$$

and the inequality $\mu(T') > 0$, we get $\mu(Y) > 0$. On the other hand, for any two distinct natural numbers $n > n_0$ and $m > n_0$, we have the equality

$$\mu(g_n(Y) \cap g_m(Y)) = 0,$$

because the set $g_n(Y) \cap g_m(Y)$ lies in an affine linear manifold whose dimension is strictly less than $r = dim(L)$. Therefore, we see that the family of sets $\{g_n(Y) \ : \ n_0 < n < \omega\}$ is almost disjoint with respect to μ and

$$\mu(\cup\{g_n(Y) \ : \ n_0 < n < \omega\}) = +\infty.$$

But, for each integer $n > n_0$, we can write

$$g_n(Y) \subset g_n(T') \subset T \subset K$$

and, consequently,

$$\cup\{g_n(Y) \ : \ n_0 < n < \omega\} \subset K,$$

$$\mu(\cup\{g_n(Y) \ : \ n_0 < n < \omega\}) \le 1.$$

Thus, we obtained a contradiction which shows that the set Y does not belong to the domain of μ.

This ends the proof of Theorem 2.

We would like to finish the chapter with some remarks concerning the results presented above.

Remark 1. Let F be a set, Γ be a group of transformations of F and let μ be a σ-finite Γ-invariant measure on F. We shall say that the group Γ acts freely on (in) F with respect to μ if, for any two distinct transformations g and h from Γ, we have

$$\mu^*(\{x \in F \ : \ g(x) = h(x)\}) = 0.$$

For instance, the group of all motions of an euclidean space E acts freely on E with respect to the Lebesgue measure λ.

Some generalizations of the Vitali classical theorem were obtained for various groups of transformations acting freely with respect to a given nonzero σ-finite invariant measure (see [208], [209] and Chapter 11 of this book). The free action of a group of transformations with respect to a given invariant measure seems to be rather natural. However, we wish to point out that, even in the case of a two-dimensional euclidean space E, an example of a group G of motions of this space can be constructed such that:

(a) G has the Vitali property;

(b) for some G-measure ν given on E, the group G does not act freely with respect to ν.

In order to present such an example, let us take

$$E = \mathbf{R}^2 = \mathbf{R} \times \mathbf{R}$$

and let us define a group G as follows. First of all, we put

$$g_1 = (0, 1/2), \qquad G_2 = \mathbf{R} \times \{0\}.$$

Let G_1 denote the group generated by g_1. Obviously, G_1 is a discrete group of translations of E. Denote also by s the symmetry of E with respect to the straight line $\mathbf{R} \times \{0\}$. Finally, let G be the group of transformations of E, generated by $G_1 \cup G_2 \cup \{s\}$. Evidently, G is not discrete and

$$\cup\{g(K_2) \ : \ g \in G\} = E,$$

where

$$K_2 = \{(x, y) \ : \ 0 < x < 1, \ \ 0 < y < 1\}$$

is the open unit cube in $E = \mathbf{R}^2$. According to Theorem 1, G possesses the Vitali property.

On the other hand, consider the set

$$P = \cup\{g(\mathbf{R} \times \{0\}) \ : \ g \in G\}.$$

Clearly, we have

$$P = G(P) = G_1(P) = \cup\{g(\mathbf{R} \times \{0\}) \ : \ g \in G_1\}.$$

Finally, for each Borel subset X of E, let us put

$$\nu(X) = \sum_{g \in G_1} \lambda_1(X \cap g(\mathbf{R} \times \{0\})),$$

where λ_1 is the standard one-dimensional Lebesgue measure. It can easily be checked that ν is a σ-finite G-measure singular with respect to the usual two-dimensional Lebesgue measure λ_2 on $E = \mathbf{R}^2$. Also, for the two distinct transformations Id_E and s from G, the set

$$\{z \in E \ : \ Id_E(z) = s(z)\} = \mathbf{R} \times \{0\}$$

is ν-measurable and, for this set, we have

$$\nu(\{z \in E \ : \ Id_E(z) = s(z)\}) = +\infty.$$

Thus, G does not act freely on E with respect to ν.

Note that a similar example can be constructed for any euclidean space E with $dim(E) \geq 2$.

Remark 2. It is easy to see that the corresponding analogues of Theorems 1 and 2 hold true for the euclidean unit sphere $\mathbf{S}_n$ $(n \geq 1)$ equipped with a group of its isometric transformations.

Remark 3. Let G be a discrete group of motions of an euclidean space E with $dim(E) \geq 1$. As shown above, there are G-measures on E defined on the family of all subsets of E. Consequently, we cannot assert that for any G-measure there exist nonmeasurable sets in E. But for certain G-measures, we are able to establish the existence of nonmeasurable subsets of E and even the existence of nonmeasurable G-selectors. Namely, let G be a discrete group of motions of E, containing at least two distinct elements, and let λ denote the Lebesgue measure on E. Then, starting with Lemma 1 and using an argument similar to the classical Bernstein construction (see Chapter 2), it can be proved that there exists a G-selector Z satisfying the equalities

$$\lambda_*(Z) = \lambda_*(E \setminus Z) = 0.$$

In other words, both of the sets Z and $E \setminus Z$ are λ-thick in E and, consequently, they are not measurable with respect to λ (considered as a G-measure on E).

EXERCISES

1. Let E be a finite-dimensional euclidean space. Suppose that an injective sequence $\{g_m : m < \omega\}$ of motions of E is given, relatively compact in the topological group of all motions of E. Denoting by G the group generated by $\{g_m : m < \omega\}$, show that there exists an injective sequence

$$\{h_0, h_1, ..., h_n, ...\} \subset G$$

such that

$$lim_{n \to +\infty} h_n = Id_E.$$

Applying this result, complete the proof of Theorem 2.

2. Let E be a finite-dimensional euclidean space and let G be a group of translations of E. Deduce from Theorem 1 that the following two statements are equivalent:

(a) G possesses the Vitali property;

(b) G is not discrete and $\cup\{g(K) : g \in G\} = E$.

3. Let E be a finite-dimensional euclidean space and let G be a group of translations of E. Deduce from Theorem 2 that the following two assertions are equivalent:

(a) G possesses the weak Vitali property;

(b) G is not discrete.

4. Formulate and prove the corresponding analogues of Theorems 1 and 2 for the euclidean unit sphere $\mathbf{S}_n$ $(n \geq 1)$ endowed with a group G of its isometric transformations.

5. For a finite-dimensional euclidean space E with $dim(E) > 2$, construct a group G of motions of E and a G-measure ν on E, analogous to those ones presented in Remark 1.

6. Let E be a finite-dimensional euclidean space equipped with the Lebesgue measure λ. Let G be a discrete group of motions of E, such that

$$card(G) \geq 2.$$

Construct, by using the method of transfinite recursion, a G-selector Z for which the equalities

$$\lambda_*(Z) = \lambda_*(E \setminus Z) = 0$$

are valid. In particular, Z turns out to be nonmeasurable with respect to the G-measure λ on E.

Chapter 9
Selectors associated with countable subgroups

Here we continue our discussion of analogues and generalizations of the classical Vitali construction producing a non Lebesgue-measurable set on the real line $\mathbf{R}$. We shall consider in this chapter several general constructions of nonmeasurable sets in uncountable groups. In fact, those sets will be various selectors associated with some countable subgroup of a given uncountable group. Naturally, the latter group is assumed to be equipped with a nonzero σ–finite invariant (more generally, quasi-invariant) measure.

The aim of this chapter is twofold.

On the one hand, we shall show that a number of questions arising in connection with the Vitali theorem can be completely solved for uncountable commutative groups. Namely, we shall establish, in our further considerations, the corresponding analogue of the Vitali theorem for uncountable commutative groups endowed with nonzero σ–finite quasi-invariant measures.

On the other hand, it will also be shown that nonmeasurable sets obtained by the methods developed in this chapter are not absolutely nonmeasurable with respect to the class of all nonzero σ-finite quasi-invariant measures (compare Chapter 11 devoted to absolutely nonmeasurable sets).

First of all, we wish to recall some notation and terminology which is systematically used below.

Let $\{X_\alpha \ : \ \alpha \in A\}$ be a family of nonempty pairwise disjoint sets and let

$$E = \cup\{X_\alpha \ : \ \alpha \in A\}.$$

We say that a set $X \subset E$ is a partial selector of the family $\{X_\alpha : \alpha \in A\}$ if for each index $\alpha \in A$, we have the inequality

$$card(X \cap X_\alpha) \leq 1.$$

Accordingly, we say that a set $X \subset E$ is a selector of $\{X_\alpha : \alpha \in A\}$ if for each $\alpha \in A$, we have the equality

$$card(X \cap X_\alpha) = 1.$$

Obviously, we can state (with the aid of the Axiom of Choice) that any partial selector of the family $\{X_\alpha : \alpha \in A\}$ is extendible to a selector of $\{X_\alpha : \alpha \in A\}$.

Let E be a nonempty set and let G be a group of transformations of E. Consider an arbitrary subgroup H of G. We recall that a subset X of E is an H-selector if X is a selector of the family of all H-orbits in E. In this situation, we also say that X is a Vitali type subset of E (with respect to the subgroup H of the original group G).

In particular, various selectors appear naturally in the case where two abstract groups G and H are given such that $H \subset G$ (in other words, H is a subgroup of G). In this case, we have $E = G$, the group G is identified with the group of all left (right) translations of E and H is considered as a subgroup of the group of all left (right) translations of E.

For instance, let $(G, +)$ be a commutative group and let H be a subgroup of G. According to the definition formulated above, we say that a set $X \subset G$ is an H-selector (respectively, a partial H-selector) if X is a selector (respectively, a partial selector) of the family G/H consisting of all H-orbits in G. As a rule, we suppose in our considerations below that H is a nontrivial ($=$ nonzero) subgroup of G.

In particular, returning to the classical Vitali theorem [224], we recall that this theorem deals with the additive group $G = \mathbf{R}$ and its countable dense subgroup $H = \mathbf{Q}$. We know that all selectors of the family G/H turn out to be nonmeasurable with respect to the Lebesgue measure λ on $\mathbf{R}$.

Let E be a set, G be an arbitrary group of transformations of E and let μ be a measure defined on some σ-algebra of subsets of E. We recall that μ is a G-invariant measure if $dom(\mu)$ is invariant with respect to G and the equality

$$\mu(g(Y)) = \mu(Y)$$

holds for each transformation $g \in G$ and for each set $Y \in dom(\mu)$.

We have also a more general concept, namely the concept of a G-quasi-invariant measure.

For a given nonzero measure μ on E, let us denote by $\mathcal{I}(\mu)$ the σ-ideal of subsets of E, generated by the family of all μ-measure zero sets. We say that μ is a G-quasi-invariant measure if the classes of sets $dom(\mu)$ and $\mathcal{I}(\mu)$ are invariant with respect to all transformations from G. If μ is identically equal to zero, then μ is assumed to be G-quasi-invariant by definition.

Various aspects of the general theory of invariant (quasi-invariant) measures are presented in the works [27], [36], [62], [68], [70], [100], [116], [117], [133], [139], [164], [205], [225], [226], [235].

An important special case of invariant and quasi-invariant measures is the following. Let $(G, \cdot)$ be an arbitrary group. Put $E = G$ and consider G as the group of all left translations of E. Then we naturally come to the notions of left G-invariant and left G-quasi-invariant measures on G. Similarly, the notions of right G-invariant and right G-quasi-invariant measures can be introduced. It is obvious that if G is a commutative group, then the concepts of left and right G-invariant (respectively, G-quasi-invariant) measures are identical.

We would like to emphasize that the measures considered in the sequel are always assumed to be σ-finite. Consequently, the countable chain condition holds for those measures (see Exercise 10 from Chapter 1).

Let us also recall that if we have a space (E, G) equipped with a nonzero σ-finite G-quasi-invariant (G-invariant) measure μ on E, then we can easily define a probability G-quasi-invariant measure ν on E such that

$$dom(\nu) = dom(\mu), \quad \mathcal{I}(\nu) = \mathcal{I}(\mu).$$

In other words, μ and ν are equivalent measures. This simple observation will be applied many times below.

Of course, without loss of generality, we may restrict our further considerations to the case of left G-quasi-invariant (left G-invariant) measures defined on various σ-algebras of subsets of a given abstract group $(G, \cdot)$.

If the original group G is commutative, then there is no difference between left G-quasi-invariant (left G-invariant) and right G-quasi-invariant (right G-invariant) measures on G, and we can simply utilize the notion of a G-quasi-invariant (respectively, G-invariant) measure.

Now, let us fix an uncountable commutative group $(G, +)$ with a nonzero σ-finite G-quasi-invariant measure μ defined on some σ-algebra of subsets of G. Let H be a subgroup of G. In connection with the classical Vitali theorem stating the nonmeasurability (in the Lebesgue sense) of all $\mathbf{Q}$-selectors in $\mathbf{R}$, the following three questions arise very naturally (compare [105]; see also [100]).

Question 1. Let H be an infinite countable subgroup of G. Is it true that all H-selectors are nonmeasurable with respect to μ?

Question 2. Let H be a nontrivial countable subgroup of G. Is it true that there exists at least one H-selector nonmeasurable with respect to μ?

Question 3. Let H be a countable subgroup of G and let

$$G/H = \{X_\alpha \ : \ \alpha \in A\}.$$

Is it true that there exists a subset B of A such that all selectors of the partial family $\{X_\beta \ : \ \beta \in B\}$ are nonmeasurable with respect to μ?

This chapter is primarily devoted to the discussion of the questions posed above.

First of all, let us mention that the answer to Question 1 is negative even in the classical case where

$$G = \mathbf{R}, \quad H = \mathbf{Q}.$$

Moreover, in Chapter 11 a certain measure ν will be constructed, satisfying the following conditions:

1) ν is defined on some σ-algebra of subsets of $\mathbf{R}$;

2) ν is a nonzero σ-finite nonatomic measure;

3) $dom(\lambda)$ is contained in $dom(\nu)$;

4) for each Lebesgue measurable subset X of $\mathbf{R}$ with $\lambda(X) = 0$, we have $\nu(X) = 0$;

5) for each Lebesgue measurable subset X of $\mathbf{R}$ with $\lambda(X) > 0$, we have $\nu(X) = +\infty$;

6) ν is invariant under the group of all isometric transformations of $\mathbf{R}$; in particular, ν is invariant under the group of all translations of $\mathbf{R}$;

7) there exists a ν-measurable $\mathbf{Q}$-selector.

The last condition says that a certain Vitali set is measurable with respect to ν. This fact seems to be rather surprising since we know that all Vitali sets are nonmeasurable with respect to every $\mathbf{Q}$-invariant measure on $\mathbf{R}$ extending λ (compare Exercise 4 of Chapter 1).

For a detailed construction of ν, see Example 2 in Chapter 11.

Now, we are going to demonstrate that the answers to Questions 2 and 3 are positive. Moreover, a much stronger result will be established in Theorem 1 below.

We need several preliminary propositions.

Lemma 1. *Let $\{X_\alpha : \alpha \in A\}$ be a family of pairwise disjoint sets such that $card(X_\alpha) > 1$ for all indices $\alpha \in A$, and let X be a partial selector of $\{X_\alpha : \alpha \in A\}$. Then there exist two selectors Y_1 and Y_2 of $\{X_\alpha : \alpha \in A\}$ satisfying the equality*

$$Y_1 \cap Y_2 = X.$$

This lemma is trivial, but rather useful in the sequel. In particular, it immediately implies the next auxiliary statement.

Lemma 2. *Let E be a set, μ be a measure on E and let $\{X_\alpha : \alpha \in A\}$ be a partition of E such that $card(X_\alpha) > 1$ for all $\alpha \in A$. Suppose also that there exists a partial selector of $\{X_\alpha : \alpha \in A\}$ nonmeasurable with respect to μ. Then there exists a selector of $\{X_\alpha : \alpha \in A\}$ nonmeasurable with respect to μ.*

Proof. Let X be an arbitrary μ-nonmeasurable partial selector of the family $\{X_\alpha : \alpha \in A\}$. According to Lemma 1, there exist two selectors Y_1 and Y_2 of $\{X_\alpha : \alpha \in A\}$ such that

$$X = Y_1 \cap Y_2.$$

Suppose for a moment that $Y_1 \in dom(\mu)$ and $Y_2 \in dom(\mu)$. Then, in view of the closedness of $dom(\mu)$ under finite intersections, we immediately get $X \in dom(\mu)$. Since this relation is impossible, we must have

$$Y_1 \notin dom(\mu) \quad \vee \quad Y_2 \notin dom(\mu);$$

in other words, at least one of the selectors Y_1 and Y_2 is nonmeasurable with respect to μ. This finishes the proof of the lemma.

Let G_1 be a group equipped with a left G_1-quasi-invariant probability measure μ_1, let G_2 be another group, and let ϕ be an arbitrary surjective homomorphism from G_1 onto G_2. We denote

$$\mathcal{S} = \{Y \subset G_2 : \phi^{-1}(Y) \in dom(\mu_1)\}.$$

Obviously, $\mathcal{S}$ is a σ-algebra of subsets of the group G_2 and one can easily verify that $\mathcal{S}$ is invariant with respect to the group of all left translations of G_2. We define a functional μ_2 on $\mathcal{S}$ by the formula

$$\mu_2(Y) = \mu_1(\phi^{-1}(Y)) \qquad (Y \in \mathcal{S}).$$

It is not hard to see that the following proposition holds.

Lemma 3. *The functional μ_2 is a left G_2-quasi-invariant probability measure on G_2. Moreover, if the original measure μ_1 is left G_1-invariant, then μ_2 is left G_2-invariant.*

We leave an easy proof of Lemma 3 to the reader. We only wish to note, in connection with this lemma, that if μ_1 is an arbitrary σ-finite left G_1-quasi-invariant (respectively, left G_1-invariant) measure on the group G_1, then the measure μ_2 on the group G_2, defined by the same formula, is left G_2-quasi-invariant (respectively, left G_2-invariant), but we cannot assert, in general, that μ_2 is σ-finite.

Lemma 4. *Let G be an uncountable commutative group equipped with a nonzero σ-finite G-quasi-invariant measure μ. Then there exists a subgroup Γ of G nonmeasurable with respect to μ.*

The reader can find in [84] a detailed proof of this auxiliary proposition (see also Exercise 9 from Chapter 13 where a much stronger result is given). Here we only want to remark that the proof of Lemma 4 is essentially based on some combinatorial properties of an Ulam $(\omega \times \omega_1)$-matrix (see [222] or Chapter 7 of this book) and on a well-known theorem from group theory, concerning the algebraic structure of commutative groups. More precisely, the above-mentioned theorem states that every commutative group can be represented as the union of a countable family of subgroups each of which is the direct sum of cyclic groups (see [129] or Appendix 2).

Remark 1. We may assume that the μ-nonmeasurable group $\Gamma \subset G$ in Lemma 4 is also uncountable. Indeed, it is sufficient to apply Lemma 4 to a G-quasi-invariant extension μ' of μ such that $dom(\mu')$ contains all countable subsets of G. (Evidently, μ can be extended onto the family of all countable subsets of G in such a way that its extension will be G-quasi-invariant, too.)

From Lemma 4 we easily deduce the next proposition.

Lemma 5. *Let G be a commutative group with a nonzero σ-finite G-quasi-invariant measure μ and let H be a subgroup of G satisfying the inequality $card(G/H) > \omega$. Then there exists a subgroup Γ of G such that:*

1) H is contained in Γ;

2) Γ is nonmeasurable with respect to μ.

Proof. We may assume, without loss of generality, that μ is a probability G-quasi-invariant measure on G. Let us denote by ϕ the canonical homomorphism from the given group G onto the factor group G/H and put

$$S = \{Y \subset G/H : \phi^{-1}(Y) \in dom(\mu)\}.$$

Further, define a functional ν on the σ-algebra S by the formula

$$\nu(Y) = \mu(\phi^{-1}(Y)) \qquad (Y \in S).$$

According to Lemma 3, ν is a (G/H)-quasi-invariant probability measure on the uncountable group G/H. In view of Lemma 4, there exists a subgroup Γ^* of G/H nonmeasurable with respect to ν. Let us put

$$\Gamma = \phi^{-1}(\Gamma^*).$$

One can readily verify that Γ is a subgroup of G nonmeasurable with respect to the original measure μ and

$$H = ker(\phi) = \phi^{-1}(0) \subset \phi^{-1}(\Gamma^*) = \Gamma.$$

Lemma 5 has thus been proved.

Lemma 6. *Let G be an uncountable group equipped with a nonzero σ-finite left G-quasi-invariant measure μ. Then there exists a subset of G nonmeasurable with respect to μ.*

This lemma is a particular case of the following statement.

Lemma 7. *Let (E, G) be a space with a transformation group and let μ be a nonzero σ-finite G-quasi-invariant measure on E. Suppose also that G contains an uncountable subgroup Γ acting freely in E. Then there exists a subset of E nonmeasurable with respect to μ.*

Proof. We may assume, without loss of generality, that:

1) the group Γ coincides with the original group G;

2) $card(G) = card(\Gamma) = \omega_1$;

3) μ is a G-quasi-invariant probability measure on E.

Further, we denote by $\{X_\alpha \ : \ \alpha \in A\}$ the disjoint family of all G-orbits in E. Since G acts freely in E, we have

$$(\forall \alpha \in A)(card(X_\alpha) = \omega_1).$$

Let X be an arbitrary selector of the family $\{X_\alpha \ : \ \alpha \in A\}$. If X is a nonmeasurable set with respect to μ, then there is nothing to prove. So we can suppose that $X \in dom(\mu)$. Now, consider the family of sets

$$\{g(X) \ : \ g \in G\}.$$

Obviously, the following three relations are satisfied:

(a) $E = \cup\{g(X) \ : \ g \in G\}$;

(b) if $g \in G$, $h \in G$ and $g \neq h$, then $g(X) \cap h(X) = \emptyset$;

(c) $\mu(g(X)) = 0$ for each element g from G.

Note that relation (c) is implied directly by the G-quasi-invariance of our measure μ and the countable chain condition for μ.

So we have a certain partition

$$\{Z_\xi \ : \ \xi < \omega_1\} = \{g(X) \ : \ g \in G\}$$

of the space E, consisting of μ-measure zero sets. We now assert that there exists a subset Ξ of ω_1 for which the union

$$\cup\{Z_\xi \ : \ \xi \in \Xi\} \subset E$$

is not measurable with respect to μ. Suppose otherwise, that is for all subsets Ξ of ω_1 the corresponding unions $\cup\{Z_\xi \ : \ \xi \in \Xi\}$ are always μ-measurable. Then we may put

$$\nu(\Xi) = \mu(\cup\{Z_\xi \ : \ \xi \in \Xi\}) \quad\quad (\Xi \subset \omega_1).$$

In this way, we get a diffused probability measure ν defined on the family of all subsets of ω_1. But this contradicts the classical Ulam theorem stating that ω_1 is not a real-valued measurable cardinal number (see Chapter 7).

The contradiction obtained shows the existence of a subset of E nonmeasurable with respect to the original measure μ. The proof of the lemma is thus completed.

Remark 2. Lemma 7 was first established in papers [80], [38] and [188]. Actually, Lemma 7 gives an answer to one question posed by Oxtoby. At the present time, several statements are known which generalize this lemma or are closely related to it. For instance, let us mention paper [1] where a result analogous to Lemma 7 is discussed for the case of nonzero invariant measures which are not necessarily σ-finite (compare also Exercise 8 of this chapter).

Now, we can formulate and prove the following statement.

Theorem 1. *Let G be an uncountable commutative group equipped with a nonzero σ-finite G-quasi-invariant measure μ and let H be a countable subgroup of G. Denote by*

$$G/H = \{X_\alpha \ : \ \alpha \in A\}$$

the partition of G canonically associated with H. Then there exists a subset B of A such that:

1) the union of the partial family $\{X_\beta \ : \ \beta \in B\}$ is a subgroup of G nonmeasurable with respect to μ;

2) all selectors of $\{X_\beta \ : \ \beta \in B\}$ are nonmeasurable with respect to μ;

3) if H is a nontrivial subgroup of G, then there exists an H-selector nonmeasurable with respect to μ.

Proof. Applying Lemma 5, we see that there exists a subgroup Γ of G such that

$$H \subset \Gamma, \quad \Gamma \notin dom(\mu).$$

In view of the inclusion $\Gamma/H \subset G/H$, we may write

$$\Gamma/H = \{X_\beta \ : \ \beta \in B\} \subset \{X_\alpha \ : \ \alpha \in A\}$$

for some set $B \subset A$. Obviously, we have the equality

$$\Gamma = \cup\{X_\beta \ : \ \beta \in B\}.$$

Consequently, relation 1) holds for the family $\{X_\beta \ : \ \beta \in B\}$.

Further, let X be an arbitrary selector of $\{X_\beta \ : \ \beta \in B\}$. We assert that X is nonmeasurable with respect to μ. Suppose otherwise: $X \in dom(\mu)$. Then we have

$$\Gamma = \cup\{h + X : h \in H\},$$

where all sets $h + X$ $(h \in H)$ are μ-measurable. Taking into account the fact that H is a countable subgroup of G, we get $\Gamma \in dom(\mu)$, which yields a contradiction. Therefore, X does not belong to $dom(\mu)$ and relation 2) holds for $\{X_\beta : \beta \in B\}$. Finally, applying Lemma 2 to the partition G/H of G, we immediately obtain that relation 2) implies relation 3). The proof of Theorem 1 is thus completed.

Remark 3. Unfortunately, Theorem 1 cannot be generalized to the class of all uncountable groups equipped with nonzero σ-finite left quasi-invariant measures. Indeed, Shelah proved in his well-known work [191] that there exists a group G with the following properties:

(a) $card(G) = \omega_1$;

(b) G does not contain a proper uncountable subgroup.

Let us take such a group G and fix a countable subgroup H of G. Further, denote by $\mathcal{S}$ the σ-algebra of subsets of the group G, generated by the family of all countable sets in G. One can easily define a left (right) G-invariant probability measure μ on $\mathcal{S}$ such that $\mu(Y) = 0$ for each countable subset Y of G. It is clear now that for (G, μ) and H, an analogue of Theorem 1 fails to be valid.

However, we have the following result (compare [100], [105]).

Theorem 2. *Let G be an arbitrary uncountable group equipped with a nonzero σ-finite left G-quasi-invariant measure μ and let $\{X_\alpha : \alpha \in A\}$ be a partition of G such that*

$$1 < card(X_\alpha) \leq \omega$$

for all indices $\alpha \in A$. Then there exists a selector of $\{X_\alpha : \alpha \in A\}$ non-measurable with respect to the measure μ. In particular, if H is a nontrivial countable subgroup of G and $\{X_\alpha : \alpha \in A\}$ is an injective family of all left (right) H-orbits in G, then there exists a selector of $\{X_\alpha : \alpha \in A\}$ nonmeasurable with respect to μ.

Proof. According to Lemma 6, there is a subset Y of G nonmeasurable with respect to μ. Starting with the inequalities

$$card(X_\alpha) \leq \omega \qquad (\alpha \in A),$$

we easily infer that the set Y can be represented in the form

$$Y = \cup\{Y_n : n < \omega\},$$

where each set Y_n $(n < \omega)$ is a partial selector of $\{X_\alpha : \alpha \in A\}$. Since Y does not belong to $dom(\mu)$, there exists a natural number n such that Y_n

does not belong to $dom(\mu)$ either. Finally, applying Lemma 2, we conclude that there exists at least one selector of $\{X_\alpha \ : \ \alpha \in A\}$ extending Y_n and nonmeasurable with respect to μ.

Theorem 2 has thus been proved.

Remark 4. Let E be a set equipped with a measure μ. Consider any partition $\{X_\alpha \ : \ \alpha \in A\}$ of E such that

$$1 < card(X_\alpha) \leq \omega$$

for all indices $\alpha \in A$. Actually, the argument utilized in the proof of Theorem 2 shows that these two assertions are equivalent:

1) there exists a subset of E nonmeasurable with respect to μ;

2) there exists a selector of $\{X_\alpha \ : \ \alpha \in A\}$ nonmeasurable with respect to μ.

We can prove some analogues of the preceding results in a more general situation. Namely, let G be an uncountable group, let $\mathcal{S}$ be a σ-algebra of subsets of G and let $\mathcal{I}$ be a σ-ideal of subsets of G, such that $\mathcal{I} \subset \mathcal{S}$. Suppose that the following relations are valid:

(a) $\mathcal{S}$ is invariant under the group of all left translations of G;

(b) $\mathcal{I}$ is invariant under the group of all left translations of G;

(c) the pair $(\mathcal{S}, \mathcal{I})$ satisfies the Suslin condition (the countable chain condition); in other words, any disjoint family of sets belonging to $\mathcal{S} \setminus \mathcal{I}$ is at most countable.

Then a result similar to Theorem 2 holds for the group G, the pair $(\mathcal{S}, \mathcal{I})$ and a nontrivial countable subgroup H of G. In addition, if G is commutative, then a result similar to Theorem 1 holds for the group G, the pair $(\mathcal{S}, \mathcal{I})$ and a countable subgroup H of G. The proofs of these two results are based on the corresponding analogues of the lemmas presented above.

In particular, we can formulate the following topological statement.

Theorem 3. *Let G be an uncountable commutative group and let $\mathcal{T}$ be a topology on G such that:*

1) $(G, \mathcal{T})$ is a second category topological space;

2) the σ-algebra of sets having the Baire property in $(G, \mathcal{T})$ is invariant under the group of all translations of G;

3) the σ-ideal of first category sets in $(G, \mathcal{T})$ is invariant under the group of all translations of G;

4) the space $(G, \mathcal{T})$ satisfies the Suslin condition (the countable chain condition).

Furthermore, let H be a countable subgroup of G and let

$$G/H = \{X_\alpha \; : \; \alpha \in A\}$$

denote the partition of G canonically associated with H. Then there exists a subset B of A such that:

(i) the union of the partial family $\{X_\beta \; : \; \beta \in B\}$ is a subgroup of G without the Baire property in $(G, \mathcal{T})$;

(ii) no selector of $\{X_\beta \; : \; \beta \in B\}$ has the Baire property in $(G, \mathcal{T})$;

(iii) if H is a nontrivial subgroup of G, then there exists an H-selector not possessing the Baire property in $(G, \mathcal{T})$.

The proof is left to the reader.

In a similar way, one can formulate and prove a topological statement analogous to Theorem 2. We leave again to the reader the formulation and proof of this statement.

In Chapter 11 we shall consider a more strong version of the nonmeasurability of selectors associated with countable subgroups. In particular, we will be dealing there with those selectors which are nonmeasurable not only with respect to a given nonzero σ-finite invariant measure, but are also nonmeasurable with respect to any invariant extension of this measure. The corresponding notion will be introduced and investigated, namely the notion of an absolutely nonmeasurable set with respect to a given class of invariant (more generally, quasi-invariant) measures.

EXERCISES

1. Let Y be an arbitrary Lebesgue measurable subset of $\mathbf{R}$ with strictly positive measure. Show that there exists a Vitali subset X of $\mathbf{R}$ satisfying the following relations:

(a) $\lambda(X \setminus Y) = 0$;

(b) $\lambda_*(Y \setminus X) = 0$.

In particular, infer from this result that, for any nonempty open interval $U \subset \mathbf{R}$, there exists a Vitali set X such that

$$X \subset U, \quad \lambda_*(U \setminus X) = 0.$$

More generally, suppose that E is an infinite set equipped with a group G of its transformations and μ is a σ-finite G-quasi-invariant measure on E satisfying the following conditions:

(a) for each μ-measurable set Z with $\mu(Z) > 0$, we have the equality $card(Z) = card(E)$;

(b) there exists a family $\{Z_i : i \in I\}$ of μ-measurable sets, such that

$$card(I) \leq card(E),$$

$$(\forall i \in I)(\mu(Z_i) > 0),$$

$$(\forall Z \in dom(\mu) \setminus \mathcal{I}(\mu))(\exists i \in I)(Z_i \subset Z);$$

(c) there is a countable subgroup H of G such that μ is metrically transitive with respect to H.

Prove that, for each μ-measurable set Y with $\mu(Y) > 0$, there exists a selector X of the family $\{H(x) \ : \ x \in E\}$, such that

(i) $\mu^*(X \setminus Y) = 0$;
(ii) $\mu_*(Y \setminus X) = 0$.

2. Let (E, G) be a space with a transformation group and let μ be a σ-finite G-quasi-invariant measure given on E. Suppose also that G contains a subgroup Γ satisfying these two conditions:

(a) Γ is an uncountable group;

(b) for any two distinct elements $g \in \Gamma$ and $h \in \Gamma$, the equality

$$\mu^*(\{x \in E \ : \ g(x) = h(x)\}) = 0$$

is valid.

In other words, condition (b) says that the group Γ acts $\mathcal{I}(\mu)$-freely in E (or Γ acts freely in E with respect to μ).

Further, let Y be an arbitrary subset of E such that $\mu^*(Y) > 0$. Show that there exists a subset of Y nonmeasurable with respect to μ.

Clearly, the result presented in this exercise is a slight generalization of Lemma 7.

3. Let $E = \mathbf{R}^n$, where $n \geq 1$, and let G be an arbitrary group of affine transformations of E. Let μ be a measure on E such that all affine hyperplanes in E belong to $dom(\mu)$ and the values of μ on those hyperplanes are equal to zero. Verify that G acts freely in E with respect to μ.

4. Let E be a set, G be a group of transformations of E and let μ be a nonzero σ-finite G-quasi-invariant measure defined on a σ-algebra of subsets of E. Suppose also that the group G contains an uncountable subgroup Γ acting freely in E with respect to the measure μ. Finally, let H be an arbitrary countable subgroup of Γ and let $\{H(x) : x \in E\}$ be the partition of E into H-orbits. Prove that there exists a subfamily of $\{H(x) : x \in E\}$ such that its union is nonmeasurable with respect to μ.

Deduce from this result that there exists a subfamily of $\{H(x) : x \in E\}$ such that all its selectors are nonmeasurable with respect to μ.

5. Let G be an arbitrary group with $card(G) = \omega_1$. Let us put $E = G$ and identify the given group G with the group of all left translations of E. Show that there exist a family of sets $\{X_\alpha : \alpha < \omega_1\}$ and a measure μ, satisfying the following conditions:

(a) $\{X_\alpha : \alpha < \omega_1\}$ is a partition of E;

(b) $card(X_\alpha) = \omega$ for all ordinals $\alpha < \omega_1$;

(c) μ is a complete diffused left G-invariant probability measure defined on a σ-algebra of subsets of E;

(d) there is a selector X of $\{X_\alpha : \alpha < \omega_1\}$ such that $X \in dom(\mu)$ and $\mu(X) = 0$.

In particular, condition (d) implies that there does not exist a subfamily of $\{X_\alpha : \alpha < \omega_1\}$ whose all selectors are nonmeasurable with respect to the measure μ.

We thus see that it is essential, for the validity of the result presented in Exercise 4, that the partition $\{H(x) : x \in E\}$ consists of H-orbits, where H is a countable subgroup of the group Γ.

6. Formulate and prove a topological analogue of Theorem 2.

7. Give a proof of Theorem 3.

8. Generalize Lemma 6, Lemma 7 and Exercise 2 to the case of quasi-invariant measures satisfying the countable chain condition.

9. Let E be a set, G be a group of transformations of E, and let μ be a complete σ-finite G-invariant measure on E. Suppose that G contains an uncountable subgroup acting freely on E with respect to μ. Prove that the following two assertions are equivalent:

(a) μ is metrically transitive with respect to G;

(b) μ has the uniqueness property, that is for any σ-finite G-invariant measure ν with $dom(\nu) = dom(\mu)$, there exists a real number $t = t(\nu)$ such that $\nu = t \cdot \mu$.

10. Let E be a set and let μ be a measure defined on some σ-algebra of subsets of E. We recall that μ is semifinite if, for any set $X \in dom(\mu)$ with $\mu(X) > 0$, there exists a set $Y \in dom(\mu)$ such that

$$Y \subset X, \quad 0 < \mu(Y) < +\infty.$$

Evidently, every σ-finite measure is semifinite. The converse assertion is not true (give an example).

Let us denote

$$\kappa_0 = \text{ the first real-valued measurable cardinal.}$$

Demonstrate that, for any group $(G, \cdot)$ satisfying the relation

$$card(G) \geq \kappa_0,$$

there exists a measure μ on G such that:

(a) $dom(\mu) = \mathcal{P}(G)$;

(b) μ is nonzero, diffused and semifinite;

(c) μ is left G-invariant.

In order to establish this fact, identify κ_0 with the first ordinal of the same cardinality and fix a subgroup H of G with $card(H) = \kappa_0$. Further, construct by transfinite recursion a κ_0-sequence

$$\{H_\alpha : \alpha < \kappa_0\}$$

of subgroups of H, having the following properties:

(i) $\cup\{H_\beta : \beta < \alpha\} \subset H_\alpha$ and $H_\alpha \setminus \cup\{H_\beta : \beta < \alpha\} \neq \emptyset$ for any ordinal number $\alpha < \kappa_0$;

(ii) $card(H_\alpha) \leq card(\alpha) + \omega$ for any ordinal number $\alpha < \kappa_0$;

(iii) $\cup\{H_\alpha : \alpha < \kappa_0\} = H$.

Now, for each $\alpha < \kappa_0$, put

$$D_\alpha = H_\alpha \setminus \cup\{H_\beta : \beta < \alpha\}$$

and take a selector Z of the disjoint family of nonempty sets $\{D_\alpha : \alpha < \kappa_0\}$.

Check that $card(Z) = \kappa_0$ and that

$$card(Z \cap gZ) < \kappa_0$$

for all $g \in G \setminus \{e\}$, where e denotes the neutral element of G.

Fix a diffused probability measure ν on Z with $dom(\nu) = \mathcal{P}(Z)$. Then $\nu(T) = 0$ for all sets $T \subset Z$ with $card(T) < \kappa_0$. Finally, for each set $X \subset G$, define

$$\mu(X) = \sum_{g \in G} \nu(gX \cap Z).$$

Verify that μ is the required measure on G; in other words, μ satisfies all conditions (a), (b) and (c).

Thus, we see that Lemma 6 of this chapter cannot be generalized to semifinite invariant measures on groups.

The result of the last exercise is essentially due to Pelc [168] and Zakrzewski [236].

Chapter 10
Selectors associated with uncountable subgroups

The previous chapter was devoted to measurability properties of selectors associated with countable subgroups. Here we wish to consider analogous properties of those selectors which are associated with uncountable subgroups of a given group.

First, let us recall (see Theorem 1 of Chapter 1) that according to one generalized version of Vitali's theorem, if Γ is a countable dense subgroup of the additive group $\mathbf{R}$, then all Γ-selectors are nonmeasurable in the Lebesgue sense (and, respectively, they do not have the Baire property).

We thus see that only two assumptions on the group $\Gamma \subset \mathbf{R}$ (namely, the countability of Γ and its density in $\mathbf{R}$) imply the classical result of Vitali. In other words, under these assumptions every Γ-selector is not Lebesgue measurable and does not possess the Baire property. Equivalently, the above-mentioned assumptions are sufficient for the validity of Vitali's result.

In this connection, the following question arises naturally: what is the situation for other subgroups of the real line? The present chapter is primarily devoted to the investigation of this problem (compare also [101]).

Let $\lambda = \lambda_1$ denote the standard Lebesgue measure on $\mathbf{R} = \mathbf{R}^1$.

First of all, we are going to discuss the most simple case where a given subgroup Γ of $\mathbf{R}$ is not everywhere dense in $\mathbf{R}$. It can easily be shown that such a group Γ is always closed and discrete. We leave the proof of this fact to the reader, as an exercise. Actually, Γ can be represented in the form

$$\Gamma = \{na \ : \ n \in \mathbf{Z}\},$$

where $\mathbf{Z}$ denotes the set of all integers and a is a fixed point of $\mathbf{R}$.

If $a = 0$, then $\Gamma = \{0\}$. Here we have the unique Γ-selector which coincides with $\mathbf{R}$ and, obviously, is λ-measurable.

Suppose now that $a \neq 0$. We may assume, without loss of generality, that $a > 0$. Then it is easy to check that the interval $[0, a[$ is one of the Γ-selectors, and this interval is Lebesgue measurable as well.

Consequently, we obtain that if Γ is not dense in $\mathbf{R}$, then there exist Lebesgue measurable Γ-selectors; hence, the assertion of the Vitali theorem does not hold for Γ.

At the same time, it is not difficult to show that if Γ is a nontrivial (nonzero) discrete subgroup of $\mathbf{R}$, then there are some λ-nonmeasurable Γ-selectors. In order to establish this fact, we may apply a construction very similar to the classical Bernstein argument (see, for instance, well-known textbooks [155], [165] or Chapter 2 of the present book).

Let α denote the least ordinal number with $card(\alpha) = \mathbf{c}$, where $\mathbf{c}$ is the cardinality of the continuum, and let us consider an injective family

$$\{P_\xi \ : \ \xi < \alpha\}$$

consisting of all nonempty perfect subsets of $\mathbf{R}$. We fix an arbitrary nonzero discrete subgroup Γ of $\mathbf{R}$ and, using the method of transfinite recursion, define two α-sequences

$$\{x_\xi \ : \ \xi < \alpha\}, \quad \{y_\xi \ : \ \xi < \alpha\}$$

of points of $\mathbf{R}$, satisfying the following three conditions:

1) $x_\xi \in P_\xi$ and $y_\xi \in P_\xi$ for each ordinal $\xi < \alpha$;

2) $(x_\xi + \Gamma) \cap (x_\zeta + \Gamma) = \emptyset$ and $(y_\xi + \Gamma) \cap (y_\zeta + \Gamma) = \emptyset$ for all ordinals $\xi < \alpha, \zeta < \alpha, \xi \neq \zeta$;

3) $(x_\xi + \Gamma) \cap (y_\zeta + \Gamma) = \emptyset$ for all ordinals $\xi < \alpha$ and $\zeta < \alpha$.

The construction of such α-sequences is not hard and we leave the corresponding details to the reader.

Now, let us put

$$X = \{x_\xi \ : \ \xi < \alpha\},$$

$$Y = \{y_\xi \ : \ \xi < \alpha\}.$$

It follows directly from our construction that

$$X \cap Y = \emptyset$$

and both of the sets X and Y are totally imperfect in $\mathbf{R}$. Also, each of these two sets can be regarded as a partial Γ-selector. In other words, for any point $z \in \mathbf{R}$, we have the relations

$$card((z + \Gamma) \cap X) \leq 1,$$

$$card((z + \Gamma) \cap Y) \leq 1.$$

Taking into account the fact that

$$(\forall z \in \mathbf{R})(card(z + \Gamma) = \omega),$$

we can extend the partial Γ-selector X to a Γ-selector X' such that

$$X' \cap Y = \emptyset.$$

Obviously, X' is a Bernstein subset of $\mathbf{R}$ and, therefore, it is nonmeasurable in the Lebesgue sense (respectively, it does not possess the Baire property).

We thus see that any nontrivial discrete group $\Gamma \subset \mathbf{R}$ admits a non Lebesgue-measurable Γ-selector (compare Exercise 6 from Chapter 8).

Let us return to the case of an arbitrary subgroup Γ of the additive group $\mathbf{R}$. We have already mentioned that if all Γ-selectors are nonmeasurable in the Lebesgue sense, then Γ must be everywhere dense in $\mathbf{R}$. Moreover, we also know that if a group Γ is countable and dense in $\mathbf{R}$, then the Vitali theorem holds for Γ, that is all Γ-selectors are nonmeasurable with respect to λ and do not possess the Baire property.

Consequently, the density in $\mathbf{R}$ of a given group $\Gamma \subset \mathbf{R}$ is a necessary condition for the validity of the Vitali theorem. In this context, the following question can be posed: is the countability of Γ necessary for the validity of the Vitali theorem? In other words, suppose that for a given subgroup Γ of $\mathbf{R}$, all Γ-selectors are nonmeasurable in the Lebesgue sense. Does this property of Γ imply that Γ is necessarily countable? It is not difficult to show that by assuming Martin's Axiom with the negation of the Continuum Hypothesis, the answer to this question is negative. Our goal is to establish a stronger result stating that the answer to this question is also negative, by assuming only Martin's Axiom. More precisely, we shall prove below that, under Martin's Axiom, there exists a subgroup Γ of the real line such that $card(\Gamma) = \mathbf{c}$ and every Γ-selector is nonmeasurable with respect to λ.

It should be noted that the question of the existence of such a subgroup Γ of $\mathbf{R}$ was first raised by Ryll-Nardzewski (personal communication).

The following result was obtained in [79]. Several related facts and statements can be found in works [18], [83], [163], [208] and [209].

Theorem 1. *Assume Martin's Axiom. Then there exists a subset Γ of* $\mathbf{R}$ *satisfying the following conditions:*

1) $card(\Gamma) = \mathbf{c}$;

2) Γ *is a vector space over the field* $\mathbf{Q}$ *of all rational numbers (in particular,* Γ *is a subgroup of the additive group* $\mathbf{R}$*);*

3) all Γ*-selectors are nonmeasurable in the Lebesgue sense.*

Proof. Let α denote the first ordinal number with $card(\alpha) = \mathbf{c}$. Let $\{X_\xi : \xi < \alpha\}$ be an injective family of all Lebesgue measure zero Borel subsets of $\mathbf{R}$. We are going to construct (using the method of transfinite recursion) two α-sequences

$$\{V_\xi : \xi < \alpha\}, \qquad \{X'_\xi : \xi < \alpha\}$$

of subsets of $\mathbf{R}$, such that:

(a) $card(V_\xi) \leq card(\xi) + \omega$ for any ordinal $\xi < \alpha$;

(b) if $\xi < \zeta < \alpha$, then V_ξ is a proper subset of V_ζ;

(c) V_ξ is a vector space over $\mathbf{Q}$ for each $\xi < \alpha$;

(d) the set X'_ξ is a translate of X_ξ for each $\xi < \alpha$; in other words, there exists an element h_ξ of $\mathbf{R}$ satisfying the equality $X'_\xi = h_\xi + X_\xi$;

(e) for any two ordinals $\xi < \alpha$ and $\zeta < \alpha$, we have $V_\xi \cap X'_\zeta = \emptyset$.

In order to construct the above-mentioned α-sequences of sets, we first put

$$V_0 = \mathbf{Q}, \qquad X'_0 = h_0 + X_0,$$

where h_0 is an element of $\mathbf{R}$ such that

$$\mathbf{Q} \cap (h_0 + X_0) = \emptyset.$$

The existence of h_0 is evident, so the definition of X'_0 is correct. Suppose now that, for a nonzero ordinal number $\beta < \alpha$, two partial families

$$\{V_\xi : \xi < \beta\}, \qquad \{X'_\xi : \xi < \beta\}$$

have already been defined. Then we put

$$V = \cup\{V_\xi : \xi < \beta\},$$

$$X' = \cup\{X'_\xi : \xi < \beta\}.$$

Obviously, the following three relations are satisfied:

$$V \cap X' = \emptyset, \quad \lambda(X') = 0, \quad card(V) \leq card(\beta) + \omega < \mathbf{c}.$$

In addition, V is a vector space over the field $\mathbf{Q}$. Now, we can choose elements $y \in \mathbf{R} \setminus V$ and $z \in \mathbf{R}$ in such a way that

$$(\mathbf{Q}y + V) \cap X' = \emptyset,$$

$$(\mathbf{Q}y + V) \cap (z + X_\beta) = \emptyset.$$

The possibility of such a choice of y and z follows directly from the fact that under Martin's Axiom, the union of an arbitrary family of cardinality strictly less than $\mathbf{c}$, consisting of Lebesgue measure zero sets, is of Lebesgue measure zero, too (see Appendix 1). Thus, we can put

$$V_\beta = \mathbf{Q}y + V,$$

$$h_\beta = z,$$

$$X'_\beta = z + X_\beta.$$

Continuing the process in this manner, we are able to construct the desired families $\{V_\xi : \xi < \alpha\}$ and $\{X'_\xi : \xi < \alpha\}$. Finally, we define

$$\Gamma = \cup\{V_\xi : \xi < \alpha\}.$$

Evidently, Γ is a vector space over $\mathbf{Q}$ and

$$card(\Gamma) = \mathbf{c}.$$

Also, we assert that all Γ-selectors are nonmeasurable in the Lebesgue sense. Indeed, let Z be an arbitrary Γ-selector and suppose, for a while, that Z belongs to $dom(\lambda)$. Since our group Γ is uncountable, there exists an uncountable disjoint family of translates of Z. Consequently, the equality

$$\lambda(Z) = 0$$

must be valid. Remembering the definition of $\{X_\xi : \xi < \alpha\}$, we see that there exists an ordinal $\xi < \alpha$ such that $Z \subset X_\xi$. But, for the same ordinal, we simultaneously have the relation

$$\Gamma \cap X'_\xi = \Gamma \cap (h_\xi + X_\xi) = \emptyset$$

and, therefore,

$$(-h_\xi + \Gamma) \cap X_\xi = \emptyset,$$

$$(-h_\xi + \Gamma) \cap Z = \emptyset.$$

On the other hand, since Z is a Γ-selector, we must have

$$card((-h_\xi + \Gamma) \cap Z) = 1,$$

which yields a contradiction. This contradiction completes the proof of the theorem.

Let us emphasize that we do not need the full power of Martin's Axiom for the validity of Theorem 1. The previous argument relies only on the following well-known consequence of this axiom: for any family $\{Y_j : j \in J\}$ of Lebesgue measure zero subsets of the real line, with $card(J) < \mathbf{c}$, the set $\cup\{Y_j : j \in J\}$ is of Lebesgue measure zero, too.

Also, it is easy to see that an analogous argument yields the corresponding result for the Baire property (instead of the Lebesgue measure). In this connection, let us point out that a more general result (containing both cases of measure and category) is formulated in Exercise 2.

It is not difficult to observe that the group Γ constructed above has the following interesting property: for every set $Y \subset \mathbf{R}$ of Lebesgue measure zero, we have

$$Y + \Gamma \neq \mathbf{R}.$$

Indeed, suppose to the contrary that there exists a Lebesgue measure zero set $Y \subset \mathbf{R}$ for which

$$Y + \Gamma = \mathbf{R}.$$

Then, for each $x \in \mathbf{R}$, we get

$$Y \cap (x - \Gamma) = Y \cap (x + \Gamma) \neq \emptyset,$$

from which it follows that there exists a Γ-selector Z being a subset of Y. Therefore, Z has Lebesgue measure zero and, in particular, is measurable in the Lebesgue sense. But this is impossible for our group Γ in view of Theorem 1.

Conversely, if Γ is an uncountable subgroup of $\mathbf{R}$ such that for any Lebesgue measure zero set $Y \subset \mathbf{R}$, the relation

$$Y + \Gamma \neq \mathbf{R}$$

holds true, then all Γ-selectors turn out to be nonmeasurable in the Lebesgue sense. An easy checking of this fact is left to the reader.

Notice that any uncountable group $\Gamma \subset \mathbf{R}$ with the above-mentioned property is necessarily totally imperfect; in other words, Γ does not contain a nonempty perfect set. The latter fact follows directly from one deep result of Mokobodzki [150] which states that if P is an arbitrary nonempty perfect set in $\mathbf{R}$, then there always exists a set $Y \subset \mathbf{R}$ such that

$$\lambda(Y) = 0, \quad Y + P = \mathbf{R}.$$

Returning to the question of Lebesgue measurability of selectors associated with subgroups of $\mathbf{R}$ and summarizing all said before, we conclude that the countability and density (in $\mathbf{R}$) of a given group $\Gamma \subset \mathbf{R}$ cannot be described in terms of the Lebesgue nonmeasurability of all Γ-selectors. On the other hand, we have

Theorem 2. *Let Γ be a subgroup of $\mathbf{R}$. Then the following two assertions are equivalent:*

1) Γ is countable and nondiscrete (thus, everywhere dense in $\mathbf{R}$);

2) any Γ-selector is nonmeasurable with respect to each Γ-invariant measure on $\mathbf{R}$ extending λ.

Proof. The first implication 1) $\Rightarrow$ 2) is, in fact, the generalized version of Vitali's theorem (compare Theorem 1 from Chapter 1). Consequently, we only need to establish the converse implication 2) $\Rightarrow$ 1). Suppose that 2) is true. Then, applying 2) to λ, we see that all Γ-selectors are nonmeasurable in the Lebesgue sense and, as shown above, the group Γ must be nondiscrete (equivalently, dense in $\mathbf{R}$). It remains to prove that Γ is also at most countable.

Suppose for a moment that $card(\Gamma) \geq \omega_1$. Take an arbitrary Γ-selector Z and denote by $\mathcal{I}$ the Γ-invariant σ-ideal of subsets of $\mathbf{R}$, generated by the one-element family $\{Z\}$. It can easily be checked that for each set $X \in \mathcal{I}$, we have $\lambda_*(X) = 0$ where λ_* denotes, as usual, the inner measure associated with λ.

Therefore, applying the standard method of extending invariant measures (see [82], [212] or [214]), we can construct a complete Γ-invariant measure μ on $\mathbf{R}$ such that μ extends λ and

$$\mathcal{I} \subset dom(\mu), \quad (\forall X \in \mathcal{I})(\mu(X) = 0).$$

In particular, for the taken Γ-selector Z, we have $Z \in dom(\mu)$, which contradicts our assumption that 2) is true. This contradiction yields the inequality

$$card(\Gamma) \leq \omega$$

and, consequently, the implication 2) $\Rightarrow$ 1). Thus, Theorem 2 is proved.

In addition, let us point out that if Γ is an arbitrary uncountable subgroup of $\mathbf{R}$, then every Γ-selector is a Γ-negligible subset of $\mathbf{R}$. Moreover, it turns out that every Γ-selector is a Γ-absolutely negligible subset of $\mathbf{R}$. (The notion of an absolutely negligible set, with respect to a given class of σ-finite quasi-invariant measures, is introduced in Exercise 4 of this chapter.)

It is well known, in general, that various nice descriptive properties of subsets of $\mathbf{R}$ are not preserved under the operation of vector sum. In other words, if some sets $A \subset \mathbf{R}$ and $B \subset \mathbf{R}$ have a good descriptive structure, then we cannot assert that the set

$$A + B = \{a + b : a \in A, \ b \in B\}$$

has the same good descriptive structure. The following two examples are typical in this context.

Example 1. There exist two Borel subsets X and Y of $\mathbf{R}$ such that the set $X + Y$ is not Borel. (Obviously, $X + Y$ is an analytic subset of $\mathbf{R}$.)

Example 2. There exist two sets $X \subset \mathbf{R}$ and $Y \subset \mathbf{R}$, both of Lebesgue measure zero, such that the set $X + Y$ is not measurable in the Lebesgue sense.

In connection with Example 1, see [43], [182], [206] and [207].

Recall that Example 2 is due to Sierpiński (see Theorem 5 in Chapter 3 of this book).

Now, let us consider a rather general situation where the operation of vector sum plays an essential role for the descriptive structure of selectors associated with subgroups of $\mathbf{R}$ (compare [99]).

Theorem 3. *Let $\mathcal{K}$ be some class of subsets of the real line, satisfying the following three conditions:*

1) if $X \in \mathcal{K}$ and $Y \in \mathcal{K}$, then $X \cap Y \in \mathcal{K}$;

2) if $X \in \mathcal{K}$ and $Y \in \mathcal{K}$, then $X + Y \in \mathcal{K}$;

3) the σ-algebra generated by $\mathcal{K}$ contains the Borel σ-algebra of the real line.

Then at least one of the next two relations is valid:

(a) there exists a set belonging to $\mathcal{K}$ and nonmeasurable in the Lebesgue sense;

(b) for each proper dense subgroup Γ of $\mathbf{R}$ belonging to $\mathcal{K}$, no Γ-selector belongs to $\mathcal{K}$.

Proof. Suppose that relation (a) is not true; in other words, suppose that all sets from the given class $\mathcal{K}$ are Lebesgue measurable. Let us show that, in such a case, relation (b) is fulfilled. In order to do this, take an arbitrary proper dense subgroup Γ of $\mathbf{R}$ belonging to $\mathcal{K}$ and fix an arbitrary Γ-selector Z. Our aim is to establish the relation $Z \notin \mathcal{K}$. Suppose for a while that $Z \in \mathcal{K}$. Obviously, we can define a function

$$f \ : \ \mathbf{R} \to \mathbf{R}$$

in the following manner: for each point $z \in \mathbf{R}$, let the value $f(z)$ be equal to a unique element of the set $(z+\Gamma) \cap Z$. From the definition of f and from conditions 1) and 2) we can immediately derive that for any set $X \in \mathcal{K}$, the preimage

$$f^{-1}(X) = (X \cap Z) + \Gamma$$

belongs to $\mathcal{K}$, too. Hence, according to our assumption, the set $f^{-1}(X)$ is measurable in the Lebesgue sense. Taking condition 3) into account, we conclude that f is a Lebesgue measurable function. In addition, f simultaneously is a Γ-invariant function; in other words, for any $z \in \mathbf{R}$ and $h \in \Gamma$, we have

$$f(z) = f(z + h).$$

But our Γ is everywhere dense in $\mathbf{R}$. So, applying the metrical transitivity of the Lebesgue measure with respect to Γ, we obtain that f is equivalent to a constant function. This means that there exists an element $t \in \mathbf{R}$ for which

$$\lambda(\mathbf{R} \setminus f^{-1}(t)) = 0.$$

At the same time, it is clear that

$$f^{-1}(t) = t + \Gamma.$$

Since Γ is a Lebesgue measurable subgroup of $\mathbf{R}$ and the relation

$$\Gamma \neq \mathbf{R}$$

holds, we have $\lambda(\Gamma) = 0$ (we recall that the last equality follows directly from the Steinhaus property for λ). Thus, we come to the relation

$$\lambda(f^{-1}(t)) = \lambda(t + \Gamma) = \lambda(\Gamma) = 0,$$

which contradicts the equality $\lambda(\mathbf{R} \setminus f^{-1}(t)) = 0$.

This contradiction completes the proof of Theorem 3.

Example 3. Let us take as $\mathcal{K}$ the class of all projective subsets of $\mathbf{R}$. Evidently, conditions 1), 2) and 3) of Theorem 3 are satisfied for such a $\mathcal{K}$. Hence, at least one of the following two assertions is true:

(a) there exists a projective subset of $\mathbf{R}$ nonmeasurable in the Lebesgue sense;

(b) if Γ is a proper dense subgroup of $\mathbf{R}$ and Γ is simultaneously a projective set in $\mathbf{R}$, then all Γ-selectors are not projective subsets of $\mathbf{R}$.

In this connection, we wish to point out that none of assertions (a) and (b) presented above can be proved within the theory **ZFC**.

Indeed, on the one hand, it was established by Solovay [210] that there is a model of set theory in which all projective subsets of the real line are Lebesgue measurable and possess the Baire property.

On the other hand, in the Constructible Universe of Gödel, there exist **Q**-selectors belonging to the projective class $\Sigma_2^1(\mathbf{R})$. (Recall that, according to the theorem of Vitali, all such selectors are nonmeasurable in the Lebesgue sense and do not have the Baire property.)

Now, let us fix an uncountable commutative group $(G, +)$ with a nonzero σ-finite G-quasi-invariant measure μ defined on some σ-algebra of subsets of G. Let H be a subgroup of G. In Chapter 9 we have considered the three questions concerning the measurability of various H-selectors in the case where

$$card(H) \leq \omega.$$

Here we will discuss an analogous question for an uncountable subgroup H of G satisfying a natural condition. Namely, let H be an uncountable subgroup of G such that

$$card(G/H) = card(G).$$

Is it true that there exists at least one H-selector nonmeasurable with respect to μ?

It is not hard to show that the answer to the question just formulated is negative. Indeed, let us put

$$G = \mathbf{R} \times \mathbf{R} = \mathbf{R}^2$$

and take as H the subgroup $\{0\} \times \mathbf{R}$ of G. Evidently, we have the relations

$$card(G/H) = card(\mathbf{R}) = card(G) = \mathbf{c}.$$

Denote by λ_2 the standard two-dimensional Lebesgue measure on G. Further, let $\mathcal{I}$ be the σ-ideal of subsets of G, generated by the family of all H-selectors (or, equivalently, by the family of all graphs of functions acting from $\mathbf{R}$ into $\mathbf{R}$). It can readily be checked that $\mathcal{I}$ possesses the following properties:

(a) $\mathcal{I}$ is invariant under the group of all translations of G;

(b) for each set $X \in \mathcal{I}$, the inner λ_2-measure of X is equal to zero.

Starting with these two properties of $\mathcal{I}$ and applying the standard argument, we can establish the existence of a measure μ on G such that:

(1) μ is an extension of λ_2;

(2) μ is invariant under G;

(3) $\mathcal{I}$ is contained in $dom(\mu)$;

(4) $\mu(X) = 0$ for each set $X \in \mathcal{I}$.

In particular, we easily obtain that all H-selectors are measurable with respect to μ.

Remark 1. Let $n \geq 1$ be a natural number, $\mathbf{R}^n$ be the n-dimensional euclidean space, and let λ_n be the standard n-dimensional Lebesgue measure on $\mathbf{R}^n$. Finally, let H denote an arbitrary uncountable subgroup of the additive group $\mathbf{R}^n$. Suppose that Martin's Axiom with the negation of the Continuum Hypothesis hold. Then it can be proved (see [83]) that there exists a measure ν on $\mathbf{R}^n$ satisfying the following conditions:

1) ν is an extension of λ_n;

2) ν is invariant under the group of all translations and central symmetries of $\mathbf{R}^n$;

3) all H-selectors are measurable with respect to ν;

4) if X is an arbitrary H-selector, then $\nu(X) = 0$.

For the euclidean space $\mathbf{R}^n$, where $n \geq 2$, the question on the measurability of selectors associated with uncountable subgroups of $\mathbf{R}^n$ sometimes turns out to be trivial because we have a constructive representation of the space $\mathbf{R}^n$ in the form of the direct sum of its two uncountable vector subspaces. For example, we may write

$$\mathbf{R}^n = \Gamma + H \qquad (\Gamma \cap H = \{0\}),$$

where

$$\Gamma = \mathbf{R}, \quad H = \mathbf{R}^{n-1}$$

and, therefore, the subspace $\mathbf{R}^{n-1}$ can be regarded as a Lebesgue measurable selector associated with the uncountable subgroup Γ of the whole space $\mathbf{R}^n$.

In this connection, it should be noted that for $n = 1$, we do not have an analogous representation of the space $\mathbf{R}^n = \mathbf{R}$ in the form of a direct sum of its two proper subgroups with a nice descriptive structure. Indeed, take any representation

$$\mathbf{R} = G + H \qquad (G \cap H = \{0\}),$$

where G and H are some Borel (or, more generally, analytic) subgroups of $\mathbf{R}$. We assert that this decomposition is trivial; in other words, either $G = \{0\}$ or $H = \{0\}$. To see this, assume to the contrary that $G \neq \{0\}$ and $H \neq \{0\}$ and observe at once that, in this case, we must have the inequalities

$$card(G) > \omega, \quad card(H) > \omega.$$

Indeed, suppose for a while that $card(G) \leq \omega$. Then the equality

$$\mathbf{R} = \cup\{g + H : g \in G\}$$

leads to the claim that H is not a first category set in $\mathbf{R}$. Taking into account the fact that H has the Baire property and applying the Banach-Kuratowski-Pettis theorem (see Exercise 6 from Chapter 1), we obtain that H has nonempty interior and, consequently, is a clopen subgroup of $\mathbf{R}$. This immediately implies $H = \mathbf{R}$ and $G = \{0\}$ which contradicts our assumption.

So we necessarily come to the uncountability of the groups G and H, from which it follows that both these groups are also dense in $\mathbf{R}$. For our further purposes, it suffices to assume that H is everywhere dense in $\mathbf{R}$. Since, for any $x \in \mathbf{R}$, we have the unique representation

$$x = y + z \qquad (y = pr_G(x) \in G, \ z = pr_H(x) \in H),$$

it is natural to consider a function

$$f \; : \; \mathbf{R} \to \mathbf{R}$$

defined by the formula:

$$f(x) = pr_G(x) \qquad (x \in \mathbf{R}).$$

Obviously,

$$f(x + x') = f(x) + f(x') \qquad (x \in \mathbf{R}, \; x' \in \mathbf{R}).$$

We claim that f turns out to be an additive functional on $\mathbf{R}$ (or, in other words, f turns out to be a group homomorphism from $\mathbf{R}$ into itself). Also, an easy verification shows that f is measurable in the Lebesgue sense. This implies (see Chapter 3) the linearity of f, that is there exists a real number $a \in \mathbf{R}$ for which

$$f(x) = ax \qquad (x \in \mathbf{R}).$$

Then $H = f^{-1}(0)$ and we get either $H = \{0\}$ (if $a \neq 0$), or $H = \mathbf{R}$ (if $a = 0$).

In both these cases a contradiction is obtained, which yields the required result.

Remark 2. For subgroups of $\mathbf{R}$ belonging to higher projective classes, a result analogous to the one presented above fails to be any longer true because in the Gödel Constructible Universe we have a representation

$$\mathbf{R} = \mathbf{Q} + \Gamma \qquad (\mathbf{Q} \cap \Gamma = \{0\}),$$

where Γ is some vector hyperplane in $\mathbf{R}$ (considered as a vector space over $\mathbf{Q}$) belonging to the class $\Sigma_2^1(\mathbf{R})$. Since Γ is a Vitali subset of $\mathbf{R}$, it is not measurable in the Lebesgue sense and does not possess the Baire property.

Remark 3. Some interesting additive properties of subsets of $\mathbf{R}$ are discussed in [41]. (See also [95] and [107].)

EXERCISES

1. Modifying the construction of an uncountable universal measure zero subset of $\mathbf{R}$, given in Chapter 4, demonstrate the existence of an uncountable universal measure zero subgroup of $\mathbf{R}$ (for a much stronger result, see [172]).

2. Assuming Martin's Axiom, prove that there exists a subgroup Γ of $\mathbf{R}$ such that:

(a) $card(\Gamma) = \mathbf{c}$;

(b) every Γ-selector is nonmeasurable in the Lebesgue sense;

(c) every Γ-selector does not possess the Baire property.

3. Let G be a group and let H be an arbitrary uncountable subgroup of G. We identify G with the group of all left translations of G. Show that each H-selector is an H-negligible (and a G-negligible) subset of G.

4. Let E be a set, G be a group of transformations of E and let X be a subset of E. We shall say that X is G-absolutely negligible in the space (E, G) if for any σ-finite G-invariant measure μ on E, there exists a G-invariant measure μ' on E extending μ and such that

$$X \in dom(\mu'), \quad \mu'(X) = 0.$$

The notion of an absolutely negligible set plays an essential role in some questions concerning invariant extensions of σ-finite invariant measures. See, for instance, [82] or [100].

Demonstrate the following properties of absolutely negligible sets:

(a) the class of all G-absolutely negligible sets is hereditary;

(b) the same class is closed under finite unions of its members;

(c) if X is G-absolutely negligible and $\{g_n : n < \omega\}$ is an arbitrary countable family of transformations from G, then the set

$$Y = \cup\{g_n(X) : n < \omega\}$$

is also G-absolutely negligible;

(d) every G-absolutely negligible set is G-negligible as well.

In connection with property (b), it should be mentioned that in general the class of absolutely negligible sets is not closed under countable unions of its members. (See [82], [100].)

In connection with property (d), let us recall that every selector associated with an uncountable subgroup H of $\mathbf{R}$ is $\mathbf{R}$-negligible in $\mathbf{R}$. At the same time, it is not true that every selector associated with an uncountable subgroup H of $\mathbf{R}$ is $\mathbf{R}$-absolutely negligible.

In order to establish this, start with the group $\mathbf{R}^2 = \mathbf{R} \times \mathbf{R}$ and its subgroup $\{0\} \times \mathbf{R}$. Denote by T_2 the group of all translations of $\mathbf{R}^2$. Assuming the Continuum Hypothesis and applying the method of transfinite induction, show that there exists a partial function

$$f : \mathbf{R} \to \mathbf{R} \quad (dom(f) \subset \mathbf{R})$$

such that:

(i) the graph Γ_f of this partial function is λ_2-thick in $\mathbf{R}^2$;

(ii) for some countable family $\{h_n : n < \omega\} \subset T_2$, the set

$$Y = \cup\{h_n(\Gamma_f) : n < \omega\}$$

is almost invariant with respect to T_2; in other words, we have

$$card(h(Y) \triangle Y) \leq \omega$$

for any translation h of $\mathbf{R}^2$.

Infer from (ii) that the set Γ_f, being a partial selector for the family of all translates of the uncountable subgroup $H = \{0\} \times \mathbf{R}$ of $\mathbf{R}^2$, is T_2-negligible but is not T_2-absolutely negligible.

Construct an analogous example for the real line $\mathbf{R}$ and for some uncountable subgroup H of $\mathbf{R}$. (Use the fact that $\mathbf{R}$ and $\mathbf{R}^2$ are isomorphic as vector spaces over $\mathbf{Q}$.)

5. Let G be a commutative group and let H be a subgroup of G such that

$$card(G/H) \geq \omega_1.$$

Show that H is a G-absolutely negligible set in G.

For additional information about negligible and absolutely negligible sets, see [82], [89], [94], [96], [97], [100] and Chapter 14 of this book.

6. Prove the existence of a measure ν described in Remark 1 of this chapter.

7. Deduce from the Mokobodzki theorem [150] that for every nonempty perfect set $P \subset \mathbf{R}$ there exists a set $X \subset \mathbf{R}$ of Lebesgue measure zero, such that

$$P + X = \mathbf{R}.$$

8. Generalize Theorem 1 of this chapter to the case of the euclidean space $\mathbf{R}^n$ where $n \geq 1$.

9. Assume Martin's Axiom. Let Γ be a subgroup of the additive group **R**, satisfying the relation

$$\omega_1 \leq card(\Gamma) < \mathbf{c},$$

and let X be an arbitrary Γ-selector. Show that:
(a) X is not measurable in the Lebesgue sense;
(b) X does not possess the Baire property.
Prove an analogous result for the euclidean space $\mathbf{R}^n$ where $n \geq 2$.

Chapter 11
Absolutely nonmeasurable sets in groups

In this chapter we wish to discuss some specific analogues and generalizations of the classical Vitali construction producing a non Lebesgue-measurable subset of the real line. More precisely, we will be dealing here with several constructions of so-called absolutely nonmeasurable sets in abstract groups. Obviously, every uncountable group can be equipped with various nonzero diffused σ-finite left invariant measures. Considering subsets of a group which are nonmeasurable with respect to all above-mentioned measures, we naturally come to the notion of absolutely nonmeasurable sets. Such sets have extremely bad properties from the viewpoint of the general theory of invariant (quasi-invariant) measures.

Notice that the material presented in this chapter is essentially based on results obtained by the author (see [78], [82], [85], [96], [99]). In our opinion, the concept of absolute nonmeasurability is rather important for measure theory and deserves to be studied in more details.

We have already indicated a number of works devoted to the Vitali theorem and its different versions. Namely, the theorem was generalized in [16], [17], [18], [38], [69], [80], [103], [208], and [209]. Also, interesting generalizations and analogues of this theorem were obtained for a locally compact topological group equipped with a left invariant Haar measure (see, for instance, [25] and [68]).

Some other questions related to the Vitali theorem were considered in [22], [55], [99], [143], [155], [173], [197], [226], [228] and in the previous chapters of this book.

Of course, it is impossible to mention here all papers, monographs, or textbooks in which this theorem is discussed. Recall that in the well-known monograph by Morgan [155] a rather complete list of references to various works (closely connected with the Vitali theorem) is presented.

The aim of this chapter is to investigate the concept of an absolutely nonmeasurable set with respect to a given class of nonzero σ-finite quasi-invariant (invariant) measures. We shall establish, in our further considerations, the existence of absolutely nonmeasurable sets in any uncountable commutative group. Also, a generalization of this result to an uncountable solvable group will be obtained and several related open problems and questions will be posed.

First of all, we wish to formulate some auxiliary facts and statements which will be utilized systematically throughout the chapter.

Let E be a nonempty set and let μ be a measure defined on a σ-algebra of subsets of E. We recall that μ satisfies the countable chain condition (or the Suslin condition) if for any disjoint family of sets

$$\{X_i : i \in I\} \subset dom(\mu),$$

the relation

$$(\forall i \in I)(\mu(X_i) > 0)$$

implies the relation

$$card(I) \leq \omega.$$

It can easily be observed that every σ-finite measure satisfies the Suslin condition, and simple examples show that the converse assertion is not true.

In the sequel, we will employ one advantage of measures satisfying the Suslin condition over σ-finite measures. This advantage plays an essential role in some technical tricks below. Namely, the following simple lemma turns out to be very useful for our purposes.

Lemma 1. *Let $(E_1, \mathcal{S}_1, \mu_1)$ and $(E_2, \mathcal{S}_2, \mu_2)$ be two measure spaces and let*

$$\phi : (E_1, \mathcal{S}_1, \mu_1) \to (E_2, \mathcal{S}_2, \mu_2)$$

be a homomorphism of the first measure space into the second one; in other words, suppose that

$$(\forall X \in dom(\mu_2))(\phi^{-1}(X) \in dom(\mu_1) \ \& \ \mu_1(\phi^{-1}(X)) = \mu_2(X)).$$

If $(E_1, \mathcal{S}_1, \mu_1)$ satisfies the countable chain condition, then $(E_2, \mathcal{S}_2, \mu_2)$ satisfies this condition, too.

Therefore, the Suslin condition is preserved under homomorphisms.

An easy proof of Lemma 1 is left to the reader.

In general topology we have an analogous Suslin condition for topological spaces (see [37]). It is well known that this condition is preserved under surjective continuous mappings.

Simple examples show that the homomorphic image of a σ-finite measure can be a non-σ-finite measure (compare Exercise 1 of this chapter).

Let us also recall one simple fact from measure theory: if we have a space (E, G) equipped with a nonzero σ-finite G-quasi-invariant (G-invariant) measure μ on E, then we can define a probability G-quasi-invariant measure ν on E such that $dom(\nu) = dom(\mu)$ and $\mathcal{I}(\nu) = \mathcal{I}(\mu)$; in other words, μ and ν are equivalent measures. This observation will be applied many times in the following constructions.

Of course, without loss of generality, we may restrict our further considerations to the case of left G-quasi-invariant (left G-invariant) measures defined on various σ-algebras of subsets of a given abstract group $(G, \cdot)$. If an original group G is commutative, then the situation is significantly simplified since the concept of a left G-quasi-invariant (left G-invariant) measure is identical with the concept of a right G-quasi-invariant (right G-invariant) measure.

The following definition is basic for this chapter.

Let (E, G) be a space with a transformation group and let M be some class of G-quasi-invariant (G-invariant) measures on E. We say that a set $X \subset E$ is absolutely nonmeasurable with respect to M if

$$(\forall \mu \in M)(X \notin dom(\mu)).$$

It immediately follows from this definition that if M and M' are any two classes of G-quasi-invariant (G-invariant) measures on E, satisfying the inclusion $M \subset M'$, and $X \subset E$ is absolutely nonmeasurable with respect to M', then X is also absolutely nonmeasurable with respect to M.

Example 1. Let $E = \mathbf{R}$ and let G be the group of all rational translations of $\mathbf{R}$; in other words, we put $G = \mathbf{Q}$. Let M denote the class of all those G-invariant measures μ on $\mathbf{R}$ which satisfy the conditions

$$]0, 1[\ \in dom(\mu), \quad 0 < \mu(]0, 1[) < +\infty.$$

Then, according to the Vitali theorem, we can assert that each Vitali subset of $\mathbf{R}$ is absolutely nonmeasurable with respect to M.

Example 2. Let again $E = \mathbf{R}$ and let G be the group of all motions (isometric transformations) of $\mathbf{R}$. Denote by M the class of all those G-invariant measures μ on $\mathbf{R}$ which satisfy the conditions:

1) μ is nonzero and σ-finite;

2) $dom(\lambda) \subset dom(\mu)$ where $\lambda = \lambda_1$ stands, as usual, for the Lebesgue measure on $\mathbf{R}$.

Let us demonstrate that there exists a Vitali subset of $\mathbf{R}$ which is not absolutely nonmeasurable with respect to M.

For this purpose, consider $E = \mathbf{R}$ as a vector space over the field $\mathbf{Q}$. Take $e = 1$. In view of the fundamental theorem of the theory of vector spaces (over arbitrary fields), the one-element set $\{e\}$ can be extended to a basis of E, that is there exists a Hamel basis $\{e_i : i \in I\}$ for E, containing e. The latter means that $\{e_i : i \in I\}$ is a maximal (with respect to inclusion) linearly independent (over $\mathbf{Q}$) family of elements of E and $e \in \{e_i : i \in I\}$. Now, find the index $i_0 \in I$ for which $e_{i_0} = e$, and consider the vector subspace V of E generated by the family $\{e_i : i \in I \setminus \{i_0\}\}$. It is obvious that V turns out to be a vector hyperplane in E complementary to the vector "line" $\mathbf{Q}$. In other words, we have the representation

$$E = V + \mathbf{Q} \qquad (V \cap \mathbf{Q} = \{0\})$$

of the space E in the form of a direct sum of its two vector subspaces. In particular, for each $x \in \mathbf{R}$, the relation

$$card(V \cap (x + \mathbf{Q})) = 1$$

is true, from which it follows that V is a Vitali subset of $\mathbf{R}$.

Further, a simple argument shows that V is everywhere dense in $\mathbf{R}$. Indeed, starting with the formula

$$\mathbf{R} = V + \mathbf{Q} = \cup\{V + q : q \in \mathbf{Q}\}$$

and applying the classical Baire theorem on category, we claim that for some rational number q, the set $cl(V + q)$ has nonempty interior. This fact immediately implies that $cl(V)$ has nonempty interior, too, and since V is a subgroup of $\mathbf{R}$, we conclude that V is everywhere dense in $\mathbf{R}$. Moreover, we can even assert that V is a λ-thick set in $\mathbf{R}$. To see this, let us fix any λ-measurable set $X \subset \mathbf{R}$ with $\lambda(X) > 0$ and suppose, for a moment, that

$$X \cap V = \emptyset.$$

Let $\{v_n : n < \omega\}$ be a sequence of elements from V, everywhere dense in $\mathbf{R}$. Obviously, we have

$$(\cup\{X + v_n : n < \omega\}) \cap V = \emptyset.$$

In view of the metrical transitivity of λ (see Exercise 7 from Chapter 1), the set $\cup\{X + v_n : n < \omega\}$ is of full λ-measure; in other words,

$$\lambda(\mathbf{R} \setminus \cup\{X + v_n : n < \omega\}) = 0,$$

from which it follows that
$$\lambda(V) = 0.$$

But the last equality is impossible for a Vitali set. In this manner, we have shown that V is a λ-thick subset of the real line.

Denote by $\{V_n : n < \omega\}$ the Vitali partition of $\mathbf{R}$ produced by the Vitali set V. It is not hard to check that:

(a) each set V_n $(n < \omega)$ of this partition is some translate of V;

(b) if g is any motion of $\mathbf{R}$, then the set $g(V)$ coincides with some V_n.

We thus see that $\{V_n : n < \omega\}$ forms a G-invariant partition of $\mathbf{R}$ into λ-thick sets.

Consider now the family $\mathcal{S}$ of all those subsets Y of $\mathbf{R}$ which can be represented in the form

$$Y = \cup\{V_n \cap X_n : n < \omega\},$$

where $\{X_n : n < \omega\}$ is a countable family of λ-measurable sets in $\mathbf{R}$ (of course, depending on Y). A straightforward verification shows that the following relations hold:

(c) $\mathcal{S}$ is a G-invariant family of sets;

(d) $\mathcal{S}$ is a σ-algebra of subsets of $\mathbf{R}$;

(e) $dom(\lambda) \subset \mathcal{S}$.

Now, for any set $Y = \cup\{V_n \cap X_n : n < \omega\}$ belonging to $\mathcal{S}$, we put

$$\nu(Y) = \sum_{n<\omega} \lambda(X_n).$$

Notice that the value $\nu(Y)$ depends only on Y and does not depend on the representation of Y in the above-mentioned form. Indeed, if we have any two such representations:

$$Y = \cup\{V_n \cap X_n : n < \omega\} = \cup\{V_n \cap X'_n : n < \omega\},$$

then, taking into account the λ-thickness of all V_n, we get

$$V_n \cap X_n = V_n \cap X'_n,$$

$$V_n \cap (X_n \triangle X'_n) = \emptyset,$$

$$\lambda(X_n \triangle X'_n) = 0,$$

$$\lambda(X_n) = \lambda(X'_n)$$

for each $n < \omega$. Therefore, the functional

$$\nu : \mathcal{S} \to \mathbf{R} \cup \{+\infty\}$$

is well defined. Utilizing once more the λ-thickness of all V_n, we easily infer that ν is countably additive; in other words, ν turns out to be a nonzero measure on $\mathcal{S}$. The σ-finiteness of ν can be observed immediately: if $\{X_k : k < \omega\}$ is any countable covering of $\mathbf{R}$ such that

$$(\forall k < \omega)(X_k \in dom(\lambda) \,\&\, \lambda(X_k) < +\infty),$$

then the family of ν-measurable sets

$$\{V_n \cap X_k : n < \omega, \ k < \omega\}$$

is a countable covering of $\mathbf{R}$, too, and for all $n < \omega$ and $k < \omega$, we have

$$\nu(V_n \cap X_k) = \lambda(X_k) < +\infty.$$

Also, the G-invariance of the Vitali partition $\{V_n : n < \omega\}$ and the definition of our measure ν readily imply the G-invariance of ν. Further, if X is an arbitrary λ-measurable subset of $\mathbf{R}$, then we can write

$$X = \cup\{V_n \cap X_n : n < \omega\}$$

where $X_n = X$ for all $n < \omega$. This yields the relation

$$\nu(X) = \sum_{n<\omega} \lambda(X_n) = \lambda(X) + \lambda(X) + \dots \in \{0, +\infty\}.$$

Finally, there exists a natural number m such that $V_m = V$. So we have the representation

$$V = \cup\{V_n \cap X_n : n < \omega\},$$

where $X_n = \emptyset$ if $n \neq m$, and $X_n = \mathbf{R}$ if $n = m$. This shows that the Vitali set V is measurable with respect to ν.

Example 3. Let again $E = \mathbf{R}$ and let G be the group of all motions (isometric transformations) of $\mathbf{R}$. Take the measure ν constructed in the previous example and denote by ν' a probability measure equivalent to ν. In other words, ν' is a probability measure defined on $dom(\nu)$ and having the property that for each set $Y \in dom(\nu)$, the relation

$$\nu(Y) = 0 \Leftrightarrow \nu'(Y) = 0$$

is valid. Notice, by the way, that ν' is not a G-invariant measure but is a G-quasi-invariant measure. Thus, we see that there exists a probability measure ν' on $\mathbf{R}$ which is G-quasi-invariant and whose domain contains all Lebesgue measurable subsets of $\mathbf{R}$ and some Vitali set.

In a similar way, we can define a G-quasi-invariant measure ν'' on $\mathbf{R}$ which extends λ and whose domain includes some Vitali set (compare Theorem 1 from Chapter 13).

Let G be again the group of all motions of $\mathbf{R}$. Example 2 shows us that there exist Vitali sets which are not absolutely nonmeasurable with respect to the class of all nonzero σ-finite G-invariant measures on $\mathbf{R}$. Analogously, Example 3 shows that there exist Vitali sets which are not absolutely nonmeasurable with respect to the class of all probability G-quasi-invariant measures on $\mathbf{R}$.

In view of the above demonstrated, we can conclude that the classical Vitali construction is not able to provide us with absolutely nonmeasurable sets with respect to the following two natural classes of measures: the class of all nonzero σ-finite invariant measures on $\mathbf{R}$ and the class of all probability quasi-invariant measures on $\mathbf{R}$. Now, we are going to discuss the question of the existence of absolutely nonmeasurable sets with respect to these classes of measures. Of course, it is not reasonable to restrict considerations only to the case of the real line $\mathbf{R}$ endowed with some canonical group of its transformations (for example, translations or motions). We may also consider the question of the existence of absolutely nonmeasurable sets in more general situations.

The central problems of this topic can be formulated in the following manner (compare [82], [99], [100]).

Problem 1. Let (E, G) be a space with a transformation group. Find necessary and sufficient conditions (in terms of the pair (E, G)) for the existence in E of an absolutely nonmeasurable set with respect to the class $M^1_{(E,G)}$ of all probability G-quasi-invariant measures on E.

Simple examples show that in some cases, for a given space (E,G), there are not absolutely nonmeasurable subsets of E with respect to the class $M^1_{(E,G)}$ (see, for instance, Exercise 3 of this chapter).

Naturally, an analogue of Problem 1 for invariant measures can be formulated as follows.

Problem 2. Let (E,G) be a space with a transformation group. Find necessary and sufficient conditions (in terms of the pair (E,G)) for the existence in E of an absolutely nonmeasurable set with respect to the class $M^\sigma_{(E,G)}$ of all nonzero σ-finite G-invariant measures on E.

Problems 1 and 2 presented above remain unsolved and, in our opinion, are of certain interest for the theory of invariant (quasi-invariant) measures.

We now formulate the next problem concerning a geometric description of absolutely nonmeasurable sets.

Problem 3. Let (E,G) be again a space with a transformation group. Give some characterization (in terms of (E,G)) of subsets of E which are absolutely nonmeasurable with respect to the class $M^1_{(E,G)}$ (with respect to the class $M^\sigma_{(E,G)}$).

In other words, necessary and sufficient geometric conditions must be found for a set $X \subset E$, under which X turns out to be absolutely nonmeasurable with respect to $M^1_{(E,G)}$ ($M^\sigma_{(E,G)}$). In this connection, let us note that some simple necessary conditions can be found without any difficulty (compare Exercise 5).

In addition to this, let us notice that if a set $X \subset E$ is absolutely nonmeasurable with respect to the class of all probability G-quasi-invariant measures on E, then X is also absolutely nonmeasurable with respect to the class of all nonzero σ-finite G-invariant measures on E.

Of course, the formulation of Problem 1 seems to be rather general. It should be mentioned that even particular cases of this problem are of interest from the point of view of the theory of quasi-invariant measures. One of more concrete versions of Probem 1 is the next question.

Problem 4. Let E be an uncountable group and let G be the group of all left translations of E. Does there exist a subset of E which is absolutely nonmeasurable with respect to $M^1_{(E,G)}$?

In other words, identifying an uncountable group E with the group G of all its left translations, we are interested whether there exist absolutely nonmeasurable subsets of G with respect to the class $M_G^1 = M_{(E,G)}^1$. Up to now, Problem 4 remains unsolved. Let us point out that this problem was posed by the author many years ago (see [82] where several related open questions are formulated).

Our further argument is connected with some particular cases of Problem 4. Namely, we are going to demonstrate that a solution to this problem is positive if a given uncountable group E is solvable. Consequently, we also get a positive solution for any uncountable commutative group E.

First, let us consider the case of an uncountable commutative group E which is identified below with the group G of all its translations.

We need several preliminary statements.

Lemma 2. *Let $(G, +)$ be an uncountable commutative group. Then there exist two subgroups G_0 and G_1 of G, such that:*
1) $card(G_0) = \omega_0 = \omega$;
2) $card(G_1) = \omega_1$;
3) $G_0 \cap G_1 = \{0\}$ where 0 denotes the neutral element (zero) in G.

Proof. We apply one deep theorem from the general theory of commutative groups, which yields an algebraic description of these groups. Namely (see [129] and Appendix 2), if Γ is an arbitrary commutative group, then it can be represented in the form

$$\Gamma = \cup\{\Gamma_n : n < \omega\},$$

where the family of sets $\{\Gamma_n : n < \omega\}$ is increasing by inclusion and each Γ_n is a subgroup of Γ representable as a direct sum of cyclic groups. Hence, in our case we may write

$$G = \cup\{H_n : n < \omega\},$$

where $\{H_n : n < \omega\}$ is a sequence of subgroups of G and each H_n is a direct sum of cyclic groups. Since our G is uncountable, at least one H_n is uncountable, too; actually, there are infinitely many uncountable groups in the above-mentioned sequence. We may suppose, without loss of generality, that

$$card(H_0) \geq \omega_1$$

and, in fact, we may assume that

$$card(H_0) = \omega_1.$$

Now, we can express H_0 as a direct sum

$$H_0 = \sum_{\xi < \omega_1} [g_\xi],$$

where g_ξ ($\xi < \omega_1$) are some nonzero elements of G and the symbol $[g_\xi]$ denotes the cyclic group generated by g_ξ. Finally, let us put

$$G_0 = \sum_{\xi < \omega} [g_\xi],$$

$$G_1 = \sum_{\omega \leq \xi < \omega_1} [g_\xi].$$

Then it is not hard to check that

$$G_0 \cap G_1 = \{0\},$$

$$card(G_0) = \omega, \quad card(G_1) = \omega_1.$$

This shows that the groups G_0 and G_1 are the required ones.

Lemma 2 has thus been proved.

A direct proof of Lemma 2 is sketched in Exercise 4.

We recall that if E is an arbitrary set, then Id_E stands for the identity transformation of E.

The next lemma (see [96]) plays the key role in further constructions.

Lemma 3. *Let E be a set of cardinality ω_1, let G, G_0 and G_1 be some groups of transformations of E, such that:*
1) G acts freely and transitively in E;
2) $G_0 \subset G$, $G_1 \subset G$, $G_0 \cap G_1 = \{Id_E\}$;
3) $card(G_0) = \omega$ and $card(G_1) = \omega_1$.
Then there exists a subset X of E possessing the following properties:
(a) $G_0(X) = E$;
(b) for any two distinct transformations $g \in G_1$ and $h \in G_1$, we have

$$card(g(X) \cap h(X)) \leq \omega.$$

In particular, the set X turns out to be absolutely nonmeasurable with respect to the class of all nonzero G-quasi-invariant measures on E satisfying the countable chain condition.

Proof. First, let us observe that relation 1) of our lemma immediately implies the equality

$$card(G) = \omega_1.$$

Take now an arbitrary ω_1-sequence $(\Gamma_\xi)_{\xi<\omega_1}$ of subgroups of G, such that:

(1) $\Gamma_0 = G_0$;

(2) for all ordinals $\xi < \omega_1$, we have $card(\Gamma_\xi) = \omega$;

(3) for each ordinal $\xi < \omega_1$, the set $\cup\{\Gamma_\zeta : \zeta < \xi\}$ is a proper subset of Γ_ξ (in particular, this ω_1-sequence of subgroups of G is strictly increasing by inclusion);

(4) $\cup\{\Gamma_\xi : \xi < \omega_1\} = G$.

The proof of the existence of such an ω_1-sequence is not difficult and is left to the reader (compare Exercise 10 from Chapter 9).

Further, fix a point $y \in E$ and, for any $\xi < \omega_1$, put

$$Y_\xi = \Gamma_\xi(y) \setminus \cup\{\Gamma_\zeta(y) : \zeta < \xi\}.$$

A straightforward verification shows that the family of sets $\{Y_\xi : \xi < \omega_1\}$ forms a partition of E and each Y_ξ is a Γ'_ξ-invariant subset of E, where the group Γ'_ξ is defined by the equality

$$\Gamma'_\xi = \cup\{\Gamma_\zeta : \zeta < \xi\}.$$

We recall that, according to relation (3), the group Γ'_ξ is a proper subgroup of Γ_ξ. Also, in virtue of the free action of G in E, it is not hard to see that

$$card(Y_\xi) = \omega \qquad (\xi < \omega_1).$$

Now, for each $\xi < \omega_1$, let us introduce the group

$$G_{1,\xi} = G_1 \cap \Gamma'_\xi.$$

Obviously, the ω_1-sequence of groups $(G_{1,\xi})_{\xi<\omega_1}$ is increasing by inclusion and

$$\cup\{G_{1,\xi} : \xi < \omega_1\} = G_1.$$

Fix for a while an ordinal $\xi < \omega_1$ and consider the two partitions of Y_ξ associated with the groups G_0 and $G_{1,\xi}$, respectively. Taking into account the free action of G in E and the relation

$$G_0 \cap G_1 = \{Id_E\},$$

we infer that the above-mentioned two partitions of Y_ξ are mutually transversal; in other words, any equivalence class of the first partition has at most one common point with any equivalence class of the second partition. Starting with this fact, we are going to define by recursion an ω-sequence

$$x_{\xi,0}, \ x_{\xi,1}, \ ..., \ x_{\xi,k}, \ ...$$

of points from Y_ξ, such that:

(i) $G_0(\{x_{\xi,k} : k < \omega\}) = Y_\xi$;

(ii) for any two distinct natural numbers k and m, the point $x_{\xi,k}$ does not belong to the orbit $G_{1,\xi}(x_{\xi,m})$.

Indeed, let $\{Z_{\xi,k} : k < \omega\}$ denote an injective family of all those G_0-orbits which are contained in Y_ξ. Suppose that, for a natural number k, the elements

$$x_{\xi,0} \in Z_{\xi,0}, \ x_{\xi,1} \in Z_{\xi,1}, ..., x_{\xi,k-1} \in Z_{\xi,k-1}$$

have already been defined and that they lie in pairwise distinct $G_{1,\xi}$-orbits. Consider the set

$$P_k = G_{1,\xi}(x_{\xi,0}) \cup G_{1,\xi}(x_{\xi,1}) \cup ... \cup G_{1,\xi}(x_{\xi,k-1}).$$

Clearly, we have

$$card(P_k \cap Z_{\xi,k}) \le k, \quad card(Z_{\xi,k}) = \omega.$$

Consequently, there exists an element

$$x \in Z_{\xi,k} \setminus P_k,$$

and we can put $x_{\xi,k} = x$.

Therefore, for each ordinal $\xi < \omega_1$, we get the corresponding ω-sequence $\{x_{\xi,k} : k < \omega\}$ of points from Y_ξ, fulfilling conditions (i) and (ii).

Finally, we define

$$X = \{x_{\xi,k} : \xi < \omega_1, \ k < \omega\}.$$

Let us check that the set X is the required one. In other words, we must check that X satisfies relations (a) and (b) of the lemma.

On the one hand, we may write

$$G_0(X) = \cup\{G_0(\{x_{\xi,k} : k < \omega\}) : \xi < \omega_1\} = \cup\{Y_\xi : \xi < \omega_1\} = E,$$

which shows that relation (a) holds.

On the other hand, let us take an arbitrary g such that

$$g \in G_1, \quad g \neq Id_E.$$

Then there exists an ordinal $\xi_0 < \omega_1$ for which $g \in G_{1,\xi_0}$. Further, for any $\xi < \omega_1$, let us denote

$$X_\xi = \{x_{\xi,k} : k < \omega\}.$$

Evidently,

$$(\forall \xi < \omega_1)(card(X_\xi) = \omega).$$

Also, we have the equality

$$X = \cup\{X_\xi : \xi < \omega_1\}$$

and the inclusion

$$g(X) \cap X \subset \cup\{g(X_\zeta) \cap X_\eta : \zeta < \omega_1, \ \eta < \omega_1\}.$$

If $\zeta < \omega_1$ and $\eta < \omega_1$ satisfy the relations $\xi_0 < \zeta$ and $\xi_0 < \eta$, then

$$g(X_\zeta) \cap X_\eta = \emptyset.$$

In addition to this, if $\zeta < \xi_0$ and $\eta > \xi_0$, or, respectively, $\zeta > \xi_0$ and $\eta < \xi_0$, then

$$g(X_\zeta) \cap X_\eta = g(X_\zeta \cap g^{-1}(X_\eta)) \subset g(Y_\zeta \cap Y_\eta) = \emptyset,$$

or, respectively,

$$g(X_\zeta) \cap X_\eta \subset Y_\zeta \cap Y_\eta = \emptyset,$$

from which it follows that

$$g(X) \cap X \subset (\cup\{g(X_\zeta) : \zeta \leq \xi_0\}) \cup (\cup\{X_\eta : \eta \leq \xi_0\})$$

and, therefore,

$$card(g(X) \cap X) \leq \omega.$$

Finally, suppose that g and h are any two distinct elements of G_1. Then

$$h^{-1} \circ g \neq Id_E, \quad h^{-1} \circ g \in G_1,$$

and, according to the said above, we may write

$$card((h^{-1} \circ g)(X) \cap X) \leq \omega$$

which implies at once

$$card(g(X) \cap h(X)) \leq \omega.$$

Thus, we have shown that relation (b) of the lemma holds, too.

It remains to demonstrate that X is an absolutely nonmeasurable set with respect to the class of all nonzero G-quasi-invariant measures on E satisfying the countable chain condition.

To see this, let us suppose to the contrary that there exists a nonzero G-quasi-invariant measure μ on E which satisfies the countable chain condition and for which $X \in dom(\mu)$. Without loss of generality, we may assume that μ vanishes on all countable subsets of E. In view of the equalities

$$E = G_0(X) = \cup\{g(X) : g \in G_0\}$$

and of the countability of G_0, we must have $\mu(X) > 0$. Consequently, the G-quasi-invariance of μ yields

$$(\forall g \in G_1)(\mu(g(X)) > 0).$$

But, considering the family of sets $\{g(X) : g \in G_1\}$ and remembering that this family is almost disjoint with respect to any diffused measure on E and that all members of this family are of positive μ-measure, we get a contradiction with the countable chain condition for μ. The contradiction obtained establishes the absolute nonmeasurability of X with respect to the class of all nonzero G-quasi-invariant measures on E satisfying the countable chain condition.

This completes the proof of Lemma 3.

The next statement gives the existence of absolutely nonmeasurable sets in an arbitrary uncountable commutative group. Actually, a much stronger result is contained in this statement.

Theorem 1. *Let $(\Gamma, +)$ be a commutative group and let Γ' be an uncountable subgroup of Γ. Then there exists a subset Y of Γ which is absolutely nonmeasurable with respect to the class of all nonzero Γ'-quasi-invariant measures on Γ satisfying the Suslin condition.*

Proof. In virtue of Lemma 2, we can find two subgroups G_0 and G_1 of Γ', such that:

1) $card(G_0) = \omega$;

2) $card(G_1) = \omega_1$;

3) $G_0 \cap G_1 = \{0\}$ where 0 denotes the neutral element (zero) in Γ.

Let G be the subgroup of Γ generated by the set $G_0 \cup G_1$. Obviously, we have

$$G \subset \Gamma', \quad card(G) = \omega_1.$$

Applying Lemma 3 to the groups G, G_0 and G_1, we can find a set $X \subset G$ satisfying the relations:

(a) $G_0(X) = G$;

(b) $card((g + X) \cap (h + X)) \leq \omega$ for any two distinct elements $g \in G_1$ and $h \in G_1$.

Consider now the partition of Γ associated with G. In other words, we are dealing with the partition of Γ into G-orbits. Denote by $\{h_i : i \in I\}$ any selector of this partition and put

$$Y = \cup\{h_i + X : i \in I\}.$$

We assert that the set Y is the required one. First of all, let us establish some properties of Y.

On the one hand, we may write

$$G_0(Y) = \cup\{h_i + G_0(X) : i \in I\} = \cup\{h_i + G : i \in I\} = \Gamma.$$

On the other hand, let g and g' be any two distinct elements from G_1. Then it can easily be verified that for some countable set $Z_{g,g'} \subset G$, the inclusion

$$(g + Y) \cap (g' + Y) \subset \cup\{h_i + Z_{g,g'} : i \in I\}$$

holds true (compare the proof of Lemma 3 and use a similar argument).

At last, we are ready to demonstrate that Y is absolutely nonmeasurable with respect to the class of all nonzero G-quasi-invariant measures on Γ satisfying the countable chain condition. (Since $G \subset \Gamma'$, we get in this way the desired result.) Suppose to the contrary that there exists a measure μ from the above-mentioned class, for which $Y \in dom(\mu)$. Then, taking into account the relation

$$\Gamma = \cup\{g + Y : g \in G_0\},$$

we claim that

$$\mu(Y) > 0.$$

Consequently, in view of the G_1-quasi-invariance of μ, we obtain that all sets from the family $\{g + Y : g \in G_1\}$ are of strictly positive μ-measure.

Further, for any two distinct elements g and g' from G_1, the μ-measurable set

$$(g + Y) \cap (g' + Y)$$

has the property that there are uncountably many pairwise disjoint translates of this set. (Indeed, the set $\cup\{h_i + Z_{g,g'} : i \in I\}$ obviously has the above-mentioned property and contains $(g + Y) \cap (g' + Y)$.) This yields at once the equality

$$\mu((g + Y) \cap (g' + Y)) = 0,$$

which shows that our family $\{g + Y : g \in G_1\}$ turns out to be almost disjoint (in the sense of μ). However, the last fact contradicts the countable chain condition for μ. The contradiction obtained establishes the absolute nonmeasurability of Y with respect to the class of all nonzero G-quasi-invariant measures on Γ satisfying the Suslin condition.

Theorem 1 has thus been proved.

Let G_1 be a group equipped with a left G_1-quasi-invariant measure μ_1, let G_2 be another group, and let ϕ be an arbitrary surjective homomorphism from G_1 onto G_2. Denote

$$\mathcal{S} = \{Y \subset G_2 : \phi^{-1}(Y) \in dom(\mu_1)\}.$$

Clearly, $\mathcal{S}$ is a σ-algebra of subsets of the group G_2 and one can readily verify that $\mathcal{S}$ is invariant with respect to the group of all left translations of G_2. We define a functional μ_2 on $\mathcal{S}$ by the formula

$$\mu_2(Y) = \mu_1(\phi^{-1}(Y)) \qquad (Y \in \mathcal{S}).$$

It is not difficult to see that the following proposition is valid.

Lemma 4. *The functional μ_2 is a left G_2-quasi-invariant measure on the group G_2.*

If μ_1 is a left G_1-invariant probability measure, then μ_2 is a left G_2-invariant probability measure.

Also, if μ_1 satisfies the countable chain condition, then μ_2 satisfies this condition, too.

We omit a simple proof of Lemma 4 and leave it to the reader. Actually, this lemma is a more complete version of Lemma 3 from Chapter 9.

We only wish to underline once more, in connection with Lemma 4, that if μ_1 is an arbitrary σ-finite left G_1-quasi-invariant (respectively, left G_1-invariant) measure on the group G_1, then the measure μ_2 on the group G_2, as defined by the same formula, is left G_2-quasi-invariant (respectively, left G_2-invariant). However, we cannot assert that μ_2 is σ-finite, and this is a weak side of the class of σ-finite measures.

A straightforward application of Lemma 4 leads us to the next auxiliary proposition.

Lemma 5. *Let $(G_1, \cdot)$ and $(G_2, \cdot)$ be any two groups and let*

$$\phi : G_1 \to G_2$$

be a surjective group homomorphism from G_1 onto G_2. If a set Z in G_2 is absolutely nonmeasurable with respect to the class of all nonzero (left) G_2-quasi-invariant measures on G_2 satisfying the Suslin condition, then the set $\phi^{-1}(Z)$ is absolutely nonmeasurable with respect to the class of all nonzero (left) G_1-quasi-invariant measures on G_1 satisfying the same condition.

A simple proof of Lemma 5 is also left to the reader.

In order to formulate the main result of this chapter, it is reasonable to recall the notion of a solvable group, which generalizes the notion of a commutative group. The fundamental role of finite solvable groups in the classical theory of algebraic equations is well known (see, for instance, [223]).

Let Γ be a group. According to the standard definition, this group is called solvable if there exists a finite sequence

$$(\Gamma_0, \Gamma_1, ..., \Gamma_n)$$

of subgroups of Γ, such that:

(a) $\Gamma_0 = \Gamma$;

(b) Γ_n is a trivial group (i.e. Γ_n is a one-element group);

(c) for each natural number $i \in [0, n-1]$, we have $\Gamma_{i+1} \subset \Gamma_i$ and Γ_{i+1} is a normal subgroup of Γ_i for which the factor group Γ_i/Γ_{i+1} is commutative.

The above-mentioned sequence of subgroups of Γ is usually called a composition series for Γ.

Note that any subgroup of a solvable group is also solvable. (An easy proof of this fact is left to the reader.)

The group of all motions (isometric transformations) of the euclidean plane $\mathbf{R}^2$ yields a classical example of a solvable group.

For $n \geq 3$, the group of all motions of $\mathbf{R}^n$ is not solvable because it contains a free subgroup of rank 2. (For more details, see [226] and also Chapter 14 of this book.)

Theorem 2. *Let Γ be an uncountable solvable group. Then there exists a subset of Γ absolutely nonmeasurable with respect to the class of all nonzero left Γ-quasi-invariant measures on Γ satisfying the Suslin condition.*

Proof. Since our Γ is a solvable group, we can find a finite composition series

$$(\Gamma_0, \Gamma_1, ..., \Gamma_n)$$

for this group. Now, it is natural to use the induction on n in order to demonstrate the existence of an absolutely nonmeasurable subset of Γ.

If $n = 1$, then Γ is a commutative group and we are able to utilize Theorem 1, which gives at once the required result.

Suppose now that our assertion is already established for a natural number $n - 1 \geq 1$. Take the group Γ_1 and introduce the canonical surjective homomorphism

$$\phi : \Gamma_0 \to \Gamma_0/\Gamma_1.$$

Only two cases are possible.

1. $card(\Gamma_0/\Gamma_1) \leq \omega$.

In this case, we obviously have

$$card(\Gamma_1) \geq \omega_1.$$

According to the inductive assumption, there is a set $Z \subset \Gamma_1$ absolutely nonmeasurable with respect to the class of all nonzero (left) Γ_1-quasi-invariant measures on Γ_1 satisfying the countable chain condition. Then, for some countable family $\{h_i : i \in I\}$ of elements of Γ_1, we get

$$\cup\{h_i + Z : i \in I\} = \Gamma_1.$$

This circumstance is thoroughly explained in Exercise 5 and also implies, in view of the inequality $card(\Gamma_0/\Gamma_1) \leq \omega$, that there exists a countable family $\{g_j : j \in J\}$ of elements of Γ_0, for which

$$\cup\{g_j + Z : j \in J\} = \Gamma_0 = \Gamma.$$

Now, it is not difficult to deduce from these facts that the same set Z is absolutely nonmeasurable with respect to the class of all nonzero (left) Γ-quasi-invariant measures on Γ satisfying the Suslin condition.

2. $card(\Gamma_0/\Gamma_1) \geq \omega_1$.

In this case, we can apply Theorem 1 to the uncountable commutative group Γ_0/Γ_1. Hence, there exists a set $Z \subset \Gamma_0/\Gamma_1$ absolutely nonmeasurable with respect to the class of all nonzero (Γ_0/Γ_1)-quasi-invariant measures on Γ_0/Γ_1 satisfying the countable chain condition. Finally, utilizing Lemma 5, we conclude that the set

$$Z' = \phi^{-1}(Z) \subset \Gamma_0 = \Gamma$$

is absolutely nonmeasurable with respect to the class of all nonzero (left) Γ-quasi-invariant measures on Γ satisfying the countable chain condition.

Theorem 2 has thus been proved.

As an immediate consequence of the above demonstrated, we get the following statement.

Theorem 3. *Let Γ be an uncountable solvable group. Then there exists a subset of Γ absolutely nonmeasurable with respect to the class of all nonzero σ-finite left Γ-quasi-invariant measures on Γ.*

Now, we wish to turn our attention to σ-finite invariant measures given on an arbitrary (not necessarily commutative) uncountable group. Namely, for such measures we formulate (without proof) a result very similar to the classical Vitali theorem. This result is due to Solecki (see [208]).

Theorem 4. *Let (E, G) be a space equipped with a transformation group and let μ be a σ-finite G-invariant measure on E. Suppose also that:*
1) the group G is uncountable;
2) G acts freely in E with respect to μ, that is for any two distinct transformations $g \in G$ and $h \in G$, we have

$$\mu^*(\{x \in E : g(x) = h(x)\}) = 0.$$

Then, for each μ-measurable set X with $\mu(X) > 0$, there exists a subset of X nonmeasurable with respect to any G-invariant measure on E extending μ. In other words, there exists a subset of X absolutely nonmeasurable with respect to the class $M_{(E,G,\mu)}$ of all G-invariant extensions of μ.

The next statement is an important particular case of Theorem 4.

Theorem 5. *Let G be an arbitrary uncountable group equipped with a σ-finite left G-invariant measure μ. Then, for each μ-measurable set X*

with $\mu(X) > 0$, there exists a subset of X nonmeasurable with respect to all left G-invariant measures on G extending μ.

For the proof of Theorems 4 and 5, see [208] and also Chapter 4 in monograph [100].

It is reasonable to remark, in connection with the two theorems above, that their proofs are essentially based on the assumption that the measure μ participating in their formulation is G-invariant (hence, G-quasi-invariant). One can easily observe that the argument used in the proofs of these theorems does not work in the case of σ-finite G-quasi-invariant measures. In other words, the following problem posed by the author (see [99]) remains open.

Problem 5. Let (E, G) be a space with a transformation group and let μ be a σ-finite G-quasi-invariant measure on E. Suppose that:
  1) G is an uncountable group;
  2) G acts freely in E with respect to μ;
  3) X is an arbitrary μ-measurable set with $\mu(X) > 0$.
Does there exist a subset V of X such that V is nonmeasurable with respect to every G-quasi-invariant measure on E extending μ?

Finally, we would like to formulate another result of Solecki [209] which shows, for an arbitrary uncountable group G, the existence of an infinite countable subgroup H of G with properties analogous to the corresponding properties of the subgroup $\mathbf{Q}$ of $\mathbf{R}$. More precisely, the above-mentioned subgroup H of G is such that all H-selectors in G turn out to be nonmeasurable with respect to every left G-invariant measure extending a given nonzero σ-finite left G-invariant measure μ on G.

Namely, the second result of Solecki is formulated as follows.

Theorem 6. *Let (E, G) be a space with a transformation group and let μ be a nonzero σ-finite G-invariant measure on E. Suppose that the group G is uncountable and acts freely in E with respect to μ. Then there exists a subgroup H of G satisfying the following two conditions:*
  1) $card(H) = \omega$;

  2) all H-selectors are nonmeasurable with respect to any G-invariant measure on E extending μ; in other words, all H-selectors turn out to be absolutely nonmeasurable sets with respect to the class $M_{(E,G,\mu)}$.

The next statement is a particular (but important) case of the preceding theorem.

Theorem 7. *Let G be an uncountable group equipped with a nonzero σ-finite left G-invariant measure μ. Then there exists an infinite countable subgroup H of G such that all H-selectors are nonmeasurable with respect to every left G-invariant measure on G extending μ.*

The proof of Theorem 6 can be found in [209] (see also Chapter 4 in monograph [100]).

Actually, the following slightly more general result can be obtained by using the same argument as in the proof of Theorem 6. Namely, let (E, G) be a space equipped with a transformation group and let μ be a nonzero σ-finite G-invariant measure on E. Suppose also that G is uncountable and acts freely in E with respect to μ. Then there exists a subgroup H of G satisfying these two conditions:

(a) $card(H) = \omega$;

(b) all H-selectors are nonmeasurable with respect to any H-invariant measure on E which extends μ.

The following example provides an application of the above-mentioned fact to the classical situation where E coincides with a finite-dimensional euclidean space and G is a certain (essentially noncommutative) group of isometric transformations of E, acting transitively in E.

Example 4. Let $E = \mathbf{R}^n$ where $n \geq 3$. It can be proved (see, for instance, [99]) that there exists a free group G of isometric transformations of E, acting transitively in E. Notice that the proof of this fact is based on the existence of a large free group Γ of rotations of E about its origin. ("Large" means here that the cardinality of Γ is equal to $\mathbf{c}$.) Evidently, for the group G, we also have

$$card(G) = \mathbf{c}.$$

In particular, G is an uncountable group. Further, let μ be an arbitrary nonzero σ-finite G-invariant measure on E vanishing on all affine hyperplanes of E. Then it is not hard to observe that G acts freely in E with respect to μ. Hence, according to the Solecki result, there exists an infinite countable subgroup H of G such that all H-selectors in E are nonmeasurable with respect to any H-invariant extension of μ.

We wish to emphasize that H, being a subgroup of a free group, is also a free group; actually, this fact is a direct consequence of a well-known general theorem from the theory of groups (see [129]).

In addition, let us remark that the problem of finding a characterization of all those groups of isometric transformations (or, more generally, affine transformations) of the space $\mathbf{R}^n$, which act transitively in $\mathbf{R}^n$, remains open and seems to be a nontrivial one.

EXERCISES

1. Give an example of two measures μ_1 and μ_2, such that μ_1 is σ-finite, μ_2 is a homomorphic image of μ_1 and μ_2 is not σ-finite.

2. Starting with the λ-thickness of all sets V_n $(n < \omega)$ of Example 2, verify the countable additivity of the functional ν introduced there.

3. Let E be a nonempty set and let G be a group of transformations of E such that $card(G) \leq \omega$. Show that there exists no absolutely nonmeasurable set in E with respect to the class $M^1_{(E,G)}$.

Let now E be an uncountable set and let G be the family of all those transformations g of E for which we have

$$card(\{x \in E : g(x) \neq x\}) \leq \omega.$$

Verify that G is a group (with respect to the usual composition operation) which acts transitively on E.

Show that E does not contain an absolutely nonmeasurable set with respect to the class $M^1_{(E,G)}$.

4. Give a direct proof of Lemma 2 from this chapter by using the method of transfinite induction or by applying the Kuratowski-Zorn lemma. Namely, for an uncountable commutative group $(G, +)$, construct two subgroups G_0 and G_1, such that

$$G_0 \cap G_1 = \{0\},$$

$$card(G_0) = \omega, \quad card(G_1) = \omega_1,$$

where 0 denotes the neutral element (zero) in G.

Moreover, prove that there are uncountably many subgroups of G which are uncountable and have the property that the intersection of any two of them coincides with $\{0\}$.

5. Let (E, G) be a space equipped with a transformation group, let $\mathcal{I}$ be a G-invariant σ-ideal of subsets of E and let X be a set in E. Suppose that, for any countable family $\{g_n : n < \omega\}$ of transformations from G, the relation

$$E \setminus \cup\{g_n(X) : n < \omega\} \notin \mathcal{I}$$

holds true. Denote by $\mathcal{I}^*$ the δ-filter of subsets of E, dual to $\mathcal{I}$, and define on the σ-algebra

$$\mathcal{S} = \mathcal{I} \cup \mathcal{I}^*$$

a functional μ by the formula:

$\mu(Z) = 0$ if $Z \in \mathcal{I}$;

$\mu(Z) = 1$ if $Z \in \mathcal{I}^*$.

Obviously, μ turns out to be a G-invariant probability measure on E. Prove that there exists a measure μ' on the same E, satisfying the following conditions:

(a) μ' is G-invariant;

(b) μ' is an extension of μ;

(c) $X \in dom(\mu')$ and $\mu'(X) = 0$.

In particular, suppose that G is a transitive group of transformations of an uncountable set E and put

$$\mathcal{I} = \{Z : Z \subset E, \ card(Z) \leq \omega\}.$$

Let X be a subset of E such that for any countable family $\{g_n : n < \omega\}$ of transformations from G, we have

$$\cup\{g_n(X) : n < \omega\} \neq E.$$

Show that there exists a G-invariant probability measure ν on E for which

$$X \in dom(\nu), \quad \nu(X) = 0.$$

Conclude from this fact that X is not absolutely nonmeasurable with respect to the class $M^1_{(E,G)}$.

6. Let μ be a σ-finite G-quasi-invariant (G-invariant) measure on a space (E, G) and let X be a subset of E absolutely nonmeasurable with respect to the class $M_{(E,G,\mu)}$ of all G-quasi-invariant (G-invariant) measures on E extending μ. Demonstrate that there exists a countable family $\{g_n : n < \omega\}$ of transformations from G, for which

$$\mu_*(\cup\{g_n(X) : n < \omega\}) > 0,$$

where μ_* denotes the inner measure associated with μ.

In particular, if our measure μ is metrically transitive (see the corresponding definition in Exercise 7 of Chapter 1), then, for the same set X, there exists a countable family $\{h_n : n < \omega\}$ of transformations from G, such that

$$\mu^*(E \setminus \cup\{h_n(X) : n < \omega\}) = 0,$$

where μ^* denotes the outer measure associated with μ.

7. Let H be an infinite-dimensional separable Hilbert space (over the field $\mathbf{R}$) and let T be a closed ball in H. Denote by $\mathcal{S}_T$ the smallest H-invariant σ-algebra of subsets of H, containing T. Prove the equality

$$\mathcal{S}_T = \mathcal{B}(H),$$

where $\mathcal{B}(H)$ is the Borel σ-algebra of H.

Further, let K be any σ-compact set in H. Show that there exists an ω_1-sequence $\{h_\xi : \xi < \omega_1\}$ of elements of H, such that the family of sets

$$\{h_\xi + K : \xi < \omega_1\}$$

is disjoint. Deduce from this result that K is H-absolutely negligible in H and that there is no nonzero σ-finite H-quasi-invariant Borel measure on H. (Take into account the fact that H is a Radon topological space; in other words, every σ-finite Borel measure on H is Radon.)

Finally, conclude from the above stated that the ball T is an H-absolutely nonmeasurable subset of H and check that the absolute nonmeasurability of T in H can be established within the theory **ZF** & **DC**.

The latter circumstance is of some interest from a logical viewpoint because, as we know, even the existence of subsets of $\mathbf{R}^n$ ($0 < n < \omega$) nonmeasurable in the Lebesgue sense needs rather strong versions of the Axiom of Choice.

8. Let n be a nonzero natural number. Consider the n-dimensional unit sphere

$$\mathbf{S}_n = \{x : \|x\| - 1\}$$

in the $(n+1)$-dimensional euclidean space $\mathbf{R}^{n+1}$. Let G be some group of rotations of $\mathbf{S}_n$ about its centre (in other words, $G \subset O_{n+1}^+$).

We denote by $M_n^1(G)$ the class of all G-quasi-invariant diffused probability measures on $\mathbf{S}_n$.

It follows directly from Theorem 1 that for $n = 1$, the sphere $\mathbf{S}_n$ (regarded as a commutative group) has a subset absolutely nonmeasurable with respect to $M_n^1(G)$ whenever G is uncountable.

By starting with this result, give an example of an uncountable commutative group G of rotations of $\mathbf{S}_n$ ($n \geq 2$) for which there exist absolutely nonmeasurable sets in $\mathbf{S}_n$ with respect to the class $M_n^1(G)$.

As a rule, a group G of rotations of $\mathbf{S}_n$ about its centre is not commutative in the case $n \geq 2$. Moreover, in this case there are many free groups of rotations of $\mathbf{S}_n$, whose cardinalities are equal to $\mathbf{c}$. This circumstance implies the existence of extremely paradoxical subsets of $\mathbf{S}_n$ from the viewpoint of equidecomposability theory (for more details, see [226]). At the same time, there is a nice property of the rotation group O_3^+, which is frequently useful in applications. Namely, the action of the rotation group on $\mathbf{S}_2$ is locally commutative. The last means that for any two rotations g and h of $\mathbf{S}_2$, the existence of a point $x \in \mathbf{S}_2$ satisfying the relation

$$g(x) = h(x) = x$$

necessarily implies the equality $g \circ h = h \circ g$ (that is g and h commute whenever they have a common fixed point).

9. Let $n \geq 1$ be a natural number. We denote

$$G = \mathbf{R}^n, \quad \mu = \lambda_n.$$

Demonstrate that there exists a set $X \subset G$ with $\mu^*(X) > 0$ such that there is no subset of X absolutely nonmeasurable with respect to the class of all G-invariant extensions of μ.

This fact shows us that in the formulation of Theorem 4 the assumption $\mu(X) > 0$ cannot be replaced by the weaker assumption $\mu^*(X) > 0$.

On the other hand, we know that any set $Z \subset G$ with $\mu^*(Z) > 0$ contains at least one subset nonmeasurable with respect to μ. In this connection, see an essentially more general result presented in Exercise 2 from Chapter 9, which is valid not only for σ-finite invariant measures but also for σ-finite quasi-invariant measures.

Chapter 12
Ideals producing nonmeasurable unions of sets

The present chapter is concerned with those descriptive properties of a σ-ideal of sets, which are implied by the existence of a projective base for this ideal and are closely connected with the existence of nonmeasurable sets. The main result of the chapter (namely, Theorem 1) generalizes the result of [16] and can successfully be applied to some questions of measure theory and set-theoretical topology (compare also [20], [49] and Exercise 2 from Chapter 2).

The notation utilized throughout this chapter is fairly standard. However, we shall recall some definitions and concepts (especially, those ones which belong to classical descriptive set theory).

As usual, we denote by ω $(= \mathbf{N})$ the set of all natural numbers.

If X is any set, then $[X]^{\leq \omega}$ is the family of all countable subsets of X.

The symbol $\mathcal{P}(X)$ denotes the family of all subsets of X.

If X is an arbitrary topological space, then $\mathcal{B}(X)$ is the family of all Borel subsets of X. Respectively, the well-known classes of projective subsets of X are denoted by

$$\Sigma_1^1(X), \quad \Pi_1^1(X), \quad \Sigma_2^1(X), \quad \Pi_2^1(X), \dots \quad .$$

It is also convenient to put

$$\Sigma_0^1(X) = \Pi_0^1(X) = \mathcal{B}(X).$$

As usual, we define

$$\Delta_n^1(X) = \Sigma_n^1(X) \cap \Pi_n^1(X).$$

A Σ_n^1–space is any Σ_n^1–subset of a Polish space X, equipped with the topology induced by the topology of X. In particular, according to this definition, a Suslin space is any Σ_1^1–space.

For more detailed information about the projective hierarchy, see [64], [65], and [125]. These books also contain the standard proof of the classical fact from descriptive set theory, stating that

$$\Pi_n^1(X) \setminus \Sigma_n^1(X) \neq \emptyset \qquad (n \geq 1)$$

for every uncountable Polish space X.

If $\mathcal{S}$ and $\mathcal{T}$ are any two families of subsets of a given set X, then we denote by $\mathcal{S} \triangle \mathcal{T}$ the family of sets

$$\{Y \triangle Z : Y \in \mathcal{S} \,\&\, Z \in \mathcal{T}\},$$

where $\triangle$ stands for the operation of symmetric difference of two sets:

$$Y \triangle Z = (Y \setminus Z) \cup (Z \setminus Y).$$

It is clear that if $\mathcal{S}$ is a σ-algebra of sets (in X) and $\mathcal{I}$ is a σ-ideal of sets (in the same X), then $\mathcal{S} \triangle \mathcal{I}$ is a σ-algebra of sets, too.

If X is a metric space and $x \in X$, then $B(x, \varepsilon)$ denotes the open ball with centre x and with radius ε.

Let X be a topological space and let $\mathcal{I}$ be an ideal of subsets of X. We recall that $\mathcal{I}$ has a Borel base if for every set $Y \in \mathcal{I}$, there exists a set $Z \in \mathcal{I} \cap \mathcal{B}(X)$ such that $Y \subset Z$. In an analogous manner, we say that $\mathcal{I}$ has a Π_n^1–base if for every set $Y \in \mathcal{I}$, there exists a set $Z \in \mathcal{I} \cap \Pi_n^1(X)$ such that $Y \subset Z$. Finally, we say that $\mathcal{I}$ possesses a projective base if it has at least one base consisting of projective subsets of X.

As mentioned above, in this chapter we investigate those σ–ideals of subsets of Σ_n^1–spaces, which possess Π_n^1–bases. We are especially interested in those situations where such σ-ideals produce nonmeasurable (in the Lebesgue sense) unions of their members.

Note first that if A is a subset of a Polish space X and

$$A \in \Pi_n^1(X) \setminus \Sigma_n^1(X),$$

then the σ–ideal $\mathcal{I}$ generated by the family of sets $\mathcal{P}(A) \cup [X \setminus A]^{\leq \omega}$ has some Π_n^1–base and, on the other hand, $\mathcal{I}$ possesses no Σ_n^1–base. We now give a slightly more elaborated example.

Example 1. Let P be a nonempty perfect subset of the real line $\mathbf{R}$, consisting of linearly independent elements (over the field $\mathbf{Q}$ of all rational numbers). It is well known that such a set P exists (in this connection, see [155] or Exercise 5 from Chapter 3). Let us fix a natural number $n \geq 1$ and consider an arbitrary subset A of P belonging to the class $\Pi_n^1(\mathbf{R}) \setminus \Sigma_n^1(\mathbf{R})$. It is also well known (see, for instance, [155], [186]) that, for any two distinct real numbers t and q, we have

$$card((P + t) \cap (P + q)) \leq 1$$

and, hence,

$$card((A + t) \cap (A + q)) \leq 1.$$

The reader can check this simple fact without any difficulty. Now, let us consider the σ–ideal $\mathcal{J}$ of subsets of the real line, generated by the family of all translates of the set A. Then $\mathcal{J}$ is a σ–ideal invariant under the group of all translations of the real line. Evidently, $\mathcal{J}$ has a Π_n^1–base. On the other hand, $\mathcal{J}$ does not possess a Σ_n^1–base. To see this point, suppose to the contrary that $\mathcal{J}$ has a Σ_n^1–base. Then there exist a set $B \in \Sigma_n^1(\mathbf{R})$ and a sequence $\{t_n : n \in \omega\}$ of reals, such that

$$A \subset B \subset \cup\{A + t_n : n \in \omega\}.$$

From these inclusions we get

$$card((B \cap P) \setminus A) \leq \omega$$

and, therefore, $A \in \Sigma_n^1(\mathbf{R})$, which is impossible.

In addition, let us remark that if a σ–ideal $\mathcal{I}$ possesses a projective base, then for some $n \in \omega$, it also possesses a Π_n^1–base. This fact is an immediate consequence of the following simple set-theoretical proposition. To formulate it, let us recall that a partially ordered set $(X, \leq)$ is upward σ-centered if for any countable family $\{x_n : n \in \omega\}$ of elements from X, there exists an element $x \in X$ satisfying the relation

$$(\forall n \in \omega)(x_n \leq x);$$

in other words, each countable subset of X is bounded from above.

Lemma 1. *Suppose that $(X, \leq)$ is an upward σ–centered partially ordered set, B is a cofinal subset of X, and suppose that*

$$B = \cup\{B_n : n \in \omega\}.$$

Then, for some $n \in \omega$, the set B_n is also cofinal in X.

We omit an easy proof of this proposition.

We shall say that a class of sets in a Polish space has the perfect subset property if every uncountable set from this class contains a nonempty perfect subset.

Let us recall that, in the theory **ZFC**, the classes Σ_0^1 and Σ_1^1 have the perfect subset property. At the same time, the statement

$$\text{the class } \Pi_1^1 \text{ has the perfect subset property}$$

is independent of the theory **ZFC** (see [64], [65]). Moreover, it is known that for each natural number $n > 0$, the theory

$$\textbf{ZFC} \ \& \ \Sigma_n^1 \text{ has the perfect subset property } \&$$

$$\Pi_n^1 \text{ has not the perfect subset property}$$

is relatively consistent.

Let $\mathcal{A}$ and $\mathcal{S}$ be any two families of sets. We say that $\mathcal{A}$ is $\mathcal{S}$–summable if for every $\mathcal{A}' \subset \mathcal{A}$, we have $\cup \mathcal{A}' \in \mathcal{S}$ (see [49]).

A family $\mathcal{A}$ of sets is called point–finite if the subfamily $\{A \in \mathcal{A} : a \in A\}$ is finite for each point $a \in \cup \mathcal{A}$.

The following result can be regarded as a generalized version of the main theorem from [16].

Theorem 1. *Suppose that the class Σ_n^1 has the perfect subset property. Let X be an arbitrary Σ_n^1–space and let $\mathcal{I}$ be a σ-ideal of subsets of X with a Π_n^1–base. Suppose also that:*

1) $\mathcal{A}$ is a point–finite family of sets;

2) $\mathcal{A}$ is $(\Sigma_n^1(X) \triangle \mathcal{I})$–summable family.

Then there exists a family $\mathcal{B} \in [\mathcal{A}]^{\leq \omega}$ such that $(\cup \mathcal{A} \setminus \cup \mathcal{B}) \in \mathcal{I}$.

Proof. Without loss of generality, we may assume that X is an uncountable Σ_n^1-space and that the cardinality of the given family $\mathcal{A}$ is less than or equal to the cardinality of the continuum (denoted, as usual, by **c**). Let T be a subset of X which contains no perfect subset and satisfies the equality $card(T) = card(\mathcal{A})$. (Notice that the set T can be realized as a subset of some Bernstein set in the space X.) Obviously, we may identify T

with the set of indices of the given family $\mathcal{A}$. In other words, we can write $\mathcal{A} = \{A_t : t \in T\}$. Furthermore, we put

$$\Gamma = \{(x,t) \in X \times T : x \in A_t\}.$$

Let D be a countable dense subset of X. It is easy to check that

$$\Gamma = \bigcap_{k \in \omega} \bigcup_{d \in D} \Gamma^{-1}(B(d, \frac{1}{k+1})) \times B(d, \frac{1}{k+1}).$$

For any $k \in \omega$ and for any $d \in D$, let $S_{k,d} \in \Sigma_n^1(X)$ and $A_{k,d} \in \mathcal{I}$ be such subsets of X that

$$\Gamma^{-1}(B(d, \frac{1}{k+1})) = S_{k,d} \triangle A_{k,d}.$$

Define

$$A = \bigcup_{k \in \omega} \bigcup_{d \in D} A_{k,d}$$

and, taking into account that $A \in \mathcal{I}$, fix some Π_n^1–set $A' \in \mathcal{I}$ such that $A \subset A'$. Then the set

$$\Gamma' = \Gamma \cap ((X \setminus A') \times X)$$

is a Σ_n^1–subset of the product space $X \times X$. Hence, the set

$$T_1 = \{t \in T : (\exists x)((x,t) \in \Gamma')\}$$

is a Σ_n^1–subset of T and, consequently, T_1 must be countable. Thus, we obtain

$$\bigcup_{t \in T_1} A_t \supset \bigcup_{t \in T} A_t \setminus A'.$$

In virtue of the relation $A' \in \mathcal{I}$, this inclusion completes the proof.

Remark 1. Suppose that $\Gamma \subset A \times B$ is a binary relation with finite vertical sections; in other words, suppose that

$$card(\{y \in B : (x,y) \in \Gamma\}) < \omega$$

for each $x \in A$. Let $\{V_n : n \in \omega\}$ be a countable family of subsets of B, which separates the points in B, that is for any two distinct points $b \in B$ and $b' \in B$, there exists $n \in \omega$ such that

$$card(V_n \cap \{b, b'\}) = 1.$$

Then we have

$$\Gamma = \bigcup_{f \in 2^\omega} \bigcap_n \Gamma^{-1}(V_0^{f(0)} \cap \ldots \cap V_n^{f(n)}) \times (V_0^{f(0)} \cap \ldots \cap V_n^{f(n)}),$$

where $V^0 = V$ and $V^1 = B \setminus V$ for each set $V \subset B$. This fact enables us to develop some analogues of the above theorem for a wider class of topological spaces which, in general, are not assumed to be separable. In this connection, recall that a typical example of a nonseparable Banach space with a countable family of Borel sets separating the points is the classical space l^∞ consisting of all bounded real-valued sequences.

It is possible to apply directly Theorem 1 to the family of all analytic subsets of a Polish space X (putting $\mathcal{I} = \{\emptyset\}$ in this case), to the σ–algebra of subsets of X with the Baire property, to the σ–algebra of Lebesgue measurable subsets of the real line $\mathbf{R}$, and so on. Below, we shall give some other (more interesting) applications of this theorem.

Theorem 2. *Suppose that the class Σ_n^1 has the perfect subset property. Let $\mathcal{A}$ be an uncountable family of nonempty pairwise disjoint Σ_n^1–sets. Then there exists a subfamily $\mathcal{C}$ of $\mathcal{A}$ such that $\cup\mathcal{C}$ is not a Σ_n^1–set.*

Proof. Indeed, let us put $X = \cup\mathcal{A}$. If X is not a Σ_n^1-space, then there is nothing to prove. Assume now that X is a Σ_n^1-space. Then we can consider the given family $\mathcal{A}$ with the σ-ideal

$$\mathcal{I} = [X]^{\leq\omega}.$$

Applying Theorem 1 to $\mathcal{A}$ and $\mathcal{I}$, we easily get the required result.

Theorem 3. *Suppose that X is a Polish space and $\mathcal{I}$ is a σ–ideal of subsets of X with a Borel base. Suppose also that $\mathcal{A} \subset \mathcal{I}$ is a point–finite family of sets, such that $\cup\mathcal{A} = X$. Then there exists a subfamily $\mathcal{C}$ of $\mathcal{A}$ for which*

$$\bigcup\mathcal{C} \notin B(X)\triangle\mathcal{I}.$$

Proof. Indeed, since the class

$$B(X) = \Sigma_0^1(X) = \Pi_0^1(X)$$

has the perfect subset property, we may directly apply Theorem 1 to the family $\mathcal{A}$ and to the σ-ideal $\mathcal{I}$.

Let us mention two immediate consequences of Theorem 3.

Example 2. Let X be a Polish space and let $\mathcal{I}$ denote the σ-ideal of all first category subsets of X. Suppose also that $\mathcal{A} \subset \mathcal{I}$ is a point-finite covering of X. Then, according to Theorem 3, there exists a family $\mathcal{C} \subset \mathcal{A}$ such that the set $\bigcup \mathcal{C}$ does not have the Baire property.

In particular, if our space X contains no isolated points, then considering the partition of X into its singletons, we come to the existence of a subset of X without the Baire property.

Example 3. Let X be a Polish space and let μ be a nonzero σ-finite Borel measure on X. Denote by μ' the completion of μ and let $\mathcal{I}$ be the σ-ideal of all μ'-measure zero subsets of X. Suppose also that $\mathcal{A} \subset \mathcal{I}$ is a point-finite covering of X. Then, in view of Theorem 3, there exists a family $\mathcal{C} \subset \mathcal{A}$ such that the set $\bigcup \mathcal{C}$ is not measurable with respect to μ'.

In particular, if the original measure μ is diffused, then considering the partition of our space X into its singletons, we come to the existence of a subset of X nonmeasurable with respect to μ'.

Now, for a given family $\mathcal{S}$ of sets, we define

$$\mathcal{S}^- = \{Z : (\forall Z' \subset Z)(Z' \in \mathcal{S})\}.$$

Using this notation, we can formulate the following result.

Theorem 4. *If X is a Polish space and $\mathcal{I}$ is a σ-ideal of subsets of X covering X and possessing a Borel base, then*

$$(B(X)\triangle\mathcal{I})^- = \mathcal{I}.$$

This result easily follows from Theorem 3 and seems to be rather natural.

In order to present some other consequences of Theorem 1, we need to recall the concept of a discrete family of sets (in a topological space). Since we deal here only with metrizable topological spaces, this concept will be introduced for metric spaces (see [37] or [125]).

A family $\mathcal{F}$ of subsets of a metric space E is called discrete if there exists a nonzero $m \in \omega$ such that

$$(\forall F_1 \in \mathcal{F})(\forall F_2 \in \mathcal{F})(F_1 \neq F_2 \Rightarrow dist(F_1, F_2) > \frac{1}{m}),$$

where $dist(F_1, F_2)$ denotes the distance between F_1 and F_2.

Notice that if $\mathcal{F}$ is a discrete family of closed sets and $\mathcal{S} \subset \mathcal{F}$, then $\cup \mathcal{S}$ is a closed set, too. Moreover, if $Z \subset E$ is compact and $\mathcal{F}$ is discrete, then the family

$$\{Y \in \mathcal{F} : Y \cap Z \neq \emptyset\}$$

is necessarily finite.

In our further considerations we need the following corollary from the well-known Montgomery lemma (see [125] and [152]).

Lemma 2. *Let E be a metric space and let S be any family of open sets in E, such that $\cup S = E$. Then there exists a sequence $\{\mathcal{F}_n : n \in \omega\}$ of discrete families of closed subsets of E, satisfying the relations:*

1) $(\forall n \in \omega)(\forall F \in \mathcal{F}_n)(\exists U \in \mathcal{S})(F \subset U)$;

2) $\bigcup_{n \in \omega}(\bigcup \mathcal{F}_n) = E$.

The proof of this lemma is presented, for example, in [125], [152] and [165].

Remark 2. For any infinite cardinal number τ, let us consider the topological sum of the family of spaces $\{[0, 1] \times \{\xi\} : \xi < \tau\}$. Let us identify in this sum all points $(0, \xi)$, where $\xi < \tau$, and denote the obtained space by E_τ. It is well known that any metric space E with weight τ can be embedded into the countable product of copies of E_τ (see [37]). Starting with this fact, the previous auxiliary proposition (that is Lemma 2) can be proved without using the Montgomery lemma.

Let E be a metric space. As usual, we denote by $Comp(E)$ the family of all nonempty compact subsets of E and we equip this family with the Vietoris topology (see [37], [77], [125]). Further, let S be a family of subsets of a given set X, closed under countable unions and countable intersections.

We shall say that a (set-valued) mapping

$$\Phi \ : \ X \to Comp(E)$$

is lower $\mathcal{S}$–measurable if for every open set $Y \subset E$, the relation

$$\{x \in X : \Phi(x) \cap Y \neq \emptyset\} \in \mathcal{S}$$

is valid. It can easily be verified that, in our case, a mapping Φ is lower $\mathcal{S}$–measurable if and only if it is upper $\mathcal{S}$–measurable; in other words, for every closed set $Z \subset E$, we have

$$\{x \in X : \Phi(x) \cap Z \neq \emptyset\} \in \mathcal{S}.$$

Of course, it is essential here that, for each point $x \in X$, the set $\Phi(x)$ is compact and nonempty.

Lemma 3. *Let the class Σ_n^1 have the perfect subset property, let X be an arbitrary Σ_n^1-space, and let $\mathcal{I}$ be a σ-ideal of subsets of X with a Π_n^1-base. Suppose also that E is a metric space and*

$$\Phi \; : \; X \to Comp(E)$$

is a lower $(\Sigma_n^1(X)\triangle\mathcal{I})$-measurable mapping. Finally, let $\mathcal{F}$ be a discrete family of closed subsets of E. Then there exist a countable subfamily $\mathcal{F}'$ of $\mathcal{F}$ and a set $A \in \mathcal{I}$, such that

$$(\forall x \in X \setminus A)(\Phi(x) \cap (\cup\mathcal{F}) \neq \emptyset \Rightarrow \Phi(x) \cap (\cup\mathcal{F}') \neq \emptyset).$$

Proof. For each set $Z \in \mathcal{F}$, we put

$$A_Z = \{x \in X : \Phi(x) \cap Z \neq \emptyset\}.$$

The discreteness of $\mathcal{F}$ and the compactness of all values of Φ imply that the family $\{A_Z : Z \in \mathcal{F}\}$ is point–finite. Furthermore, the discreteness of $\mathcal{F}$ implies that $\{A_Z : Z \in \mathcal{F}\}$ is a $(\Sigma_n^1(X)\triangle\mathcal{I})$-summable family. Hence, by Theorem 1, there exist a countable subfamily $\mathcal{F}'$ of $\mathcal{F}$ and a set $A \in \mathcal{I}$, such that

$$\cup\{A_Z : Z \in \mathcal{F}'\} \cup A = \cup\{A_Z : Z \in \mathcal{F}\}.$$

Suppose now that $x \in X \setminus A$ and that $\Phi(x) \cap (\cup\mathcal{F}) \neq \emptyset$. Then we have

$$x \in \cup\{A_Z : Z \in \mathcal{F}\}.$$

Consequently, $x \in \cup\{A_Z : Z \in \mathcal{F}'\}$ and, therefore,

$$\Phi(x) \cap (\cup\mathcal{F}') \neq \emptyset.$$

This completes the proof.

Lemma 4. *Let the class Σ_n^1 have the perfect subset property, let X be an arbitrary Σ_n^1-space, and let $\mathcal{I}$ be a σ-ideal of subsets of X with a Π_n^1-base. Suppose also that E is a metric space and*

$$\Phi \; : \; X \to Comp(E)$$

is a lower $(\Sigma_n^1(X)\triangle\mathcal{I})$–measurable mapping. Finally, let $\mathcal{F}$ be a discrete family of closed subsets of E. Then there exist a countable subfamily $\mathcal{F}'$ of $\mathcal{F}$ and a set $A \in \mathcal{I}$, such that

$$(\forall x \in X \setminus A)(\forall Z \in \mathcal{F})(\Phi(x) \cap Z \neq \emptyset \Rightarrow Z \in \mathcal{F}').$$

Proof. Using Lemma 3, we can construct (by recursion) three sequences

$$(A_n)_{n\in\omega}, \quad (B_n)_{n\in\omega}, \quad (\mathcal{F}_n)_{n\in\omega}$$

satisfying the following properties:
(1) $A_0 = \{x \in X : \Phi(x) \cap (\cup\mathcal{F}) \neq \emptyset\}$;
(2) $A_0 \supset B_0 \supset A_1 \supset B_1 \supset \ldots$;
(3) $\mathcal{F}_n \cap \mathcal{F}_m = \emptyset$ for all distinct natural numbers n and m;
(4) $\mathcal{F}_n \in [\mathcal{F}]^{\leq\omega}$ for each $n \in \omega$;
(5) $A_n \setminus B_n \in \mathcal{I}$ for each $n \in \omega$;
(6) if $x \in B_n$, then $\Phi(x) \cap (\cup\mathcal{F}_n) \neq \emptyset$;
(7) for any $n \in \omega$, we have

$$A_{n+1} = \{x \in B_n : \Phi(x) \cap (\cup(\mathcal{F} \setminus (\mathcal{F}_0 \cup \ldots \cup \mathcal{F}_n))) \neq \emptyset\}.$$

The details of this construction are left to the reader.
Observe now that

$$\cap_{n\in\omega} A_n = \cap_{n\in\omega} B_n = \emptyset.$$

Indeed, supposing $x \in \cap_{n\in\omega} A_n = \cap_{n\in\omega} B_n$, we see that the set

$$\{Z \in \mathcal{F} : \Phi(x) \cap Z \neq \emptyset\}$$

is infinite. But this is impossible in view of the compactness of $\Phi(x)$. Therefore, $\cap_{n\in\omega} A_n = \emptyset$ and $\cap_{n\in\omega} B_n = \emptyset$. Let us put

$$A = \cup_{n\in\omega}(A_n \setminus B_n), \quad \mathcal{F}' = \cup_{n\in\omega}\mathcal{F}_n.$$

Note that $A \in \mathcal{I}$. Now, if $x \in X \setminus A$, $Z \in \mathcal{F}$ and $\Phi(x) \cap Z \neq \emptyset$, then we have

$$x \in (B_0 \setminus A_1) \cup (B_1 \setminus A_2) \cup \ldots.$$

This yields $x \in B_n \setminus A_{n+1}$ for some $n \in \omega$. Therefore, $x \in B_n$ and $x \notin A_{n+1}$. The last two relations imply at once that if $Z \in \mathcal{F}$ and $\Phi(x) \cap Z \neq \emptyset$, then $Z \in \mathcal{F}'$. Thus, the lemma is proved.

Lemma 5. *Let the class Σ_n^1 have the perfect subset property, let X be an arbitrary Σ_n^1-space and let $\mathcal{I}$ be a σ-ideal of subsets of X with a Π_n^1-base. Suppose also that E is a metric space and*

$$\Phi : X \to Comp(E)$$

is a lower $(\Sigma_n^1(X)\triangle\mathcal{I})$-measurable mapping. Then there exist a set $A \in \mathcal{I}$ and a closed separable subset F of E, such that

$$(\forall x \in X \setminus A)(\Phi(x) \subset F).$$

Proof. Applying Lemma 2, we can find a double sequence

$$(\mathcal{F}_{n,m})_{n\in\omega,m\in\omega}$$

of families of closed subsets of the space E such that:

1) $(\forall n \in \omega)(\forall m \in \omega)(\forall Z \in \mathcal{F}_{n,m})(diam(Z) < 1/(1+m))$;
2) $(\forall m \in \omega)(\cup_{n\in\omega}(\cup\mathcal{F}_{n,m}) = E)$;
3) $(\forall n \in \omega)(\forall m \in \omega)(\mathcal{F}_{n,m}$ is discrete$)$.

Using Lemma 4, for any two natural numbers n and m, we can find a set $A_{n,m}$ and a family $\mathcal{H}_{n,m}$ such that:

(i) $\mathcal{H}_{n,m} \in [\mathcal{F}_{n,m}]^{\leq\omega}$;
(ii) $A_{n,m} \in \mathcal{I}$;
(iii) $(\forall x \in X \setminus A_{n,m})(\forall Z \in \mathcal{F}_{n,m})(\Phi(x) \cap Z \neq \emptyset \Rightarrow Z \in \mathcal{H}_{n,m})$.

Let us consider the subspace

$$F = cl(\cap_{m\in\omega}(\cup_{n\in\omega}(\cup\mathcal{H}_{n,m})))$$

of the space E and let us put

$$A = \bigcup_{n\in\omega,m\in\omega} A_{n,m}.$$

Then it is easy to check that F and A are the required sets, which completes the proof of Lemma 5.

Now, we are able to formulate and prove several consequences of the preceding results concerning some kinds of measurable functions and selectors (compare [49], [52], [128]).

Theorem 5. *Suppose again that the class Σ_n^1 has the perfect subset property. Let X be an arbitrary Σ_n^1-space and let $\mathcal{I}$ be a σ-ideal of subsets of X with a Π_n^1-base. If E is a metric space and*

$$f : X \to E$$

is a $(\Sigma_n^1(X)\triangle\mathcal{I})$-measurable mapping, then there exists a set $A \in \mathcal{I}$ such that $f(X \setminus A)$ is a separable subspace of E.

Proof. Let us consider a set-valued mapping Φ defined by

$$\Phi(x) = \{f(x)\} \qquad (x \in X).$$

Then we obviously have $\Phi : X \to Comp(E)$ and it is easy to see that Φ is lower $(\Sigma_n^1(X)\triangle\mathcal{I})$-measurable. Consequently, we may utilize Lemma 5 to the mapping Φ. In this way, we get the desired result.

Evidently, we can also apply Theorem 5 to the class Σ_1^1 and to the ideal $\mathcal{I} = \{\emptyset\}$. So we obtain the following theorem due to Frolik (see [52]).

Theorem 6. *Suppose that X is a Suslin space, E is a metric space and $f : X \to E$ is a Borel mapping. Then the range of f is a separable subspace of E.*

Let us notice that this result of Frolik can also easily be deduced from Theorem 2.

Remark 3. Let us consider a topology $\mathcal{T}$ on the real line $\mathbf{R}$, defined by

$$\mathcal{T} = \{U \setminus D \ : \ U \text{ is open in } \mathbf{R} \ \& \ D \text{ is at most countable}\}.$$

Then a function $f : \mathbf{R} \to \mathbf{R}$ given by the formula $f(x) = x$ and treated as a mapping from the real line equipped with the standard topology into the real line equipped with the topology $\mathcal{T}$ is a Borel isomorphism, but the range of f is nonseparable. This simple example shows us that in Theorem 6 the assumption of metrizability of E is essential.

Theorem 7. *Suppose that X is a Polish space, E is a metric space and $f \ : \ E \to X$ is a mapping satisfying the following conditions:*
(a) $(\forall x \in X)(f^{-1}(x) \in Comp(E) \cup \{\emptyset\})$;
(b) for each closed set $Z \subset E$, we have $f(Z) \in \Sigma_1^1(X)$.
Then E is a separable space.

Proof. Indeed, let us define a set-valued mapping Φ by

$$\Phi(x) = f^{-1}(x) \qquad (x \in f(E))$$

and put $\mathcal{I} = \{\emptyset\}$. Then a straightforward application of Lemma 5 to Φ and $\mathcal{I}$ yields the desired result.

Let $\mathcal{S}$ be a family of subsets of a given set X, let E be another set and let $f : X \to E$ be a function. We shall say that f is an $\mathcal{S}$-step function (whose values are in the set E) if there exist a partition $\{A_n : n \in \omega\}$ of X into sets from $\mathcal{S}$ and a sequence $\{e_n : n \in \omega\}$ of elements from E, such that

$$\Gamma_f = \cup_{n \in \omega}(A_n \times \{e_n\}),$$

where Γ_f denotes the graph of f.

Suppose that the class Σ_n^1 has the perfect subset property. Let X be a Σ_n^1–space, $\mathcal{I}$ be a σ–ideal of subsets of X with a Π_n^1–base, E be an arbitrary metric space, and let $f : X \to E$ be a function. Then the following two conditions are equivalent:

(a) f is a $(\Delta_n^1(X)\triangle\mathcal{I})$–measurable function;

(b) f is equal ($\mathcal{I}$-almost everywhere) to the pointwise limit of a sequence of $(\Delta_n^1(X)\triangle\mathcal{I})$–step functions.

This equivalence directly follows from Theorem 5. Analogously the next statement can be obtained.

Theorem 8. *Suppose again that the class Σ_n^1 has the perfect subset property. Let X be a Σ_n^1–space, $\mathcal{I}$ be a σ–ideal of subsets of X with a Π_n^1–base, $(E, +)$ be any metrizable topological group, and let*

$$f : X \to E, \qquad g : X \to E$$

be two arbitrary $(\Delta_n^1(X)\triangle\mathcal{I})$–measurable functions. Then the sum $f + g$ is a $(\Delta_n^1(X)\triangle\mathcal{I})$–measurable function, too.

Proof. It suffices to reduce Theorem 8 to the case where E is a separable metrizable topological group. But this can easily be done with the aid of Theorem 5. (Here we can also directly apply the equivalence of the conditions (a) and (b) above.)

Theorem 9. *Suppose again that the class Σ_n^1 has the perfect subset property. Let X be a Σ_n^1–space and let $\mathcal{I}$ be a σ-ideal of subsets of X with a Π_n^1–base. Suppose also that E is a metric space and let*

$$\Phi : X \to Comp(E)$$

be a lower $(\Delta_n^1(X)\triangle\mathcal{I})$–measurable mapping. Then there exists at least one $(\Delta_n^1(X)\triangle\mathcal{I})$–measurable selector of Φ.

Proof. According to Lemma 5, there are a Π_n^1-set $A \in \mathcal{I}$ and a closed separable subspace F of E such that

$$(\forall x \in X \setminus A)(\Phi(x) \subset F).$$

Let F' denote the completion of the metric space F. Then F' is a Polish space and $\Phi(x)$ is a nonempty compact subset of F' for each $x \in X \setminus A$. Consequently, we may apply the classical theorem on measurable selectors (due to Kuratowski and Ryll–Nardzewski [128]) to the restriction $\Phi|(X \setminus A)$ and to the σ–algebra $(\Delta_n^1(X)\triangle\mathcal{I}) \cap \mathcal{P}(X \setminus A)$. In view of this theorem, there exists a $((\Delta_n^1(X)\triangle\mathcal{I}) \cap \mathcal{P}(X \setminus A))$–measurable selector of $\Phi|(X \setminus A)$. Then it is not hard to see that a suitable extension of this selector gives us a $(\Delta_n^1(X)\triangle\mathcal{I})$–measurable selector of the original mapping Φ.

Remark 4. Various connections between the notion of summability of families of sets and the existence of measurable selectors is thoroughly investigated in the monograph by Fremlin [49]. In that monograph the paracompactness of any metric space is utilized (instead of the Montgomery lemma) and further argument is essentially based on the assumption that an original σ–ideal of sets satisfies the countable chain condition. In this context, it is reasonable to recall that there are many natural examples of σ–ideals with a Borel base which do not satisfy the countable chain condition.

EXERCISES

1. Give a proof of Lemma 1.

2. Give an example of a σ-ideal $\mathcal{I}$ of subsets of $\mathbf{R}$, such that:
(a) $\mathcal{I}$ covers $\mathbf{R}$ and has a Borel base;
(b) the pair $(\mathcal{B}(\mathbf{R})\triangle\mathcal{I}, \mathcal{I})$ does not satisfy the countable chain condition; in other words, there exists a disjoint family of sets

$$\{X_\xi : \xi < \omega_1\} \subset (\mathcal{B}(\mathbf{R})\triangle\mathcal{I}) \setminus \mathcal{I};$$

(c) $\mathcal{I}$ is invariant under the group of all translations of $\mathbf{R}$;
(d) there exists a set $X \in \mathcal{I}$ with $card(X) = \mathbf{c}$.

3. Prove that the union of an arbitrary family of nondegenerate segments on $\mathbf{R}$ is always measurable in the Lebesgue sense.

Formulate and prove an analogous statement for the euclidean space $\mathbf{R}^n$ where $n > 1$.

4. We recall that a disc in the euclidean plane $\mathbf{R}^2$ is any subset of $\mathbf{R}^2$ homeomorphic to the closed unit circle $\mathbf{B}_2 = \{x \in \mathbf{R}^2 \ : \ ||x|| \leq 1\}$.

Give an example of a family of discs whose union is nonmeasurable with respect to the two-dimensional Lebesgue measure λ_2.

5. Demonstrate that the union of an arbitrary family of discs in $\mathbf{R}^2$ always possesses the Baire property.

6. Let E be an infinite topological space of second category and let the following two conditions be satisfied:

(a) the σ-ideal $\mathcal{K}(E)$ of all first category sets in E is $card(E)$-additive; in other words, for every cardinal number $\kappa < card(E)$ and for every family of sets

$$\{X_\xi : \xi < \kappa\} \subset \mathcal{K}(E),$$

the union $\cup\{X_\xi : \xi < \kappa\}$ belongs to $\mathcal{K}(E)$;

(b) $\mathcal{K}(E)$ possesses a base whose cardinality does not exceed $card(E)$.

Let now $\{Z_i : i \in I\}$ be a partition of E into first category sets. Prove that there exists a set $J \subset I$ for which the union $\cup\{Z_j : j \in J\}$ does not have the Baire property in E.

Consider the particular case where $card(E) = \omega_1$. (On this occasion condition (a) is, in fact, superfluous and can be omitted.)

Notice that the previous exercise is closely connected with Exercise 17 from Chapter 2. See also [92] where analogous constructions of sets lacking the Baire property are presented.

7. Let $(X, \mathcal{S})$ be a measurable space, E be a metric space, and let $\mathcal{F}(E)$ denote the family of all nonempty closed subsets of E. Suppose that a set-valued mapping

$$\Phi : X \to \mathcal{F}(E)$$

is given. We shall say that Φ is weakly $\mathcal{S}$-measurable if for each $e \in E$, the function

$$f_e : X \to \mathbf{R}$$

defined by the formula

$$f_e(x) = dist(e, \Phi(x)) \qquad (x \in X)$$

is $\mathcal{S}$-measurable in the usual sense. Show that:

(a) every lower $\mathcal{S}$-measurable Φ is weakly $\mathcal{S}$-measurable;

(b) if E is a separable metric space, then every weakly $\mathcal{S}$-measurable Φ is lower $\mathcal{S}$-measurable.

Verify also that an upper $\mathcal{S}$-measurable Φ is always lower $\mathcal{S}$-measurable.

8. Let E be a compact metric space, X be a topological space, and let μ be the completion of a σ-finite inner regular Borel measure on X. Denote $\mathcal{S} = dom(\mu)$. Let

$$\Psi : X \to \mathcal{F}(E)$$

be a set-valued mapping. Prove that the following four assertions are equivalent:

(a) Ψ is μ-measurable as a function acting from $(X, \mathcal{S}, \mu)$ into the space $\mathcal{F}(E)$ equipped with the Hausdorff metric;

(b) Ψ is lower $\mathcal{S}$-measurable;

(c) Ψ is upper $\mathcal{S}$-measurable;

(d) Ψ is weakly $\mathcal{S}$-measurable.

Chapter 13
Measurability properties of subgroups of a given group

Let λ ($= \lambda_1$) denote, as usual, the classical Lebesgue measure on the real line $\mathbf{R}$ ($= \mathbf{R}^1$). It is well known that λ is invariant under the group M_1 of all isometric transformations of $\mathbf{R}$ and, moreover, there are invariant (under the same group) measures on $\mathbf{R}$ strictly extending λ (see, for instance, [75], [82], [113], [173], [214] or Chapters 3 and 11 of the present book).

In this context, the following problem arises naturally.

Problem 1. Give a characterization of all those sets $X \subset \mathbf{R}$ for which there exists at least one M_1-invariant measure μ on $\mathbf{R}$ extending λ and satisfying the relation $X \in dom(\mu)$.

An analogous question can be posed for subsets of $\mathbf{R}$ measurable with respect to various quasi-invariant extensions of λ. More precisely, the corresponding problem is formulated as follows.

Problem 2. Give a characterization of all those sets $X \subset \mathbf{R}$ for which there exists at least one M_1-quasi-invariant measure μ on $\mathbf{R}$ extending λ and satisfying the relation $X \in dom(\mu)$.

Obviously, we can formulate analogues of Problems 1 and 2 for the n-dimensional euclidean space $\mathbf{R}^n$ and for the standard Lebesgue measure λ_n on $\mathbf{R}^n$. It will be shown that these two problems essentially differ from each other. Note that none of them is solved at the present time. Moreover, until now no reasonable approach has been found which might lead to their solution.

In connection with Problems 1 and 2, let us observe that the class of all subgroups of $\mathbf{R}$ may be regarded as a class of subsets of $\mathbf{R}$ which distinguishes these problems. Namely, we can assert that:

I. For any group $G \subset \mathbf{R}$, there exists an M_1-quasi-invariant extension μ of λ such that $G \in dom(\mu)$.

A detailed proof of this fact (and even of a more general statement) will be given later.

On the other hand, we have:

II. There exists a subgroup H of $\mathbf{R}$ such that for every $\mathbf{R}$-invariant extension ν of λ, the relation $H \notin dom(\nu)$ is valid.

Notice that H can easily be constructed by using a Hamel basis of $\mathbf{R}$. Indeed, take an arbitrary Hamel basis $\{e_i : i \in I\}$ in $\mathbf{R}$, pick an index $i_0 \in I$, and denote by H the vector subspace of $\mathbf{R}$ (over the field $\mathbf{Q}$ of rationals) generated by $\{e_i : i \in I \setminus \{i_0\}\}$. Then H is a Vitali type subset of $\mathbf{R}$ which is nonmeasurable with respect to any $\mathbf{R}$-invariant extension of λ.

In order to formulate further results, we need some preliminary facts and auxiliary propositions.

Let G be a group and let μ be a left G-quasi-invariant measure defined on some σ-algebra of subsets of G. We recall (see [62] or Exercise 7 from Chapter 1) that μ is metrically transitive if, for each set $X \in dom(\mu)$ with $\mu(X) > 0$, there exists a countable family $\{g_n : n < \omega\}$ of elements from G such that

$$\mu(G \setminus \cup\{g_n X : n < \omega\}) = 0.$$

We also recall that a topological group G is standard (see, for instance, [100]) if G coincides with some Borel subgroup of a Polish group.

For any group G and its subgroup H, the symbol G/H denotes, as usual, the set $\{gH : g \in G\}$ of all left translates of H in G.

Lemma 1. *Let G be a group equipped with a σ-finite left G-quasi-invariant measure μ and let H be a subgroup of G such that*

$$card(G/H) \geq \omega_1.$$

If μ is metrically transitive, then there exists a measure μ' on G satisfying the following relations:

1) μ' is left G-quasi-invariant;

2) μ' is metrically transitive;

3) μ' is an extension of μ;

4) H belongs to the domain of μ' and $\mu'(H) = 0$.

Proof. We may assume, without loss of generality, that the original measure μ is nonzero and complete. If $\mu(H) = 0$, then there is nothing to prove. Let us consider the case where

$$\mu^*(H) > 0$$

and let us verify that in this case, for any countable family $\{g_n : n < \omega\}$ of elements from G, the equality

$$\mu_*(\cup\{g_n H : n < \omega\}) = 0$$

is true. Indeed, suppose to the contrary that

$$\mu_*(\cup\{f_n H : n < \omega\}) > 0$$

for some countable family $\{f_n : n < \omega\} \subset G$. Then, applying the metrical transitivity of μ, we can find a countable family

$$\{f_n' : n < \omega\} \subset G$$

such that

$$\mu(G \setminus \cup\{f_n' H : n < \omega\}) = 0.$$

Since $card(G/H) > \omega$, there exists an element $f' \in G$ for which we have

$$f'H \cap (\cup\{f_n' H : n < \omega\}) = \emptyset.$$

Therefore,

$$\mu(f'H) = 0,$$

from which it also follows, in view of the left quasi-invariance of μ, that $\mu(H) = 0$ which contradicts our assumption $\mu^*(H) > 0$. The contradiction obtained shows that the inner μ-measure of any set Z of the form

$$Z = \cup\{g_n H : n < \omega\},$$

where $\{g_n : n < \omega\} \subset G$, is equal to zero.

Denote now by $\mathcal{J}$ the σ-ideal in G generated by all sets of the above-mentioned form. Let $\mathcal{S}$ be the σ-algebra in G generated by $dom(\mu) \cup \mathcal{J}$. Obviously, any set $X \in \mathcal{S}$ can be written as

$$X = (Y \cup Z_1) \setminus Z_2,$$

where $Y \in dom(\mu)$ and $Z_1 \in \mathcal{J}$, $Z_2 \in \mathcal{J}$. We put

$$\mu'(X) = \mu(Y) \quad (X \in \mathcal{S}).$$

In this way, we get the functional μ' on $\mathcal{S}$ (which is well defined). It can easily be shown, by applying the standard argument (see, for example, [82] or [100]) that μ' is a left G-quasi-invariant metrically transitive measure on E extending μ. Moreover, since the values of μ' on all sets from $\mathcal{J}$ are equal to zero, we have

$$\mu'(H) = 0.$$

The proof of Lemma 1 is thus completed.

Lemma 2. *Let G be a group equipped with a σ-finite left G-quasi-invariant measure μ and let $(H_1, H_2, ..., H_k)$ be a finite family of subgroups of G, such that $card(G/H_i) > \omega$ for each natural number $i \in [1, k]$. If μ is metrically transitive, then there exists a measure μ' on G for which the following relations are valid:*

1) μ' is left G-quasi-invariant;

2) μ' is metrically transitive;

3) μ' is an extension of μ;

4) all subgroups H_i $(i = 1, 2, ..., k)$ belong to $dom(\mu')$ and

$$\mu'(H_i) = 0 \qquad (i = 1, 2, ..., k).$$

Proof. It suffices to apply Lemma 1 and induction on k.

Lemma 3. *Let Γ be a standard group equipped with a σ-finite left Γ-quasi-invariant Borel measure ν and let G be a subgroup of Γ such that*

$$card(\Gamma/G) \leq \omega.$$

Then there exists a measure ν' on Γ satisfying the following relations:

1) ν' is left Γ-quasi-invariant;

2) ν' is metrically transitive;

3) ν' is an extension of ν;

4) G belongs to the domain of ν'.

Proof. Obviously, we can suppose that ν is not trivial (i.e. ν is not identically equal to zero).

The first part of our argument is based on the fundamental Mackey theorem [139]. Let us recall that according to this theorem, there exist a locally compact Polish topological group Γ' and a continuous group isomorphism

$$\phi : \Gamma' \to \Gamma,$$

such that the given measure ν turns out to be equivalent with the ϕ-image $\phi(\theta)$ of the left Haar measure θ on Γ'. In other words, the two measures

$\phi(\theta)$ and ν have the same σ-ideal of sets of measure zero (consequently, ν is metrically transitive).

Taking this classical result into account, we may assume without loss of generality that our Γ is a locally compact Polish topological group and the initial measure ν coincides with the left Haar measure on Γ. We preserve the same notation ν for the completion of the left Haar measure on Γ. Further, we may suppose that

$$\nu^*(G) > 0$$

and that G is everywhere dense in Γ. Otherwise, we replace Γ by $cl(G)$ (the closure of G) and deal with the restriction of ν to the Borel σ-algebra of $cl(G)$.

The second part of the proof is very similar to the argument utilized in Example 2 from Chapter 11. Since the subgroup G is of strictly positive outer ν-measure and, simultaneously, is everywhere dense in Γ, we get that G must be a ν-thick set in Γ; in other words, the equality

$$\nu_*(\Gamma \setminus G) = 0$$

must be valid. Besides, we have the inequality

$$card(\Gamma/G) \leq \omega.$$

Therefore,

$$card(\Gamma/G) = \omega \quad \vee \quad card(\Gamma/G) < \omega.$$

Let us consider only the case where

$$card(\Gamma/G) = \omega$$

since the case where $card(\Gamma/G) < \omega$ can be considered analogously and is essentially easier.

We denote by $\{Z_k : k < \omega\}$ an injective family of all left translates of G in Γ. Let $\mathcal{S}$ stand for the family of all those subsets X of Γ which can be represented in the form

$$X = \cup\{Y_k \cap Z_k : k < \omega\},$$

where Y_k $(k < \omega)$ are some ν-measurable sets in Γ. It is not hard to check that $\mathcal{S}$ is a left Γ-invariant σ-algebra of subsets of Γ and

$$\{Z_k : k < \omega\} \cup dom(\nu) \subset \mathcal{S}.$$

Let us define a functional ν' on $\mathcal{S}$ by the formula:

$$\nu'(X) = \sum_{k<\omega}(1/2)^{k+1}\nu(Y_k) \qquad (X \in \mathcal{S}).$$

Then, in view of the ν-thickness of all Z_k $(k < \omega)$, this functional is well defined and is a measure on Γ. A straightforward verification shows also that ν' satisfies relations 1), 3) and 4) of the lemma. Finally, it remains to observe that the metrical transitivity of ν implies the metrical transitivity of ν' (compare Exercise 1 from this chapter).

Lemma 3 has thus been proved.

Now, we are able to establish the following statement.

Theorem 1. *Let Γ be a standard group equipped with a σ-finite left Γ-quasi-invariant Borel measure ν and let $(G_1, G_2, ..., G_n)$ be a finite family of subgroups of Γ. Then there exists a left Γ-quasi-invariant metrically transitive measure ν' on Γ extending ν and such that*

$$\{G_1, G_2, ..., G_n\} \subset dom(\nu').$$

Proof. Without loss of generality, we may suppose that

$$card(\Gamma/G_1) \leq \omega, \quad card(\Gamma/G_2) \leq \omega, ..., \quad card(\Gamma/G_k) \leq \omega,$$

$$card(\Gamma/G_{k+1}) > \omega, \quad card(\Gamma/G_{k+2}) > \omega, ..., \quad card(\Gamma/G_n) > \omega$$

for some natural number $k \in [0, n]$. Let us put

$$G = G_1 \cap G_2 \cap ... \cap G_k.$$

Then G is a subgroup of Γ such that $card(\Gamma/G) \leq \omega$ (compare Exercise 2 of this chapter). Applying Lemma 3 to Γ and G, we see that there exists a left Γ-quasi-invariant metrically transitive measure ν_G on Γ extending ν and satisfying the relation $G \in dom(\nu_G)$. Since

$$card(G_1/G) \leq \omega, \quad card(G_2/G) \leq \omega, ..., \quad card(G_k/G) \leq \omega,$$

we also have the relations

$$G_1 \in dom(\nu_G), \quad G_2 \in dom(\nu_G), ..., \quad G_k \in dom(\nu_G).$$

Now, we can apply Lemma 2 to the measure ν_G and to the finite family

$$(G_{k+1}, G_{k+2}, ..., G_n)$$

of subgroups of Γ. In this way, we obtain the required extension ν' of ν.

Theorem 1 has thus been proved.

So far, we have been concerned with a finite family of subgroups of a standard group Γ and have been able to prove that all those subgroups can be made measurable with respect to a suitable left Γ-quasi-invariant extension of a given nonzero σ-finite left Γ-quasi-invariant Borel measure on Γ.

In dealing with countable families of subgroups of Γ, we come to a significantly different situation. For example, it is not hard to show that there exists a countable family of subgroups of $\mathbf{R}$, such that the Lebesgue measure λ cannot be extended to an $\mathbf{R}$-quasi-invariant measure whose domain includes all these subgroups. We leave to the reader the proof of this fact.

The next result generalizes the above-mentioned fact.

Theorem 2. *Let Γ be an uncountable commutative divisible group. Then there exists a countable family $\{G_i \ : \ i \in I\}$ of subgroups of Γ, such that:*

1) for each $i \in I$, we have $card(\Gamma/G_i) > \omega$;

2) $\cup\{G_i \ : \ i \in I\} = \Gamma$.

In particular, for any Γ-quasi-invariant probability measure μ on Γ, at least one group G_i is nonmeasurable with respect to μ.

Proof. Here it suffices to utilize the classical result from the theory of groups, stating that every divisible commutative group can be represented as a direct sum of a family of groups each of which is isomorphic either to $\mathbf{Q}$ (the group of all rationals) or to the quasi-cyclic group of type p^∞ where p is a prime number (see, for instance, [129] or Appendix 2). Thus, our group Γ is expressible in the form of the direct sum

$$\Gamma = \sum_{j \in J} \Gamma_j,$$

where J is some uncountable set of indices and every group Γ_j is of the above-mentioned kind. Now, it can easily be verified that, for each $j \in J$, we have

$$\Gamma_j = \cup\{H_{j,n} : n < \omega\}$$

where $\{H_{j,n} : n < \omega\}$ is an increasing (by inclusion) countable family of proper subgroups of Γ_j. For any $n < \omega$, let us put

$$G_n = \sum_{j \in J} H_{j,n}.$$

Then it is not difficult to check that the family of groups

$$\{G_i : i \in I\} = \{G_n : n < \omega\}$$

is the required one. This finishes the proof.

Remark 1. Obviously, in Theorem 2 any uncountable vector space over $\mathbf{Q}$ can be taken as Γ. In particular, we may put $\Gamma = \mathbf{R}^n$ where $n \geq 1$. Also, we may put $\Gamma = \mathbf{S}_1^\kappa$, where $\mathbf{S}_1$ denotes the one-dimensional unit torus and κ is an arbitrary nonzero cardinal.

Remark 2. Let Γ be a commutative group and let G be a subgroup of Γ such that $card(\Gamma/G) > \omega$. It can be proved that G is a Γ-absolutely negligible subset of Γ (see Exercise 5 from Chapter 10). We thus claim that each subgroup G_i in the preceding theorem turns out to be a Γ-absolutely negligible subset of Γ. Therefore, for a given $i \in I$, every σ-finite Γ-quasi-invariant measure μ on Γ can be extended to a Γ-quasi-invariant measure μ' on Γ satisfying the relation

$$G_i \in dom(\mu') \ \& \ \mu'(G_i) = 0.$$

However, there is no nonzero σ-finite Γ-quasi-invariant measure on Γ whose domain contains all subgroups G_i $(i \in I)$.

Remark 3. It would be interesting to extend Theorem 2 to a more general class of uncountable groups Γ (not necessarily divisible or commutative). In this context, let us point out that the assertion of Theorem 2 fails to be true for some uncountable groups. In particular, if Γ is uncountable and contains no proper uncountable subgroup, then the above-mentioned theorem is obviously false for Γ. On the other hand, by starting with the result formulated in Theorem 2, it is not difficult to construct an uncountable noncommutative nondivisible group Γ with a countable family $(G_i)_{i \in I}$ of its subgroups, such that each G_i $(i \in I)$ is a Γ-absolutely negligible set and for any left Γ-quasi-invariant probability measure μ on Γ, at least one G_i is nonmeasurable with respect to μ.

Example 1. Consider an arbitrary nonzero σ-finite $\mathbf{R}$-quasi-invariant measure ν on $\mathbf{R}$. In view of Theorem 2, there always exists a subgroup of $\mathbf{R}$ nonmeasurable with respect to ν. Moreover, by applying an argument similar to the proof of Theorem 2, it can be shown that there always exists a vector subspace of $\mathbf{R}$ (over the rationals) which is nonmeasurable with respect to ν (compare Remark 4 below and Exercise 9 of this chapter).

Remark 4. It is easy to see that a direct analogue of Theorem 2 is valid for vector spaces (over the field $\mathbf{Q}$ of all rational numbers) instead of commutative groups. Namely, if E is an uncountable vector space (over $\mathbf{Q}$), then there exists a countable family $(E_i)_{i \in I}$ of vector subspaces of E, such that

$$(\forall i \in I)(card(E/E_i) > \omega),$$

$$\cup \{E_i : i \in I\} = E.$$

In particular, for any nonzero σ-finite E-quasi-invariant measure ν on E, at least one subspace E_i is nonmeasurable with respect to ν. At the same time, all E_i $(i \in I)$ are E-absolutely negligible subsets of E.

In connection with the results presented above, the following problem seems to be of interest.

Problem 3. Let Γ be an uncountable commutative group. Does there exist a countable family $\{G_i : i \in I\}$ of subgroups of Γ, such that $card(\Gamma/G_i) > \omega$ for all $i \in I$ and for any nonzero σ-finite Γ-quasi-invariant measure μ on Γ, at least one subgroup G_i is nonmeasurable with respect to μ?

This problem remains open.

In preceding chapters of this book we were concerned with various nonmeasurable sets on the real line $\mathbf{R}$. Among many other results it was demonstrated, by assuming the Continuum Hypothesis, that there exists a countable family $\{X_i : i \in I\}$ of subsets of $\mathbf{R}$ such that no nonzero σ-finite diffused measure on $\mathbf{R}$ includes all these sets in its domain. However, we could not say anything about the algebraic structure of X_i and, in particular, we could not assert that at least one X_i is a subgroup of $\mathbf{R}$. In other words, the methods developed earlier were not able to solve the question whether, under $\mathbf{CH}$, for any nonzero σ-finite diffused measure μ on $\mathbf{R}$ there exists a subgroup of $\mathbf{R}$ nonmeasurable with respect to μ.

Our goal is to establish that such a subgroup always exists. Moreover, Theorem 3 formulated below contains a much stronger result stating that a nonmeasurable subgroup of a given commutative group G with $card(G) = \mathbf{c}$ can always be found in some fixed countable family of subgroups of G.

We start with a well-known Banach-Kuratowski matrix (see [8] or Chapter 7 of the present book). Let us recall that if the Continuum Hypothesis $(\mathbf{c} = \omega_1)$ is true, then every set E of cardinality $\mathbf{c}$ admits a double countable family

$$(E_{m,n})_{m<\omega,n<\omega}$$

of subsets of E, such that:

(a) $E_{m,n} \subset E_{m,n+1}$ for all natural numbers m and n;
(b) $\cup\{E_{m,n} : n < \omega\} = E$ for any natural number m;
(c) for every function $f : \omega \to \omega$, the set

$$E_{0,f(0)} \cap E_{1,f(1)} \cap ... \cap E_{m,f(m)} \cap ...$$

is at most countable.

The above-mentioned family $(E_{m,n})_{m<\omega,n<\omega}$ is usually called a Banach-Kuratowski matrix over E. It was shown in [8] (compare also Chapter 7 of this book) that, for every nonzero σ-finite diffused measure μ on E, at least one set $E_{m,n}$ is nonmeasurable with respect to μ. Applying this classical result, we can establish the following statement.

Theorem 3. *Assume that the Continuum Hypothesis is valid and let G be a commutative group of cardinality $\mathbf{c}$. Then there exists a countable family $(G_i)_{i\in I}$ of subgroups of G, such that for any nonzero σ-finite diffused measure μ on G, at least one of the groups G_i $(i \in I)$ is nonmeasurable with respect to μ.*

Proof. According to a well-known theorem of the theory of commutative groups (see, for instance, [129] or Appendix 2), the given group G can be expressed in the form

$$G = \cup\{\Gamma_k : k < \omega\},$$

where $(\Gamma_k)_{k<\omega}$ is an increasing (with respect to inclusion) countable family of subgroups of G and each group Γ_k is representable as a direct sum of cyclic groups. In addition to this, we may suppose, without loss of generality, that all groups Γ_k are of cardinality continuum. For every natural number k, we can write

$$\Gamma_k = \sum_{j\in J}[e_{k,j}],$$

where J is some set of cardinality $\mathbf{c}$ and $[e_{k,j}]$ denotes the cyclic group generated by $e_{k,j} \in \Gamma_k$. We also denote:

$$E_k = \{e_{k,j} : j \in J\} \qquad (k < \omega).$$

Further, for any natural number k, let

$$(E_{k,m,n})_{m<\omega,n<\omega}$$

be a Banach-Kuratowski matrix over the set E_k and let $\Gamma_{k,m,n}$ be the group generated by $E_{k,m,n}$. Then it is not hard to check that the family of groups

$$(\Gamma_{k,m,n})_{m<\omega,n<\omega}$$

is a Banach-Kuratowski matrix for the group Γ_k. We leave to the reader the verification of this simple fact which, however, plays an essential role in our argument.

Finally, we put

$$(G_i)_{i\in I} = \{\Gamma_k : k < \omega\} \cup \{\Gamma_{k,m,n} : k < \omega, m < \omega, n < \omega\}$$

and we assert that the countable family $(G_i)_{i\in I}$ of subgroups of G is the required one. Indeed, let μ be an arbitrary nonzero σ-finite diffused measure on G. Since every nonzero σ-finite measure is equivalent to a probability measure, we may assume, without loss of generality, that our μ is a probability diffused measure. We have to show that at least one G_i is non-measurable with respect to μ. Suppose to the contrary that all G_i $(i \in I)$ are μ-measurable. Then, taking into account the relation

$$G = \cup\{\Gamma_k : k < \omega\},$$

we infer that there exists a natural number r for which

$$\mu(\Gamma_r) > 0.$$

In other words, the corresponding restriction of μ is a nonzero finite diffused measure on the set Γ_r and all subsets $\Gamma_{r,m,n}$ $(m < \omega, n < \omega)$ of this set turn out to be μ-measurable. But this is impossible since, as mentioned above, the family $(\Gamma_{r,m,n,})_{m<\omega,n<\omega}$ forms a Banach-Kuratowski matrix over Γ_r. The contradiction obtained ends the proof of Theorem 3.

Example 2. Consider an arbitrary nonzero σ-finite diffused measure ν on the real line $\mathbf{R}$. If the Continuum Hypothesis holds, then, in view of the result presented above, there always exists a subgroup of the additive group $\mathbf{R}$, nonmeasurable with respect to ν. Moreover, by applying an argument similar to the proof of Theorem 3, it can be shown, under **CH**, that there always exists a vector subspace of $\mathbf{R}$ (over the rationals) which is nonmeasurable with respect to ν (compare Remark 7 below).

In connection with Theorem 3, let us notice that some analogue of this theorem can be established by using Martin's Axiom instead of the Continuum Hypothesis. (Recall that Martin's Axiom is much weaker than **CH** and does not essentially restrict the size of the continuum.) The corresponding result is formulated in Exercise 4 of this chapter.

In a certain sense, Theorem 3 has a weak side because it is valid under an additional set-theoretical assumption. In further considerations, we shall present an analogue of Theorem 3 for a commutative group G whose cardinality is equal to the first uncountable cardinal ω_1. Note that for this G the argument is more complicated and leads to a result in **ZFC** theory. Here the main role is played by a transfinite matrix of Ulam (see [222] and Chapter 7).

We need several lemmas.

Lemma 4. *Let E be a set, $\mathcal{S}$ be a σ-algebra of subsets of E and let $\mathcal{I}$ be a σ-ideal of subsets of E, such that $\mathcal{I} \subset \mathcal{S}$ and the pair $(\mathcal{S}, \mathcal{I})$ satisfies the countable chain condition. Further, let $\{X_\alpha : \alpha < \omega_1\}$ be an uncountable family of sets belonging to $\mathcal{S}$, let $m \geq 1$ be a fixed natural number and suppose that, for any m-element subset D of ω_1, the relation*

$$\cap\{X_\alpha : \alpha \in D\} \in \mathcal{I}$$

is true. Then there exists an uncountable subset A of ω_1 such that $X_\alpha \in \mathcal{I}$ for each ordinal α from A.

Proof. The argument is not difficult. Namely, we use induction on m.

The case $m = 1$ is trivial and the case $m = 2$ is well known. (Actually, the latter case is equivalent to the countable chain condition for the pair $(\mathcal{S}, \mathcal{I})$.)

Assume that the assertion of our lemma is valid for $m - 1 > 1$. For any set $A \subset \omega_1$, denote

$$Y_A = \cap\{X_\alpha : \alpha \in A\}.$$

Now, fix a subset A of ω_1 with $card(A) = m - 2$ and consider the family of sets

$$\{X_\xi \cap Y_A : \xi \in \omega_1 \setminus A\}.$$

It is clear that the intersection of any two distinct sets from this family belongs to the ideal $\mathcal{I}$. Consequently, there exists an ordinal $\xi(A) < \omega_1$ such that $X_\xi \cap Y_A \in \mathcal{I}$ for all ordinals ξ satisfying the inequalities

$$\xi(A) < \xi < \omega_1.$$

Starting with this fact and applying the regularity of ω_1, we can recursively define an injective subfamily

$$X_{\xi(0)}, \ X_{\xi(1)}, ..., \ X_{\xi(\alpha)}, ... \quad (\alpha < \omega_1)$$

such that, for each $(m-1)$–element subset A of ω_1, the relation

$$\cap\{X_{\xi(\alpha)} : \alpha \in A\} \in \mathcal{I}$$

holds true. Now, it remains to apply the inductive assumption. Thus, by induction, Lemma 4 is proved.

We can immediately deduce from this lemma that, actually, the inequality

$$card(\{\alpha < \omega_1 : X_\alpha \notin \mathcal{I}\}) \leq \omega$$

is valid. Notice also that, in our considerations below, we need only a particular case of Lemma 4. Namely, we will be dealing with the pair $(\mathcal{S}, \mathcal{I})$ such that

$$\mathcal{S} = dom(\mu), \quad \mathcal{I} = \mathcal{I}(\mu),$$

where μ is a nonzero σ-finite complete diffused measure given on a set of cardinality ω_1.

Let us introduce a certain type of transfinite matrices over an arbitrary set of cardinality ω_1.

Let Y be a set with $card(Y) = \omega_1$. Consider a double family

$$(Y_{n,\xi})_{n<\omega,\xi<\omega_1}$$

of subsets of Y. We shall say that this family is an admissible transfinite matrix for Y if it possesses the following two properties:

(1) for each ordinal number $\xi < \omega_1$, the partial family $(Y_{n,\xi})_{n<\omega}$ is increasing by inclusion and

$$\cup\{Y_{n,\xi} \ : \ n < \omega\} = Y;$$

(2) for each natural number n, there exists a natural number $m = m(n)$ such that for any set $D \subset \omega_1$ with $card(D) = m$, we have

$$card(\cap\{Y_{n,\xi} \ : \ \xi \in D\}) \leq \omega.$$

Lemma 5. *For any set Y with $card(Y) = \omega_1$, there exists an admissible transfinite matrix.*

Proof. The argument is essentially based on the existence of an Ulam matrix over Y. Indeed, let

$$(X_{n,\xi})_{n<\omega,\xi<\omega_1}$$

be an arbitrary Ulam matrix for Y. Then (see Chapter 7), we have:

(a) for each ordinal number $\xi < \omega_1$, the set $Y \setminus \cup\{X_{n,\xi} \ : \ n < \omega\}$ is at most countable;

(b) for each natural number n, the partial family $\{X_{n,\xi} \ : \ \xi < \omega_1\}$ is disjoint.

Let us define

$$\{x_{n,\xi} : n < \omega\} = Y \setminus \cup\{X_{n,\xi} : n < \omega\};$$

$$Y_{n,\xi} = (\cup\{X_{k,\xi} \ : \ k \leq n\}) \cup \{x_{k,\xi} : k \leq n\}$$

for all $n < \omega$ and $\xi < \omega_1$.

Then it is not hard to verify that the family $(Y_{n,\xi})_{n<\omega,\xi<\omega_1}$ is an admissible matrix of subsets of Y. (Namely, for any natural number n, we may put $m(n) = n + 2$.)

Lemma 5 has thus been proved.

Now, we present some properties of countably generated σ-algebras of subsets of ω_1, interesting from the measure-theoretical point of view. The main role here is played by the classical Sierpiński partition of $\omega_1 \times \omega_1$ (see [200] and Chapter 4 of this book).

We need the following auxiliary statement.

Lemma 6. *Let Y be a set of cardinality ω_1 and let $(Y_\alpha)_{\alpha<\omega_1}$ be a family of subsets of Y. Then there exists a countably generated σ-algebra $\mathcal{S}$ of subsets of Y, such that*

$$\{Y_\alpha : \alpha < \omega_1\} \subset \mathcal{S}.$$

In other words, any family of subsets of ω_1, having cardinality less than or equal to ω_1, can be embedded in some countably generated σ-algebra of subsets of ω_1.

Proof. We start with the Sierpiński partition $\{A, B\}$ of the product set $\omega_1 \times \omega_1$. Let us recall that this partition is defined as follows:

$$A = \{(\xi,\zeta) : \xi \leq \zeta < \omega_1\},$$

$$B = \{(\xi,\zeta) : \omega_1 > \xi > \zeta\}.$$

As already mentioned in Chapter 4, for any $\xi < \omega_1$ and $\zeta < \omega_1$, the inequalities

$$card(A^\zeta) \leq \omega, \quad card(B_\xi) \leq \omega$$

are fulfilled, where

$$A^\zeta = \{\xi : (\xi,\zeta) \in A\},$$

$$B_\xi = \{\zeta : (\xi,\zeta) \in B\}.$$

In other words, each of the sets A and B can be represented as the union of a countable family of "curves" lying in $\omega_1 \times \omega_1$. This property of the partition $\{A, B\}$ implies a number of interesting consequences. Some of them were indicated in Chapter 4. Recall that one of them is the following formula:

$$\mathcal{P}(\omega_1 \times \omega_1) = \mathcal{P}(\omega_1) \otimes \mathcal{P}(\omega_1).$$

In order to show the validity of this formula, let us take an arbitrary injection

$$\psi : \omega_1 \to \mathbf{R}$$

and identify the set ω_1 with its image $\psi(\omega_1) \subset \mathbf{R}$. Then the product set $\omega_1 \times \omega_1$ can be identified with the set $\psi(\omega_1) \times \psi(\omega_1)$ which is a subset of the euclidean plane $\mathbf{R}^2 = \mathbf{R} \times \mathbf{R}$. Evidently, any partial function acting from ω_1 into $\omega_1 \subset \mathbf{R}$ is measurable with respect to the σ-algebra $\mathcal{P}(\omega_1)$. Hence the graph of such a partial function is a measurable set with respect to the product σ-algebra $\mathcal{P}(\omega_1) \otimes \mathcal{P}(\omega_1)$. Obviously, the same is true for each subset L of $\omega_1 \times \omega_1$ such that L^{-1} is the graph of a partial function acting from ω_1 into ω_1. Briefly speaking, any "curve" lying in $\omega_1 \times \omega_1$ is measurable with respect to $\mathcal{P}(\omega_1) \otimes \mathcal{P}(\omega_1)$. Now, let X be an arbitrary subset of $\omega_1 \times \omega_1$. Then we may write

$$X = (X \cap A) \cup (X \cap B),$$

where A and B are the members of the Sierpiński partition of $\omega_1 \times \omega_1$. But, as mentioned earlier, each of the sets $X \cap A$ and $X \cap B$ can be represented as the union of a countable family of "curves" lying in $\omega_1 \times \omega_1$. Consequently, $X \cap A$ and $X \cap B$ belong to $\mathcal{P}(\omega_1) \otimes \mathcal{P}(\omega_1)$. In this way we obtain that the set X belongs to $\mathcal{P}(\omega_1) \otimes \mathcal{P}(\omega_1)$, too.

Now, let us consider an arbitrary set Y with $card(Y) = \omega_1$ and take an arbitrary family $\{Y_\alpha : \alpha < \omega_1\}$ of subsets of Y. We have to show that this family can be embedded in some countably generated σ-algebra $\mathcal{S}$ of subsets of Y. Clearly, we may identify Y with ω_1. After such an identification, let us denote

$$Z = \cup\{\{\alpha\} \times Y_\alpha \ : \ \alpha < \omega_1\}.$$

Evidently, Z is a subset of $\omega_1 \times \omega_1$ and, for each $\alpha < \omega_1$, the vertical α-section of Z is equal to Y_α. As said above, Z can be represented in the form

$$Z = \cup\{L_m \ : \ m < \omega\},$$

where L_m $(m < \omega)$ are some "curves" lying in $\omega_1 \times \omega_1$. If we identify ω_1 with a certain subset of the real line $\mathbf{R}$, then it is not difficult to demonstrate that for any "curve" L in $\omega_1 \times \omega_1$, there exists a countable family

$$\{W_k \ : \ k < \omega\} = \{U_k \times V_k \ : \ k < \omega\}$$

of rectangular subsets of $\omega_1 \times \omega_1$, such that L belongs to the σ-algebra generated by $\{W_k \ : \ k < \omega\}$. Applying this fact to each of our "curves" L_m $(m < \omega)$, we obtain that for the set Z there exists a countable family

$$\{C_n \times D_n \ : \ n < \omega\}$$

of rectangular subsets of $\omega_1 \times \omega_1$, such that Z belongs to the σ-algebra

$$\mathcal{S}' = \sigma(\{C_n \times D_n : n < \omega\})$$

generated by this family. Now, it is easy to verify that all sets Y_α $(\alpha < \omega_1)$ are in the σ-algebra

$$\mathcal{S} = \sigma(\{D_n \ : \ n < \omega\}).$$

We leave to the reader the verification of this simple fact.

Lemma 6 has thus been proved.

Remark 5. Let us return to the Sierpiński partition $\{A, B\}$ of the product set $\omega_1 \times \omega_1$. The argument used in the proof of Lemma 6 shows that there exists a countable family $\{Y_n \ : \ n < \omega\}$ of subsets of ω_1, such that

$$\{A, B\} \subset \mathcal{S} \otimes \mathcal{S},$$

where

$$\mathcal{S} = \sigma(\{Y_n : n < \omega\})$$

denotes the σ-algebra of subsets of ω_1, generated by $\{Y_n \ : \ n < \omega\}$. Obviously, we may also assume that all one-element subsets of ω_1 belong to $\mathcal{S}$ since there are countable families of subsets of ω_1, separating the points in ω_1. Then, taking into account the Fubini theorem, we readily infer that there is no nonzero σ-finite diffused measure defined on the σ-algebra $\mathcal{S}$. Consequently, there is no nonzero σ-finite diffused measure μ on ω_1 satisfying the relation

$$\{Y_n \ : \ n < \omega\} \subset dom(\mu).$$

We see again that if the Continuum Hypothesis holds, then the cardinal $\mathbf{c}$ is not real-valued measurable and, moreover, there exists a countably generated σ-algebra $\mathcal{S}$ of subsets of $\mathbf{R}$, such that all one-element subsets of $\mathbf{R}$ belong to $\mathcal{S}$ and there is no nonzero σ-finite diffused measure on $\mathcal{S}$.

Lemma 7. *Let Y be a set of cardinality ω_1 and let $(Y_{n,\xi})_{n<\omega,\xi<\omega_1}$ be an admissible transfinite matrix for Y. Then, for every nonzero σ-finite diffused measure μ on Y, at least one set $Y_{n,\xi}$ is nonmeasurable with respect to μ.*

Proof. In fact, Lemma 7 easily follows from Lemma 4. The corresponding details are left to the reader.

Lemma 8. *Let $(\Gamma, +, 0)$ be a commutative group representable in the form of a direct sum*

$$\Gamma = \sum_{\zeta < \omega_1} [e_\zeta],$$

where $(e_\zeta)_{\zeta < \omega_1}$ is an injective family of nonzero elements of Γ, and let

$$(E_{n,\xi})_{n<\omega,\xi<\omega_1}$$

be an admissible transfinite matrix over the set

$$E = \{e_\zeta : \zeta < \omega_1\}.$$

For each subset F of E, denote by $[F]$ the group generated by F. Then the family of groups

$$([E_{n,\xi}])_{n<\omega,\xi<\omega_1}$$

is an admissible transfinite matrix over Γ.

The proof of Lemma 8 can be reduced to a straightforward verification which is also left to the reader.

Now, we can formulate and prove the following statement.

Theorem 4. *Let G be any commutative group of cardinality ω_1. Then there exists a countable family $(G_i)_{i\in I}$ of subgroups of G, such that for every nonzero σ-finite diffused measure μ on G, at least one group G_i is nonmeasurable with respect to μ.*

Proof. As we know, our group G is representable in the form:

$$G = \cup\{\Gamma_k : k < \omega\},$$

where $(\Gamma_k)_{k<\omega}$ is an increasing family of subgroups of G and each Γ_k is a direct sum of cyclic groups. We may assume, without loss of generality, that all Γ_k are of cardinality ω_1. So we can write

$$\Gamma_k = \sum_{\zeta<\omega_1} [e_{k,\zeta}].$$

Let us put

$$E_k = \{e_{k,\zeta} : \zeta < \omega_1\} \qquad (k < \omega)$$

and let

$$(E_{k,n,\xi})_{n<\omega,\xi<\omega_1}$$

be an admissible transfinite matrix over E_k. (This matrix exists in view of Lemma 5.) Further, applying Lemma 6, we see that there exists a countable family S_k of subsets of E_k, such that

$$\{E_{k,n,\xi} : n < \omega, \xi < \omega_1\} \subset \sigma(S_k),$$

where $\sigma(S_k)$ denotes the σ-algebra of sets, generated by S_k. Clearly, we may suppose that S_k is an algebra of sets. Consequently, the σ-algebra $\sigma(S_k)$ coincides with the monotone class generated by S_k (see, for instance, [62]).

Now, we put

$$(G_i)_{i \in I} = \{\Gamma_k : k < \omega\} \cup \{[Z] : Z \in \cup_{k<\omega} S_k\}$$

and we assert that the family of groups $(G_i)_{i \in I}$ is the required one. To show the validity of this assertion, take any diffused probability measure μ on G and suppose to the contrary that all groups G_i $(i \in I)$ are μ-measurable. Then, for some $r < \omega$, we must have $\mu(\Gamma_r) > 0$. Consider the countable family of groups

$$\mathcal{P}_r = \{[Z] : Z \in \mathcal{S}_r\}.$$

An easy transfinite induction shows that all groups

$$[E_{r,n,\xi}] \qquad (n < \omega, \ \xi < \omega_1)$$

belong to the monotone class generated by $\mathcal{P}_r$. But, according to our assumption, all sets from $\mathcal{P}_r$ are μ-measurable. Therefore, all groups $[E_{r,n,\xi}]$ must be μ-measurable, too, which contradicts Lemmas 7 and 8. The contradiction obtained finishes the proof of Theorem 4.

Remark 6. Obviously, Theorem 3 follows from Theorem 4. But, as mentioned earlier, a certain analogue of Theorem 3 can be proved under Martin's Axiom instead of the Continuum Hypothesis. This analogue is formulated in Exercise 4 below.

Remark 7. It is easy to see that direct analogues of Theorems 3 and 4 are valid for vector spaces (over the field $\mathbf{Q}$ of all rationals) instead of commutative groups. In particular, if E is a vector space over $\mathbf{Q}$ with $card(E) = \omega_1$, then there exists a countable family $(E_i)_{i \in I}$ of vector subspaces of E, such that for any nonzero σ-finite diffused measure μ on E, at least one E_i is nonmeasurable with respect to μ.

Example 3. For noncommutative groups of cardinality ω_1, Theorem 4 fails to be true. Indeed, take any group G of the same cardinality whose all proper subgroups are at most countable. Recall that the existence of such a G was first established by Shelah in [191]. It is easy to define a G-invariant probability measure μ on G such that all countable subsets of G are of μ-measure zero. Evidently, G does not contain a subgroup nonmeasurable with respect to μ.

It would be interesting to extend Theorems 3 and 4 (respectively, their above-mentioned analogues) to a wider class of uncountable commutative groups G (respectively, to a wider class of uncountable vector spaces E). In particular, the following problem seems to be of some interest:

Problem 4. Let G be a commutative group whose cardinality is not real-valued measurable, and let μ be a nonzero σ-finite diffused measure on G. Does there exist a subgroup of G nonmeasurable with respect to μ?

This problem remains open.

EXERCISES

1. Let E be a set, G be a group of transformations of E and let μ be a σ-finite G-quasi-invariant measure on E. Suppose also that X is a subset of E satisfying the following conditions:

(i) X is μ-thick in E;

(ii) there exists a countable G-invariant partition $\{X_n : n < \omega\}$ of E into sets which are G-congruent with X.

Consider the family $\mathcal{S}$ of all those sets Z which can be represented in the form

$$Z = \cup\{Y_n \cap X_n \ : \ n < \omega\},$$

where $Y_n \in dom(\mu)$ for every $n < \omega$. Put

$$\mu'(Z) = \sum_{n<\omega} (1/2)^{n+1} \mu(Y_n)$$

and verify that:

(a) $\mathcal{S}$ is a G-invariant σ-algebra of subsets of E, containing $dom(\mu)$;

(b) the functional $\mu' : \mathcal{S} \to \mathbf{R} \cup \{+\infty\}$ is well defined on $\mathcal{S}$ and is a G-quasi-invariant measure on E extending μ;

(c) $X \in dom(\mu')$.

Suppose, in addition, that:

(iii) for each $n < \omega$, the group G contains a subgroup H_n satisfying the relation

$$(\forall h \in H_n)(h(X_n) = X_n);$$

(iv) for each $n < \omega$, the initial measure μ is metrically transitive with respect to the group H_n.

Prove that the measure μ' is metrically transitive with respect to the group G.

2. Let Γ be a group and let $(G_1, G_2, ..., G_n)$ be a finite family of subgroups of Γ. Denote

$$G = G_1 \cap G_2 \cap ... \cap G_n.$$

Show that:

(a) if $card(\Gamma/G_i) < \omega$ for all natural numbers $i \in [1, n]$, then

$$card(\Gamma/G) < \omega;$$

(b) if κ is an infinite cardinal number and $card(\Gamma/G_i) \leq \kappa$ for all natural numbers $i \in [1, n]$, then

$$card(\Gamma/G) \leq \kappa.$$

3. Let E be an arbitrary nonempty set. Show that this set can be endowed with the structure of a commutative group. Moreover, if E is infinite, then E can be equipped with the structure of a vector space over the field $\mathbf{Q}$ of all rationals.

Deduce from these facts that Theorems 3 and 4 of the present chapter generalize, respectively, the Banach–Kuratowski result [8] and the result of Ulam [222] concerning the real-valued nonmeasurability of ω_1.

4. Give a generalized version of Theorem 3 under Martin's Axiom. More precisely, assuming this axiom, prove that if G is a commutative group of cardinality $\mathbf{c}$, then there exists a countable family $\{G_n : n < \omega\}$ of subgroups of G such that for every probability measure μ on G vanishing on all subsets of G whose cardinalities are strictly less than $\mathbf{c}$, at least one subgroup G_n is nonmeasurable with respect to μ.

5. Give a detailed proof of Lemma 7 (also, of Lemma 8).

6. Reconstruct all details in the proof of Theorem 4; especially, those details which concern monotone classes of sets.

7. Let $(G, \cdot)$ be a group of cardinality ω_1. We identify the set G with ω_1 and identify each ordinal $\xi < \omega_1$ with the initial interval $[0, \xi[$. Also, we consider ω_1 as a topological space; in other words, we equip ω_1 with its order topology.

For any ordinal $\xi < \omega_1$, denote by $(\xi, \cdot)$ the substructure of $(\omega_1, \cdot)$. In general, the restriction of $\cdot$ to ξ is a partial binary operation on ξ. Let us put

$$\Xi = \{\xi < \omega_1 \ : \ (\xi, \cdot) \ is \ a \ subgroup \ of \ (\omega_1, \cdot)\}.$$

Demonstrate that $card(\Xi) = \omega_1$ and Ξ is a closed subset of ω_1. (In this connection, compare Exercise 7 from Chapter 2.)

Generalize the above result to other n-ary functional relations on ω_1, where n is a natural number.

8. Let G be a proper Borel subgroup of $\mathbf{R}$, and suppose that there exists a nonzero σ-finite G-quasi-invariant Borel measure on G. Show that

$$card(G) \leq \omega.$$

On the other hand, for $n \geq 2$, give an example of a proper uncountable Borel subgroup G of $\mathbf{R}^n$ which is everywhere dense in $\mathbf{R}^n$ and which can be equipped with a nonzero σ-finite G-invariant Borel measure.

9. Let $(\Gamma, +, 0)$ be an arbitrary uncountable commutative group. Then Γ is representable in the form

$$\Gamma = \cup\{G_n : n < \omega\},$$

where all subgroups G_n ($n < \omega$) are uncountable, too, and each of them is a direct sum of cyclic groups (see Appendix 2). Consequently, we may write

$$G_n = G_n' + G_n'' \qquad (n < \omega),$$

where the groups G_n' and G_n'' satisfy the relations

$$card(G_n') = \omega_1, \qquad G_n' \cap G_n'' = \{0\}.$$

Further, according to Theorem 4, each group G_n' admits a countable family

$$\{G_{n,k}' \ : \ k < \omega\}$$

of its subgroups, such that for any diffused probability measure μ on G_n', at least one of these subgroups is nonmeasurable with respect to μ. We define

$$G_{n,k} = G_{n,k}' + G_n'' \qquad (n < \omega, \ k < \omega).$$

Demonstrate that the countable family

$$\{G_n : n < \omega\} \cup \{G_n'' : n < \omega\} \cup \{G_{n,k} : n < \omega, \ k < \omega\}$$

of subgroups of Γ has the following property: for any nonzero σ-finite Γ-quasi-invariant measure ν on Γ, at least one subgroup from this family is nonmeasurable with respect to ν.

The result just formulated can be regarded as a generalized version of Lemma 4 from Chapter 9 and of Theorem 2 from the present chapter. (In connection with this result, see [84], [95], [106] and [108].)

10. Let Γ be an arbitrary group and let G be a subgroup of Γ. Show that there exists a nonzero σ-finite left Γ-invariant measure μ on Γ such that $G \in dom(\mu)$. In other words, G is not absolutely nonmeasurable with respect to the class of all nonzero σ-finite left Γ-invariant measures on Γ.

Chapter 14
Groups of rotations and nonmeasurable sets

As mentioned in the Preface, one of the impacts of the classical Vitali theorem was reflected in various equidecomposability paradoxes for the three-dimensional euclidean space (and, hence, for euclidean spaces of higher dimension). In this connection, it is reasonable to recall that Hausdorff [66] was the first mathematician who applied Vitali type argument to the two-dimensional euclidean sphere and, by utilizing some rather delicate properties of the group of rotations, gave a construction of a paradoxical decomposition of the sphere, and, consequently, of the three-dimensional euclidean ball. His result is extremely important from the measure-theoretical viewpoint since it directly leads to the nonexistence on the space $\mathbf{R}^n$ ($n \geq 3$) of a universal finitely additive normalized measure invariant under the group of all motions of $\mathbf{R}^n$. Some years later Banach and Tarski [9] generalized the above-mentioned Hausdorff result and formulated and proved their famous equidecomposability paradox (which now is called the Banach-Tarski paradox). Afterwards, many interesting and important theorems were obtained in this direction. A detailed account of this topic can be found in [226]. Among relatively recent achievements, the result of Dougherty and Foreman must be especially indicated (see [35]). It states that the sets participating in the Banach-Tarski paradox and being bad from the measure-theoretical viewpoint may otherwise be good from the topological point of view; namely, all of them may have the Baire property. Note that this last result solves positively an old problem formulated by Szpilrajn (Marczewski).

At the same time, it is well known that the group of all motions of the euclidean plane $\mathbf{R}^2$ is solvable. Consequently, there are no equidecomposability paradoxes in $\mathbf{R}^2$. (In this context, the classical Banach theorem on the existence of universal finitely additive motion-invariant normalized measures on $\mathbf{R}^2$ must be indicated.) However, various subsets of $\mathbf{R}^2$ were constructed with strange and, in some sense, paradoxical geometric properties. The best known example of such a set is due to Mazurkiewicz. Namely, he constructed a subset of the plane which meets every straight line of the plane in exactly two points (see Chapter 6). Other interesting examples are due to Sierpiński and Davies (see Chapters 2, 4 and 6). Let us recall that in his famous work [200], Sierpiński showed that the Continuum Hypothesis is equivalent to the existence of a partition of the plane $\mathbf{R}^2$ into two sets A and B, such that A meets every straight line parallel to the axis $\mathbf{R} \times \{0\}$ in countably many points and, similarly, B meets every straight line parallel to the axis $\{0\} \times \mathbf{R}$ in countably many points. We have already underlined in Chapter 4 that this partition became a starting point for further investigations concerning plane sets with strange properties from the measure-theoretical viewpoint. (See, for instance, Chapters 4 and 6 where such properties make contrast with the classical Fubini theorem.)

In this chapter we are going to give one more application of the above-mentioned Sierpiński partition to constructions of sets which, on the one hand, are rather good with respect to translation-invariant measures on the plane $\mathbf{R}^2$ and, on the other hand, are extremely bad with respect to motion-invariant measures on $\mathbf{R}^2$. Moreover, applying the method developed in [82] (see also [100]), we are able to extend constructions of this sort to the case of the euclidean space $\mathbf{R}^n$ where $n \geq 3$.

Throughout this chapter, it will be convenient to utilize the following notation:

T_n = the group of all translations of the space $\mathbf{R}^n$;

S_n = the group generated by all central symmetries of $\mathbf{R}^n$;

M_n = the group of all motions (isometric transformations) of $\mathbf{R}^n$.

Note that, for $n \geq 2$, we have the following proper inclusions:

$$T_n \subset S_n \subset M_n.$$

Note also that the group S_n is generated by T_n and the symmetry of the space $\mathbf{R}^n$ with respect to its origin. Moreover, T_n is a normal subgroup of M_n.

Let us briefly discuss some situations connected with the existence of nonmeasurable sets in the euclidean space $\mathbf{R}^n$. We are going to show that these situations are essentially caused by specific properties of the rotation group of $\mathbf{R}^n$ where $n \geq 3$.

We recall that the symbol O_n denotes the group of all linear orthogonal transformations of $\mathbf{R}^n$. Respectively, the symbol O_n^+ stands for the group of all proper orthogonal transformations of $\mathbf{R}^n$ which are usually called the rotations of this space (about its origin).

If $n = 1$, then O_n consists exactly of two elements: the identity transformation of $\mathbf{R}^n$ and the symmetry of $\mathbf{R}^n$ with respect to the origin. Therefore, O_n^+ is trivial: it is reduced to the identity transformation.

If $n = 2$, then O_n consists of all rotations of $\mathbf{R}^n$ (about the origin) and of all symmetries of $\mathbf{R}^n$ with respect to straight lines passing through the origin. At the same time, the group O_n^+ is commutative and is canonically isomorphic to $\mathbf{S}_{n-1}$.

For $n \geq 3$, the algebraic structure of O_n^+ becomes more complicated. Namely, it turns out that this group contains a free subgroup of cardinality continuum. This fact was first established by Sierpiński (see [198]) and implies the existence of various strong equidecomposability paradoxes in $\mathbf{R}^n$ or, respectively, in $\mathbf{S}_{n-1}$.

Let us underline that even the existence of a free subgroup of O_n^+ generated by two independent rotations is sufficient for the emergence of paradoxical sets which, therefore, turn out to be nonmeasurable in the Lebesgue sense. In fact, Hausdorff and (some years later) von Neumann discovered this phenomenon.

Indeed, take any two independent rotations f and g from O_n^+, where $n \geq 3$. The precise proof of the existence of such rotations is not difficult and can be found in many works (see, for example, [82], [87], [198] or [226]). Put $G = [\{f, g\}]$ and define:

$$H_0 = \{h \in G : h = f^{3k} \circ g^m \circ ..., \quad k \in \mathbf{Z}, \ m \in \mathbf{Z}, \ m \neq 0\},$$

$$H_1 = \{h \in G : h = f^{3k+1} \circ g^m \circ ..., \quad k \in \mathbf{Z}, \ m \in \mathbf{Z}, \ m \neq 0\},$$

$$H_2 = \{h \in G : h = f^{3k+2} \circ g^m \circ ..., \quad k \in \mathbf{Z}, \ m \in \mathbf{Z}, \ m \neq 0\}.$$

It is easy to verify that

$$H_0 \cap H_1 = H_1 \cap H_2 = H_2 \cap H_0 = \emptyset,$$

$$f H_0 = H_1, \quad f^2 H_0 = H_2,$$

$$g H_1 \cup g H_2 \subset H_0,$$

$$H_0 \cup H_1 \cup H_2 = G.$$

These relations show us that the sets H_0, H_1, H_2 form a paradoxical decomposition of the group G, which immediately implies that there exists no
nonzero finite left G-invariant finitely additive universal measure on G; in
other words, G is not amenable. Starting with this fact, it is not difficult to
obtain an analogous paradoxical decomposition for the unit sphere $\mathbf{S}_2 \subset \mathbf{R}^3$
and, respectively, for the closed unit ball

$$\mathbf{B}_3 = \{x \in \mathbf{R}^3 : ||x|| \leq 1\}.$$

Now, by applying the Banach theorem on two injections (Theorem 3 from
Chapter 1) and utilizing a simple argument, we directly come to the Banach-
Tarski paradox (see [9] and [226]; compare also Exercise 9 of this chapter).

We have already mentioned that for $n \geq 3$, the group O_n^+ contains a
free subgroup of cardinality continuum. Mycielski [156] essentially applied
this fact and was able to find much stronger paradoxical decompositions of
the sphere $\mathbf{S}_2$. From his result it follows, in particular, that there exists a
subset X of $\mathbf{S}_2$ satisfying these two conditions:

1) $\mathbf{S}_2$ can be covered by finitely many O_3^+-congruent copies of X;

2) there are uncountably many pairwise disjoint O_3^+-congruent copies of
X.

For details, see [156] or [226] where a much stronger result is formulated
and proved.

Note that, for our purposes, the existence of a set X with properties
1) and 2) is significant. In particular, we easily observe that the above-
mentioned set X turns out to be O_3^+-absolutely nonmeasurable, even with
respect to the class of all finitely additive O_3^+-quasi-invariant probability
measures on the euclidean sphere $\mathbf{S}_2$. Now, it is easy to see that if X'
denotes the union of all linear segments whose common end-point is the
origin and the second end-points range over X, then X' turns out to be
M_3-absolutely nonmeasurable, even with respect to the class of all finitely
additive M_3-quasi-invariant normalized measures on the euclidean space $\mathbf{R}^3$.

Obviously, the same results can be obtained for the groups O_n^+ and M_n,
where $n \geq 3$. (It suffices to apply the method of induction on n.)

In these constructions we essentially rely on specific properties of the
group O_n^+. Of course, the argument presented above does not work for those
groups of motions of the euclidean space $\mathbf{R}^n$ which are not paradoxical.

However, from the general result obtained in Chapter 11 we deduce that
there are T_n-absolutely nonmeasurable subsets of $\mathbf{R}^n$ where $n \geq 1$. Sup-

pose now that g is a nontrivial rotation of the space $\mathbf{R}^3$ about its origin; in other words, suppose that $g \in O_3^+$ and g differs from the identity transformation of this space. Let G stand for the group generated by g and T_3. In connection with the above said, a natural question arises whether there are G-absolutely nonmeasurable subsets of $\mathbf{R}^3$ having rather good measure-theoretical properties with respect to the group T_3. Notice that, dealing with G, we cannot appeal to equidecomposability paradoxes because G is solvable and, consequently, amenable (see Exercise 3 of this chapter). Nevertheless, by using the techniques developed in preceding chapters of the book, we are able to give a positive answer to the formulated question for many rotations g. Namely, we shall establish below the existence of G-absolutely nonmeasurable sets which, simultaneously, are T_3-negligible. The corresponding result was first obtained in paper [109] by the author.

Let us recall some definitions from the general theory of invariant (quasi-invariant) measures (compare [82], [100], Exercise 10 of Chapter 6, and Chapter 11).

Let E be a nonempty set and let G be a fixed group of transformations of E.

Let X be a subset of E. We shall say that X is G-negligible (in E) if the following two conditions are satisfied:

(a) there exists at least one nonzero σ-finite G-quasi-invariant measure μ on E such that $X \in dom(\mu)$ and $\mu(X) = 0$;

(b) for any nonzero σ-finite G-quasi-invariant measure ν on E, the relation $X \in dom(\nu)$ implies $\nu(X) = 0$.

Let Y be a subset of E. We shall say that Y is G-absolutely nonmeasurable (in E) if there exists no nonzero σ-finite G-quasi-invariant measure ν on E such that $Y \in dom(\nu)$.

Recall that some properties of absolutely nonmeasurable sets were discussed in Chapter 11.

Example 1. If X is a uniform subset of the plane $\mathbf{R}^2$ (see Chapter 6), then X is T_2-negligible (and S_2-negligible). Moreover, it was shown in [94] that if p is a straight line in $\mathbf{R}^2$ and a set $X \subset \mathbf{R}^2$ is such that every line parallel to p meets X in finitely many points, then X is T_2-negligible (and S_2-negligible). Actually, an analogous result can be formulated and proved for sets which are finite with respect to an uncountable subgroup of a given commutative group (see [89]). In particular, if every line parallel to p meets

X in one or two points, then X turns out to be T_2-negligible (S_2-negligible). This fact will be applied below.

Example 2. Assuming the Continuum Hypothesis and starting with a Sierpiński partition of $\mathbf{R}^2$, it is not hard to define two sets A' and B' in $\mathbf{R}^2$ such that:

(1) the set A' is uniform with respect to the axis $\mathbf{R} \times \{0\}$;

(2) the set B' is uniform with respect to the axis $\{0\} \times \mathbf{R}$;

(3) there exists a countable family $\{h_n : n < \omega\}$ of translations of $\mathbf{R}^2$, for which we have

$$\cup\{h_n(A' \cup B') : n < \omega\} = \mathbf{R}^2.$$

Conversely, the existence of sets A' and B' satisfying the properties (1) - (3) implies the validity of the Continuum Hypothesis. (For details, see [96].) Note also that these properties of A' and B' enable us to demonstrate the T_2-absolute nonmeasurability (consequently, S_2-absolute nonmeasurability and M_2-absolute nonmeasurability) of the set $A' \cup B'$.

Indeed, let ν be an arbitrary nonzero σ-finite T_2-quasi-invariant measure on $\mathbf{R}^2$ and suppose for a while that

$$A' \cup B' \in dom(\nu).$$

Without loss of generality, we may assume that ν is a probability measure. Now, relation (3) shows that

$$\nu(A' \cup B') > 0.$$

On the other hand, applying the classical Banach theorem (see, for instance, [82] or [226]), we can extend ν to a finitely additive T_2-quasi-invariant measure ν' defined on the family of all subsets of $\mathbf{R}^2$. Since each of the sets A' and B' is uniform, we get

$$\nu'(A') = \nu'(B') = 0.$$

On the other hand, we have

$$0 < \nu(A' \cup B') = \nu'(A' \cup B') \leq \nu'(A') + \nu'(B') = 0,$$

which yields a contradiction. The contradiction obtained establishes the T_2-absolute nonmeasurability of $A' \cup B'$.

Now, let g denote the rotation of the plane $\mathbf{R}^2$ (about its origin), which maps the axis $\mathbf{R} \times \{0\}$ onto the axis $\{0\} \times \mathbf{R}$. Let us consider the set

$$Z = g(A') \cup B'.$$

Since every straight line lying in $\mathbf{R}^2$ and parallel to the axis $\{0\} \times \mathbf{R}$ meets the set Z in one or two points, we may assert that Z is T_2-negligible in $\mathbf{R}^2$ (compare Example 1). At the same time, it can easily be observed that Z is G-absolutely nonmeasurable in $\mathbf{R}^2$, where

$$G = [\{g\} \cup T_2].$$

To see this, let us take a probability G-quasi-invariant measure ν on $\mathbf{R}^2$ and suppose for a moment that $Z \in dom(\nu)$. Without loss of generality, we may assume that ν is complete. Since $T_2 \subset G$, the measure ν is also T_2-quasi-invariant and we must have $\nu(Z) = 0$ (in view of the T_2-negligibility of Z). Therefore,

$$\nu(g^{-1}(Z)) = 0$$

by virtue of the G-quasi-invariance of ν. Consequently, the equalities

$$\nu(A') = 0, \quad \nu(B') = 0$$

are valid and yield

$$\nu(A' \cup B') = 0.$$

Now, applying (3), we readily come to the equality $\nu(\mathbf{R}^2) = 0$ which contradicts our assumption that $\nu(\mathbf{R}^2) = 1$. The contradiction obtained gives the required statement.

Note that a slight modification of this argument leads to a more general result stating (under **CH**) that there exists a uniform subset X of $\mathbf{R}^2$ which covers $\mathbf{R}^2$ with the aid of countably many transformations from G; in other words,

$$\cup\{g_k(X) \ : \ k < \omega\} = \mathbf{R}^2$$

for some countable family $\{g_k \ : \ k < \omega\} \subset G$. In particular, X is T_2-negligible and G-absolutely nonmeasurable.

Example 3. Let $\{p_k \ : \ k < \omega\}$ be an injective countable family of straight lines in the plane $\mathbf{R}^2$, passing through the origin. Recall the result of Davies [31], which states that there exists a family $\{X_k \ : \ k < \omega\}$ of subsets of $\mathbf{R}^2$ satisfying the following conditions:

(1) $\cup\{X_k : k < \omega\} = \mathbf{R}^2$;

(2) for each $k < \omega$, the set X_k is uniform with respect to the line p_k.

As we know, it is not hard to deduce from this result that there exists a uniform subset Z of $\mathbf{R}^2$ such that

$$\cup\{f_i(Z) \ : \ i < \omega\} = \mathbf{R}^2$$

for some countable family $\{f_i : i < \omega\}$ of motions of the plane. In particular, the set Z turns out to be T_2-negligible and M_2-absolutely nonmeasurable.

In the sequel, we need several auxiliary propositions.

Lemma 1. *Let g be a rotation of the plane $\mathbf{R}^2$ (about its origin 0). Then there exists a field $P \subset \mathbf{R}$ of cardinality ω_1, such that $\mathbf{R}^2$ can be represented in the form*

$$\mathbf{R}^2 = U + V \qquad (U \cap V = \{0\}),$$

where U and V satisfy the following conditions:

1) U is a two-dimensional vector subspace of $\mathbf{R}^2$ over P;

2) $g(U) = U$;

3) V is a vector subspace of $\mathbf{R}^2$ over P.

Proof. Let $\{e_1, e_2\}$ be the canonical orthonormal basis in $\mathbf{R}^2$. Obviously, we can write

$$g(e_1) = a_1 e_1 + a_2 e_2,$$
$$g(e_2) = b_1 e_1 + b_2 e_2,$$

where a_1, a_2, b_1, b_2 are some (uniquely determined) real numbers for which we have

$$1 = |a_1 b_2 - a_2 b_1| \neq 0.$$

Let P be a subfield of $\mathbf{R}$ of cardinality ω_1, such that

$$\{a_1, a_2, b_1, b_2\} \subset P.$$

This subfield can readily be constructed by the method of transfinite recursion. Now, we put

$$U = Pe_1 + Pe_2.$$

It is easy to verify that $g(U) = U$. Finally, take as V an arbitrary vector space (over the same P) satisfying the relations

$$U + V = \mathbf{R}^2, \quad U \cap V = \{0\}.$$

The existence of such a V is evident. This completes the proof.

Lemma 2. *Let G_1 and G_2 be any two groups identified, respectively, with the groups of their left translations, let*

$$\phi : G_1 \to G_2$$

be a surjective homomorphism and suppose that X is a G_2-negligible subset of G_2. Then the set $\phi^{-1}(X)$ is G_1-negligible in G_1.

We leave an easy proof of Lemma 2 to the reader (compare also Lemma 5 from Chapter 11).

Lemma 3. *Let E be a set, G be a group of transformations of E and let Y be a G-absolutely nonmeasurable set in E. Then there exists a countable family $\{g_k : k < \omega\}$ of transformations from G, such that*

$$\cup\{g_k(Y) : k < \omega\} = E.$$

Proof. Suppose to the contrary that

$$\cup\{g_k(Y) : k < \omega\} \neq E$$

for any countable family $\{g_k : k < \omega\} \subset G$. Then Y is a member of some G-invariant σ-ideal $\mathcal{I}$ of subsets of E. Obviously, there exists a G-invariant probability measure μ on E such that $\mathcal{I} = \mathcal{I}(\mu)$. In particular, $Y \in dom(\mu)$ and, consequently, Y cannot be G-absolutely nonmeasurable. The contradiction obtained finishes the proof (compare Exercise 5 from Chapter 11).

Lemma 4. *Let E be a set, G_1 and G_2 be two groups of transformations of E, such that $G_1 \subset G_2$, and let Z be a G_1-negligible set in E. Suppose also that there exists a countable family $\{g_i : i < \omega\} \subset G_2$ for which*

$$\cup\{g_i(Z) : i < \omega\} = E.$$

Then the set Z is G_2-absolutely nonmeasurable in E.

The proof of this lemma is almost trivial and is left to the reader.

Now, we can formulate and prove the following statement.

Theorem 1. *Let g be a rotation of the plane $\mathbf{R}^2$, distinct from the identity transformation of $\mathbf{R}^2$ and all central symmetries of $\mathbf{R}^2$. Then there exists a set $Z \subset \mathbf{R}^2$ such that:*

1) Z is T_2-negligible in $\mathbf{R}^2$;

2) Z is G-absolutely nonmeasurable in $\mathbf{R}^2$, where $G = [T_2 \cup \{g\}]$.

Proof. The argument below is very similar to the one given in Example 2. First of all, we may assume without loss of generality that the fixed point of g coincides with the origin 0 of $\mathbf{R}^2$; in other words, we may assume that $g \in O_2^+$. Let us represent $\mathbf{R}^2$ in the form

$$\mathbf{R}^2 = U + V \quad (U \cap V = \{0\}),$$

where U and V are vector spaces over the field $P \subset \mathbf{R}$ described in Lemma 1. For the two-dimensional vector space U, consider an analog of the Sierpiński partition, corresponding to the following two axes in U: the "line" Pe_1 and its image under the transformation g. (In fact, we are dealing here with a certain affine version of the Sierpiński partition of $\omega_1 \times \omega_1$.) Now, it is not hard to verify that the argument of Example 2 works in our situation as well, and we do not need the Continuum Hypothesis since

$$card(U) = card(P \times P) = card(P) = \omega_1.$$

Therefore, we obtain a set $X \subset U$ which is uniform in U (with respect to the "line" Pe_1) and has the property that the union $X \cup g(X)$ covers U by using countably many translations of U. In particular, we see that X is a U-negligible and G'-absolutely nonmeasurable subset of U, where G' denotes the group generated by U and $g|U$. Further, we define

$$Z = (X + V) \cup (X + g^{-1}(V)).$$

We assert that Z is the required set in $\mathbf{R}^2$. In order to demonstrate this, let us first observe that

$$g^{-1}(V) \cap U = \{0\},$$

from which it follows that every "line" in $\mathbf{R}^2$ parallel to Pe_1 meets Z in one or two points. This implies that Z is a T_2-negligible subset of $\mathbf{R}^2$ (compare Example 1). On the other hand, we have the inclusion

$$(X \cup g(X)) + V \subset Z \cup g(Z)$$

which yields that the union $Z \cup g(Z)$ covers $\mathbf{R}^2$ with the aid of countably many translations of $\mathbf{R}^2$. So we can conclude, in view of Lemma 4, that the set Z is G-absolutely nonmeasurable in $\mathbf{R}^2$. The proof is thus completed.

Remark 1. Let g be a rotation of $\mathbf{R}^2$ about its origin 0. Suppose also that g is of infinite order; in other words, suppose that g^k differs from the identity transformation of $\mathbf{R}^2$ for all strictly positive integers k. Take any straight line p_0 in $\mathbf{R}^2$ passing through 0 and put

$$p_k = g^k(p_0) \qquad (k < \omega).$$

We get the injective countable family of straight lines p_k $(k < \omega)$, to which the result of Davies (see Chapter 6) can be applied. In this manner, we obtain that there exists a uniform subset Z of $\mathbf{R}^2$ which covers $\mathbf{R}^2$ by using countably many transformations from the group G generated by T_2 and g. Therefore, Z is T_2-negligible and G-absolutely nonmeasurable in $\mathbf{R}^2$. However, this method does not work for rotations of finite order.

Remark 2. We do not know whether there exists a subset of $\mathbf{R}^2$ which is T_2-negligible and S_2-absolutely nonmeasurable.

Now, taking into account Lemmas 2, 3 and 4, we can extend Theorem 1 to euclidean spaces of higher dimension. For the sake of simplicity, we formulate and prove here the result only for the three-dimensional euclidean space $\mathbf{R}^3$.

Theorem 2. *Let g be a rotation of the space $\mathbf{R}^3$, whose corresponding angle differs from 0 and π, and let G denote the group generated by g and T_3. Then there exists a subset Z of $\mathbf{R}^3$ which is T_3-negligible and G-absolutely nonmeasurable.*

Proof. Let us represent $\mathbf{R}^3$ in the form of a direct product

$$\mathbf{R}^3 = \mathbf{R}^2 \times \mathbf{R}.$$

Without loss of generality, we may assume that the axis of fixed points of g coincides with the second factor in this product (in particular, $g \in O_3^+$). By virtue of Theorem 1, there exists a set $Y \subset \mathbf{R}^2$ which is T_2-negligible and G'-absolutely nonmeasurable, where G' stands for the group generated by T_2 and the restriction of g to the first factor in the product $\mathbf{R}^2 \times \mathbf{R}$. Now, applying Lemma 2 to the canonical surjective group homomorphism

$$pr_1 \; : \; \mathbf{R}^2 \times \mathbf{R} \to \mathbf{R}^2,$$

we infer that the set

$$Z = Y \times \mathbf{R} \subset \mathbf{R}^3$$

is T_3-negligible in $\mathbf{R}^3$. Further, applying Lemma 3 to $\mathbf{R}^2$, G' and Y, we deduce that

$$\cup \{g_i'(Y) \; : \; i < \omega\} = \mathbf{R}^2$$

for some countable family $\{g_i' : i < \omega\}$ of transformations from G' and, consequently,

$$\cup \{g_i(Z) \; : \; i < \omega\} = \mathbf{R}^2 \times \mathbf{R} = \mathbf{R}^3$$

for some countable family $\{g_i : i < \omega\}$ of transformations from G. Finally, taking into account Lemma 4, we claim that Z is G-absolutely nonmeasurable in $\mathbf{R}^3$.

Remark 3. It is not hard to verify that the group G in the preceding theorem is solvable and, consequently, amenable. Therefore, it does not admit equidecomposability paradoxes.

Remark 4. By using the same result of Davies (see Chapter 6), one can easily infer that there exists a subset of $\mathbf{R}^3$ which is T_3-negligible (or S_3-negligible) and M_3-absolutely nonmeasurable. On the other hand, there are some constructions of extremely paradoxical subsets of $\mathbf{R}^3$ whose existence is essentially caused by the existence of large free subgroups of the group O_3^+. (See [156] and [226] where strong versions of the Banach-Tarski paradox are discussed in detail.) It would be interesting to obtain a T_3-negligible and M_3-absolutely nonmeasurable set by starting with those constructions.

Remark 5. A proper subclass of negligible sets, consisting of the so-called absolutely negligible sets, is of special interest for the general theory of invariant (quasi-invariant) measures. Recall that these sets are defined as follows (see Exercise 4 from Chapter 10).

Let E be a nonempty set and let G be a group of transformations of E. (In other words, we have a space equipped with a transformation group.) Let X be a subset of E. We say that X is G-absolutely negligible if for every nonzero σ-finite G-quasi-invariant measure μ on E, there exists a G-quasi-invariant measure μ' on E extending μ and satisfying the equality $\mu'(X) = 0$.

Note that, in general, the class of negligible sets does not form an ideal of subsets of a given space (in this connection, see [96]). At the same time, it directly follows from the definition that the class of absolutely negligible sets is an ideal in the Boolean algebra of all subsets of an original space. As mentioned earlier in Example 1, any uniform subset of $\mathbf{R}^2$ is T_2-negligible. On the other hand, it was shown in [96] that there are uniform subsets of $\mathbf{R}^2$ which are not T_2-absolutely negligible (compare also Exercise 4 from Chapter 10).

In connection with this fact and with Theorems 1 and 2, the following question seems to be rather interesting:

Let a natural number n be greater than or equal to 2. Does there exist a subset of the euclidean space $\mathbf{R}^n$ which is T_n-absolutely negligible and, simultaneously, M_n-absolutely nonmeasurable?

We do not know an answer to this question even in the case $n = 2$. In other words, it is unknown whether there exists a subset of the plane $\mathbf{R}^2$ which is T_2-absolutely negligible and, simultaneously, M_2-absolutely nonmeasurable.

EXERCISES

1. We recall that a group G is amenable if G admits a universal finitely additive left G-invariant probability measure.

Show that, for an arbitrary group G, the following three assertions are equivalent:

(a) G is amenable;

(b) there exists a universal finitely additive right G-invariant probability measure on G;

(c) there exists a universal finitely additive probability measure on G which is left G-invariant and right G-invariant.

Prove also that every solvable group is amenable. (Use the induction on the length of a composition series for a given solvable group.)

2. Demonstrate that the relation $R(X, Y)$:

$$\textit{"a set } X \textit{ is equidecomposable with a set } Y \textit{"}$$

is an equivalence relation in the Boolean algebra of all subsets of an euclidean space (more generally, in the Boolean algebra of all subsets of any metric space). Show also that if X is equidecomposable to a subset of Y and Y is equidecomposable to a subset of X, then X and Y are equidecomposable. (Apply the Banach theorem from Chapter 1.)

3. Let F be a solvable group of rotations of the space $\mathbf{R}^n$ about its origin, where $n \geq 3$, and let G denote the group generated by F and T_n. Prove that the group G is solvable, too.

In particular, if $g \in O_n^+$, then the group generated by g and T_n is solvable and, consequently, amenable. Therefore, it does not admit equidecomposability paradoxes.

4. Give a proof of Lemma 2.

5. Give a detailed proof of Lemma 3.

6. Give a proof of Lemma 4.

7. Formulate and prove an analogue of Theorem 2 for the euclidean space $\mathbf{R}^n$, where $n > 3$.

8. Show that the following two assertions are equivalent in the theory **ZF & DC**:

(a) the Hahn-Banach theorem on extensions of continuous linear functionals;

(b) every Boolean algebra admits a finitely additive probability measure.

9. Let g and h be any two independent rotations of the space $\mathbf{R}^3$ about its origin. Denote by $F_2 = [\{f, g\}]$ the free group generated by these two rotations. Check that there exists a countable subset X of the two-dimensional unit sphere $\mathbf{S}_2$, such that

(a) the set $\mathbf{S}_2 \setminus X$ is F_2-invariant;

(b) the group F_2 acts freely on $\mathbf{S}_2 \setminus X$.

Deduce from this fact, by using the result presented in Exercise 2, the classical form of the Banach-Tarski paradox: if X and Y are any two bounded subsets of $\mathbf{R}^n$ $(n \geq 3)$ such that

$$int(X) \neq \emptyset, \quad int(Y) \neq \emptyset,$$

then X and Y are M_n-equidecomposable. Here $int(X)$ and $int(Y)$ denote the interiors of X and Y, respectively.

We have already mentioned in this chapter that much stronger forms of the Banach-Tarski paradox are known at the present time (see [156] and [226]).

10. Show, in the theory

$$\textbf{ZF \& DC \& } \textit{the Hahn-Banach theorem,}$$

that for every family $\{B_i : i \in I\}$ of Boolean algebras there exists a family $\{\mu_i : i \in I\}$ of finitely additive probability measures, such that

$$(\forall i \in I)(dom(\mu_i) = B_i).$$

For this purpose, consider a Boolean algebra B defined as follows. Let Q denote the weak Cartesian product $\prod_{i \in I}^{*}(B_i \setminus \{0_i\})$ endowed with the partial ordering $\preceq$ such that

$$(x_i)_{i \in I} \preceq (y_i)_{i \in I} \Leftrightarrow (\forall i \in I)(x_i \leq_i y_i).$$

Then Q can be regarded as a coinitial subset of $B \setminus \{0\}$ where B is a complete Boolean algebra obtained by the Dedekind cuts method for Q, and B has the property that for each $i \in I$, there is a canonical monomorphism ϕ_i from B_i into B. According to Exercise 8, there exists a finitely additive probability measure μ on B. It remains to put

$$\mu_i = \mu \circ \phi_i \qquad (i \in I)$$

and to check that μ_i $(i \in I)$ are the required finitely additive probability measures.

11. Let E be a set and let G be some group of transformations of E acting freely in E. Suppose also that there exists a finitely additive G-invariant probability measure μ on the Boolean algebra of all subsets of E. Demonstrate, in the same theory

$$\textbf{ZF \& DC \&} \textit{ the Hahn-Banach theorem,}$$

that the group G is amenable.

This can be done by using the following argument.

Let the symbol E/G denote the family of all G-orbits in E. For each element $x \in E$, consider the Boolean algebra

$$B_{G(x)} = \mathcal{P}(G(x))$$

of all subsets of the orbit $G(x)$. In view of the result presented in Exercise 10, there exists a family $\{\mu_T : T \in E/G\}$ of finitely additive probability measures, such that

$$dom(\mu_T) = B_T = \mathcal{P}(T)$$

for all $T \in E/G$. Further, for any set $A \subset G$, define a function

$$f_A : E \to [0,1]$$

by the formula

$$f_A(x) = \mu_{G(x)}(A(x)) \qquad (x \in E).$$

Finally, introduce a functional

$$\nu \; : \; \mathcal{P}(G) \to [0,1]$$

by putting

$$\nu(A) = \int_E f_A(x)d\mu(x)$$

for each set $A \subset G$. Show that ν is a finitely additive right G-invariant probability measure on the Boolean algebra of all subsets of G.

12. Starting with the fact that the free group F_2 (generated by any two independent rotations from O_3^+) is not amenable and applying the results of Exercises 8, 10 and 11, prove that in the theory **ZF** & **DC** the Hahn-Banach theorem implies the existence of a subset of **R** which is not measurable in the Lebesgue sense.

The result of this exercise is due to Foreman and Wehrung [46]. Developing their method, Pawlikowski was able to strengthen their result and has established that within the same theory **ZF** & **DC**, the Hahn-Banach theorem implies the Banach-Tarski paradox. (For details, see [167].)

13. Applying the result of Davies (see Chapter 6), give a direct proof of the fact that for $n \geq 2$, there exists a subset of **R**^{n} which is T_n-negligible and, simultaneously, M_n-absolutely nonmeasurable.

14. Let **S**$_{n-1}$ be the unit sphere in the euclidean space **R**^{n} where $n \geq 2$. Equip **S**$_{n-1}$ with the $(n-1)$-dimensional Lebesgue probability measure λ_{n-1} (which is invariant under the group O_n^+ of all rotations of **S**$_{n-1}$ about its centre). Let k be a strictly positive integer and let X be a Lebesgue measurable subset of **S**$_{n-1}$ with $\lambda_{n-1}(X) < 1/k$. Finally, let Y be a subset of **S**$_{n-1}$ such that $card(Y) \leq k$. Demonstrate that there exists a rotation $g \in O_n^+$ for which we have

$$g(X) \cap Y = \emptyset.$$

For this purpose, take into account a close connection between the Lebesgue measure λ_{n-1} and the Haar probability measure on the compact group O_n^+.

15. Let $n \geq 3$. By using the method of transfinite recursion and applying Lemma 1 from Chapter 6, construct a free group $G \subset O_n^+$ such that $card(G) = \mathbf{c}$.

Utilize the existence of G and show that there is a free subgroup of M_n acting transitively on the space $\mathbf{R}^n$.

Chapter 15
Nonmeasurable sets associated with filters

In Chapter 7, several individual constructions of nonmeasurable sets were considered and, among them, the construction based on the existence of a nontrivial ultrafilter in $\mathcal{P}(\omega)$ was especially underlined (See Exercise 2 from the above-mentioned chapter.) Now, we wish to develop this topic and consider some other examples of filters in $\mathcal{P}(\omega)$ leading to nonmeasurable (in the Lebesgue sense) subsets of the real line.

Our main goal in this chapter is to present the remarkable result of Shelah and Raisonnier (see [176], [192]) stating that, in the theory

$$\textbf{ZF \& DC},$$

the inequality

$$\omega_1 \leq \mathbf{c}$$

implies the existence of a non Lebesgue-measurable subset of $\mathbf{R}$. In other words, in the same theory, the existence of a non Lebesgue-measurable subset of $\mathbf{R}$ is guaranteed by the existence of an uncountable subset X of $\mathbf{R}$ which is equipped with some well-ordering $\preceq_X$. Thus, taking into account the classical result of Solovay [210], one can conclude that the relation $\omega_1 \leq \mathbf{c}$ is not provable within $\textbf{ZF \& DC}$.

In order to establish the Shelah-Raisonnier result, we first need a number of preliminary notions and facts. We begin our consideration with some auxiliary statements concerning the descriptive structure of certain filters in $\mathcal{P}(\omega)$. All these statements are due to Talagrand (see [218]).

Most constructions presented below will be carried out in the classical Cantor space which is more convenient for our further purposes. Throughout this chapter, for any natural number n, we will write

$$[0, n[\; = \{k \in \omega : 0 \leq k < n\}$$

and we will identify the interval $[0, n[$ with n. If Z is a subset of $[0, n[$, then we may canonically identify Z with the corresponding element of $\{0, 1\}^n$. Briefly speaking, the above-mentioned element is the characteristic function of Z. As usual, we denote by

$$C = 2^\omega = \{0, 1\}^\omega$$

the classical Cantor discontinuum regarded as a commutative (via the addition operation modulo 2) compact zero-dimensional topological group. Obviously, C carries a unique invariant Borel probability measure λ (the so-called Haar measure), and since C is a Polish space, this measure is Radon. Moreover, in our further considerations, we will denote by the same symbol λ the completion of the Haar measure on C.

It is a well-known fact that λ is isomorphic to the standard Lebesgue measure given on the unit segment

$$[0, 1] = \{x \in \mathbf{R} \ : \ 0 \leq x \leq 1\}$$

of the real line, and that the existence of such an isomorphism can be established within the theory **ZF** & **DC**.

Besides, it is reasonable to recall another fact stating that λ is identical with the completion of an appropriate product measure on C; more exactly, we have

$$\lambda = \prod_{n < \omega} \mu_n,$$

where each measure μ_n is defined on $\mathcal{P}(\{0, 1\})$ by

$$\mu_n(\{0\}) = \mu_n(\{1\}) = 1/2.$$

For any set $z \subset [0, n[$, we denote

$$C_z = \{x \in C : x|[0, n[\ = z\}$$

keeping in mind the identification of z with its characteristic function. Actually, C_z is the basic open (simultaneously, closed) set in C corresponding to z. It is clear that

$$\lambda(C_z) = 2^{-n}.$$

For any natural number k, let us define:

$$C_k^0 = \{x \in C \ : \ x_k = 0\},$$

$$C_k^1 = \{x \in C \ : \ x_k = 1\}.$$

The family of sets

$$(C_k^i)_{i \in \{0,1\}, k < \omega}$$

forms a canonical pre-base of the Cantor space C. Let us mention a useful probabilistic property of this pre-base. Namely, each of the two families of sets

$$(C_k^0)_{k < \omega}, \qquad (C_k^1)_{k < \omega}$$

is stochastically independent with respect to λ. (This notion can be found in any text-book devoted to probability or random processes; see, for instance, [34], [160] or Chapter 4 in [19].)

The following auxiliary proposition (due to Talagrand [218]) is an important technical tool for further constructions.

Lemma 1. *Let $A \subset C$ be a compact set with*

$$\lambda(A) = \alpha > 0.$$

Then there exist a compact set $B \subset A$ with $\lambda(B) > 0$ and a sequence of natural numbers $\{n_k : k < \omega\}$, such that:

1) $n_0 < n_1 < ... < n_k < ...$;

2) for any natural number k and for any $z \in \{0,1\}^{n_k}$, we have

$$C_z \cap B \neq \emptyset \Rightarrow \lambda(C_z \cap B) \geq (1 - 2^{-k})\lambda(C_z).$$

Proof. We define by recursion some strictly increasing sequence of natural numbers $\{n_k : k < \omega\}$ and some decreasing (with respect to inclusion) sequence $\{B_k : k < \omega\}$ of compact subsets of A. First of all, we put

$$B_0 = A, \quad n_0 = 1.$$

Suppose now that for $k < \omega$, the natural number n_k and the compact set $B_k \subset A$ have already been defined. Since

$$\lambda(B_k) < \lambda(B_k) + \alpha 2^{-n_k - 2k - 4},$$

we can choose a natural number $n_{k+1} > n_k$ so large that the relation

$$\sum \{\lambda(C_z) : z \in \{0,1\}^{n_{k+1}}, \ C_z \cap B_k \neq \emptyset\} \leq$$

$$\lambda(B_k) + \alpha 2^{-n_k - 2k - 4}$$

will be fulfilled or, equivalently,

$$m = card(\{z \in \{0,1\}^{n_{k+1}} : C_z \cap B_k \neq \emptyset\}) \leq$$

$$2^{n_{k+1}}(\lambda(B_k) + \alpha 2^{-n_k - 2k - 4}).$$

Let us put:

$$T_{k+1} = \{z \in \{0,1\}^{n_{k+1}} : \lambda(C_z \cap B_k) \geq (1 - 2^{-k-2})\lambda(C_z)\};$$

$$B_{k+1} = \cup\{C_z \cap B_k : z \in T_{k+1}\}.$$

Denoting also

$$P_{k+1} = \{z \in \{0,1\}^{n_{k+1}} : C_z \cap B_k \neq \emptyset\},$$

we obviously have

$$B_k \setminus B_{k+1} \subset \bigcup_{z \in P_{k+1} \setminus T_{k+1}} C_z \cap B_k.$$

Further, we may write

$$(1 - 2^{-k-2})(2^{-n_{k+1}} m - \lambda(B_{k+1})) =$$

$$(1 - 2^{-k-2})(\sum_{z \in P_{k+1}} \lambda(C_z) - \sum_{z \in T_{k+1}} \lambda(C_z \cap B_k)) \geq$$

$$(1 - 2^{-k-2})(\sum_{z \in T_{k+1}} (\lambda(C_z) - \lambda(C_z \cap B_k))) +$$

$$(1 - 2^{-k-2})(\sum_{z \in P_{k+1} \setminus T_{k+1}} \lambda(C_z)) \geq$$

$$\sum_{z \in P_{k+1} \setminus T_{k+1}} \lambda(C_z \cap B_k) \geq \lambda(B_k) - \lambda(B_{k+1}).$$

Taking into account the inequality

$$m \leq 2^{n_{k+1}}(\lambda(B_k) + \alpha 2^{-n_k - 2k - 4})$$

and making easy calculations, we obtain

$$\lambda(B_k) \leq \lambda(B_{k+1}) + \alpha 2^{-n_k - k - 2}.$$

Now, let us define

$$B = \cap\{B_k : k < \omega\}$$

and observe that the relation

$$\lambda(B_k) \leq \lambda(B) + \alpha 2^{-n_k - k - 1}$$

holds true for each natural number k. In particular,

$$\alpha = \lambda(B_0) \leq \lambda(B) + (1/4)\alpha,$$

from which it follows

$$0 < (3/4)\alpha \leq \lambda(B).$$

Moreover, according to our construction, for any nonzero natural number k and for any $z \in \{0,1\}^{n_k}$ such that $C_z \cap B_k \neq \emptyset$, we have

$$\lambda(C_z \cap B_k) = \lambda(C_z \cap B_{k-1}) \geq (1 - 2^{-k-1})\lambda(C_z).$$

Finally, since $\alpha \leq 1$ and $\lambda(C_z) = 2^{-n_k}$, we may write

$$\lambda(C_z \cap B) \geq (1 - 2^{-k-1})\lambda(C_z) - 2^{-n_k - k - 1} = (1 - 2^{-k})\lambda(C_z).$$

The lemma has thus been proved.

Let us introduce the notion of a rapid filter in $\mathcal{P}(\omega)$. This important notion is due to Mokobodzki [151].

Let $\mathcal{F}$ be a filter in the Boolean algebra $\mathcal{P}(\omega)$, containing all co-finite subsets of ω.

We shall say that $\mathcal{F}$ is rapid if for every increasing sequence of natural numbers $\{n_k : k < \omega\}$, there exists at least one set $X \in \mathcal{F}$ such that

$$(\forall k < \omega)(card(X \cap [0, n_k[) \leq k).$$

Clearly, $\mathcal{F}$ is rapid if and only if, for an arbitrary strictly increasing sequence $\{n_k : k < \omega\}$ of natural numbers, there exists a set $X \in \mathcal{F}$ for which the above-mentioned relation is valid.

The following auxiliary proposition yields a sufficient condition for the rapidity of a given filter of subsets of ω.

Lemma 2. *Let $\mathcal{F}$ be a filter in ω containing all co-finite subsets of ω and let*

$$\psi : \omega \to \omega$$

be an increasing mapping whose range is unbounded in ω. Suppose also that for every increasing sequence $\{n_k : k < \omega\}$ of natural numbers, there exists at least one set $X \in \mathcal{F}$ such that

$$(\forall k < \omega)(card(X \cap [0, n_k[) \leq \psi(k)).$$

Then $\mathcal{F}$ is a rapid filter.

Proof. Take an arbitrary increasing sequence $\{n_k : k < \omega\}$ of natural numbers. Obviously, there is a unique mapping

$$\phi \; : \; \omega \to \omega$$

such that $\phi(k) = n_k$ for all $k < \omega$. Let us define a mapping

$$\phi' \; : \; \omega \to \omega$$

by the formula

$$\phi'(k) = \phi(\psi(k+1)) \qquad (k < \omega).$$

Then ϕ' may be regarded as an increasing sequence of natural numbers. According to the assumption of the lemma, there exists a set $X \in \mathcal{F}$ such that

$$card(X \cap [0, \phi'(k)[) \leq \psi(k)$$

for all $k < \omega$. Let us denote

$$X' = X \cap \{n \; : \; \phi(\psi(0)) < n < \omega\}.$$

Evidently, $X' \in \mathcal{F}$. We assert that

$$(\forall m < \omega)(card(X' \cap [0, \phi(m)[) \leq m).$$

Indeed, for any natural number $m \leq \psi(0)$, we have

$$card(X' \cap [0, \phi(m)[) = 0 \leq m.$$

Suppose now that $m > \psi(0)$. Then there is a natural number k such that

$$\psi(k) \leq m \leq \psi(k+1).$$

Consequently,

$$\phi'(k) = \phi(\psi(k+1)) \geq \phi(m)$$

and we may write

$$card(X' \cap [0, \phi(m)[) \leq card(X \cap [0, \phi'(k)[) \leq \psi(k) \leq m,$$

which shows that our filter $\mathcal{F}$ is rapid. Lemma 2 has thus been proved.

Assuming some additional set-theoretical hypotheses (for example, the Continuum Hypothesis or Martin's Axiom), it is not hard to construct a rapid filter in ω. (In this connection, see Exercise 1 of the present chapter.) At the same time, the existence of rapid filters cannot be established within the theory **ZFC**.

It turns out that all rapid filters (considered as subsets of C) are non-measurable with respect to the (complete) measure λ on C. This result was first obtained by Talagrand. See his extensive work [218] where nonmeasurable filters and filters without the Baire property in C are thoroughly investigated.

Theorem 1. *Let $\mathcal{F}$ be a rapid filter in $\mathcal{P}(\omega)$. Then $\mathcal{F}$ is not λ-measurable in C.*

Proof. Since any λ-measurable filter containing all co-finite subsets of ω must be of λ-measure zero (see Exercise 2), it suffices to demonstrate that our filter $\mathcal{F}$ is a λ-thick subset of C. To show this, take any compact set $A \subset C$ with $\lambda(A) = \alpha > 0$. Let $B \subset A$ be a compact set satisfying Lemma 1 and let $\{n_k : k < \omega\}$ be a strictly increasing sequence of natural numbers, associated with B (see the formulation of Lemma 1). Clearly, there exists a set $x \in \mathcal{F}$ such that

$$(\forall k \in \omega \setminus \{0\})(card(x \cap [0, n_{k+1}[) \leq k - 1).$$

We are going to construct (by recursion) a sequence $\{z_k : k \geq 1\}$ satisfying the conditions

$$z_k \in \{0, 1\}^{n_k},$$

$$z_k = z_{k+1} \cap [0, n_k[,$$

$$x \cap [0, n_k[\subset z_k,$$

$$C_{z_k} \cap B \neq \emptyset$$

for all natural numbers $k \geq 1$. This can be done as follows.

Define $z_1 \in \{0, 1\}^{n_1}$ so that $C_{z_1} \cap B \neq \emptyset$ would be valid.

Suppose that z_k has already been defined. Since $C_{z_k} \cap B \neq \emptyset$, we may write

$$\lambda(C_{z_k} \cap B) \geq (1 - 2^{-k})\lambda(C_{z_k}) > (1 - 2^{-k+1})\lambda(C_{z_k}).$$

Taking into account the relation

$$\lambda(\{u \in C_{z_k} \; : \; x|[0, n_{k+1}[\; \subset u|[0, n_{k+1}[\}) \geq$$

$$\lambda(C_{z_k}) \cdot 2^{-card(x \cap [n_k, n_{k+1}[)} \geq \lambda(C_{z_k}) \cdot 2^{-k+1},$$

we infer that there exists an element $u \in C_{z_k} \cap B$ such that

$$z_k \subset z_{k+1}, \quad C_{z_{k+1}} \cap B \neq \emptyset, \quad x \cap [0, n_{k+1}[\; \subset z_{k+1},$$

where $z_{k+1} = u \cap [0, n_{k+1}[$. Proceeding in this manner, we are able to construct the desired sequence $\{z_k : k \geq 1\}$.

Now, there exists an element $z \in C$ such that

$$(\forall k \in \omega \setminus \{0\})(z_k = z|[0, n_k[).$$

Then it is obvious that

$$z \in B \subset A, \quad x \subset z, \quad z \in \mathcal{F}$$

and, consequently,

$$A \cap \mathcal{F} \neq \emptyset.$$

This relation establishes the λ-thickness of $\mathcal{F}$ in C and, thus, finishes the proof of Theorem 1.

The next three auxiliary propositions were formulated and proved in [176].

Lemma 3. *Let Z be a λ-measure zero subset of C. Then there exists a closed set $B \subset C$ such that:*
1) $B \cap Z = \emptyset$;
2) $\lambda(B) > 0$;
3) for each sequence $s \in 2^{<\omega}$, we have the implication

$$B \cap C_s \neq \emptyset \Rightarrow \lambda(B \cap C_s) \geq (1/8)^{|s|+1},$$

where C_s denotes again the basic open set in C corresponding to s and

$$|s| = card(dom(s))$$

is the length of s.

Proof. Construct by recursion a sequence $\{B_k \ : \ k < \omega\}$ of closed subsets of C. Let B_0 be an arbitrary closed set in C with $\lambda(B_0) > 1/2$ and such that $Z \cap B_0 = \emptyset$. Supposing that B_k has already been defined, put

$$B_{k+1} = \cup\{B_k \cap C_s \ : \ s \in 2^{k+1}, \ \lambda(B_k \cap C_s) \geq (1/8)^{k+1}\}.$$

In this way, we get the sequence of sets $\{B_k \ : \ k < \omega\}$. Finally, denote

$$B = \cap\{B_k \ : \ k < \omega\}.$$

Let us check that B is the required set. For each $k < \omega$ and for each $s \in 2^{<\omega}$ with $|s| \leq k$, we may write

$$\lambda(C_s \cap (B_k \setminus B_{k+1})) \leq 2^{k+1-|s|}/8^{k+1}.$$

Consequently, we have

$$\lambda(B) \geq 1/2 - \sum_{k \geq 0}(1/4)^{k+1} = 1/6 > 0.$$

Further, if $s \in 2^{<\omega}$ is such that $|s| > 0$ and $C_s \cap B \neq \emptyset$, then $C_s \cap B_{|s|} \neq \emptyset$ and, therefore,

$$\lambda(C_s \cap B_{|s|-1}) \geq (1/8)^{|s|}, \quad C_s \cap B_{|s|} = C_s \cap B_{|s|-1}.$$

Taking this circumstance into account, we can write

$$\lambda(C_s \cap B) \geq \lambda(C_s \cap B_{|s|}) - \sum_{k \geq |s|} \lambda(C_s \cap (B_k \setminus B_{k+1}))$$

$$\geq (1/8)^{|s|} - \sum_{k \geq |s|} 2^{k+1-|s|}/8^{k+1},$$

from which it immediately follows that

$$\lambda(C_s \cap B) \geq (1/8)^{|s|+1}.$$

Lemma 3 has thus been proved.

Let X be a subset of C of cardinality ω_1. For any binary relation $H \subset C \times C$, let us denote

$$H(X) = \cup\{H(x) \ : \ x \in X\},$$

where

$$H(x) = \{y : (x,y) \in H\}$$

stands, as usual, for the vertical section of H corresponding to x.

We shall say that H is admissible (with respect to X) if all the sections $H(x)$ ($x \in X$) are of λ-measure zero.

Lemma 4. *Suppose that there exists an admissible binary relation*

$$H \subset C \times C$$

for which the set $H(X)$ is not of λ-measure zero. Then there exists a subset of C nonmeasurable with respect to λ.

Proof. If $H(X)$ is nonmeasurable with respect to λ, then there is nothing to prove. So we may assume that $H(X) \in dom(\lambda)$ and, consequently,

$$\lambda(H(X)) > 0.$$

Let us equip the set X with a well-ordering $\preceq_X$ isomorphic to the canonical well-ordering of ω_1 and, for any element a from $H(X)$ denote by $x = x(a)$ the $\preceq_X$-least element in X such that $a \in H(x)$. Further, define a set

$$S \subset H(X) \times H(X)$$

by putting

$$S = \{(a,b) \in H(X) \times H(X) \ : \ x(a) \preceq_X x(b)\}.$$

Considering the vertical and horizontal sections of S and utilizing the classical Fubini theorem, we claim that S cannot be $(\lambda \times \lambda)$-measurable. (Here the argument is very similar to that one used for the Sierpiński partition of $\omega_1 \times \omega_1$; in this connection, compare Chapter 4 of the book.) Since λ and $\lambda \times \lambda$ are isomorphic, we conclude that there exists a non λ-measurable set in C. This ends the proof of Lemma 4.

Let us introduce some notation which will be useful in the sequel.

Let X be a fixed uncountable subset of the Cantor discontinuum C. For any two distinct elements a and b from C, denote

$$n(a,b) = inf\{n < \omega \ : \ a(n) \neq b(n)\}.$$

Let $R \subset C \times C$ be a binary relation on C. We put

$$N_R = \{n(a,b) \ : \ a \in X, \ b \in X, \ a \neq b, \ R(a,b)\}.$$

Further, we define:

$$\mathcal{R} = \{R : R \text{ is a Borel equivalence relation on } C$$

$$\text{with countably many equivalence classes}\}.$$

Note that R is a Borel equivalence relation on C if and only if its graph is a Borel subset of the product space $C \times C$.

Finally, put

$$\mathcal{F}_X = \{A \subset \omega \ : \ (\exists R \in \mathcal{R})(N_R \subset A)\}.$$

Let us establish some properties of the family of sets $\mathcal{F}_X$.

1. For any $R_1 \in \mathcal{R}$ and $R_2 \in \mathcal{R}$, we obviously have

$$R_1 \cap R_2 \in \mathcal{R}, \quad N_{R_1 \cap R_2} \subset N_{R_1} \cap N_{R_2}.$$

Therefore, $\mathcal{F}_X$ is closed under intersections of its members.

2. For each $R \in \mathcal{R}$, we have $N_R \neq \emptyset$.

This is true since X is uncountable and R has countably many equivalence classes.

In view of properties 1 and 2, $\mathcal{F}_X$ is a filter in the Boolean algebra $\mathcal{P}(\omega)$.

3. For any natural number k and for any set $A \in \mathcal{F}_X$, we have

$$A \cap \{n \ : \ k < n < \omega\} \in \mathcal{F}_X.$$

Indeed, let $R \in \mathcal{R}$ be such that $N_R \subset A$. Consider a binary relation R^* on C defined by the formula

$$R^*(a,b) \Leftrightarrow R(a,b) \ \& \ a|[0,k] = b|[0,k].$$

Then it is evident that $R^* \in \mathcal{R}$ and

$$N_{R^*} \subset A \cap \{n \ : \ k < n < \omega\}.$$

In particular, we see that the filter $\mathcal{F}_X$ contains all co-finite subsets of ω and, consequently, is nontrivial.

The next auxiliary proposition plays the key role for obtaining the main result of this chapter.

Lemma 5. *Let X be a fixed subset of C with $card(X) = \omega_1$. Suppose that, for every admissible (with respect to X) G_δ-set $H \subset C \times C$, the set $H(X)$ is of λ-measure zero. Then the filter $\mathcal{F}_X$ is rapid.*

Proof. Using the standard recursion argument, we can easily construct a family

$$(A(s,l,j))_{s \in 2^{<\omega}, l<\omega, j<\omega}$$

of basic open sets in C, such that:

1) $\lambda(A(s,l,j)) = 2^{-(l+j)}$ for all $s \in 2^{<\omega}$, $l < \omega$ and $j < \omega$;
2) for each pair (l,j) of natural numbers, the family

$$(A(s,l,j))_{s \in 2^{<\omega}}$$

is stochastically independent (with respect to λ).

Further, for every increasing function

$$\phi : \omega \to \omega,$$

let us define a set $H_\phi \subset C \times C$ by the formula

$$(a,b) \in H_\phi \Leftrightarrow (\forall j < \omega)(\exists j' \geq j)(\exists l \geq j')(b \in A(a|[0,\phi(l)[,l,j')).$$

It can easily be checked that H_ϕ is an admissible (with respect to X) G_δ-set in $C \times C$. Moreover, all vertical sections of H_ϕ are of λ-measure zero. Therefore,

$$\lambda(H_\phi(X)) = 0$$

according to our assumption. Now, for $Z = H_\phi(X)$, there exists a closed set $B = B_\phi$ satisfying Lemma 3. We define

$$U_{a,j} = (\bigcup_{j' \geq j} \bigcup_{l \geq j'} A(a|[0,\phi(l)[,l,j')) \cap B_\phi$$

for any $a \in C$ and $j < \omega$. Clearly, the sets $U_{a,j}$ are open in B_ϕ and

$$\cap\{U_{a,j} : j < \omega\} = \emptyset$$

whenever $a \in X$. Thus, if $a \in X$, we can apply the classical Baire theorem to B_ϕ and conclude that there exist $s \in 2^{<\omega}$ and $j < \omega$ for which

$$(*) \qquad B_\phi \cap C_s \neq \emptyset, \quad B_\phi \cap C_s \cap U_{a,j} = \emptyset.$$

From now on, it will be convenient to fix an injective mapping

$$\theta \ : \ \omega \times 2^{<\omega} \to \omega$$

satisfying the relations

$$\theta(j, s) \geq j, \qquad \theta(j, s) \geq |s|$$

for all pairs (j, s) from the domain of θ. We will identify $\omega \times 2^{<\omega}$ with its θ-image. Obviously, θ induces the well-ordering on the product set $\omega \times 2^{<\omega}$.

Further, for any $a \in C$, let us put:

$F(a) = $ the smallest pair (j, s) satisfying (*) if there exist such pairs;

$F(a) = \infty$ if there is no pair (j, s) satisfying (*).

Finally, define an equivalence relation R_ϕ on C by the formula:

$$R_\phi(a, b) \Leftrightarrow$$

$$(F(a) = F(b) = \infty) \vee (F(a) = F(b) \neq \infty \ \& \ a|[0, \phi(F(a))[\ = b|[0, \phi(F(b))[).$$

It can easily be verified that R_ϕ is Borel and has countably many equivalence classes. The details are left to the reader. Let

$$N_\phi = \{n(a, b) : a \in X, b \in X, a \neq b, R_\phi(a, b)\}$$

denote the corresponding set from the filter $\mathcal{F}_X$. It can be demonstrated that

$$card(N_\phi \cap [0, \phi(k)[) \leq 2^{4k} k (3k + 3)^2$$

for all natural numbers k. (For more details, see Exercise 5 of this chapter.) Now, applying Lemma 2 to the function

$$\psi(k) = 2^{4k} k (3k + 3)^2 \qquad (k < \omega),$$

we claim that $\mathcal{F}_X$ is a rapid filter.

At last, we are able to prove the famous result of Raisonnier and Shelah (see [176], [192]).

Theorem 2. *In the theory* **ZF** *&* **DC**, *the existence of a set* $Y \subset \mathbf{R}$ *with* $card(Y) = \omega_1$ *implies the existence of a subset of* $\mathbf{R}$ *nonmeasurable in the Lebesgue sense.*

Proof. The reader can check for himself (herself) that the proofs of all lemmas from this chapter were done within **ZF** & **DC**.

Suppose now that there exists a set Y in $\mathbf{R}$ with $card(Y) = \omega_1$. Then, obviously, there is a set X in C with $card(X) = \omega_1$.

Only two cases are possible.

I. For all admissible G_δ-sets H in $C \times C$, we have

$$\lambda(H(X)) = 0.$$

In this case, we apply Lemma 5 and conclude that there exists a rapid filter on ω, from which it follows (by Theorem 1) that there are subsets of C nonmeasurable with respect to λ and, consequently, there are non Lebesgue-measurable subsets of $\mathbf{R}$.

II. There is at least one admissible G_δ-set H in $C \times C$ for which

$$\lambda^*(H(X)) > 0.$$

In this case, we apply Lemma 4 and claim again that there exists a subset of C nonmeasurable with respect to λ; consequently, there exists a subset of $\mathbf{R}$ nonmeasurable in the Lebesgue sense.

The proof of Theorem 2 is completed.

Remark 1. Let us recall once more that assuming the existence of a strongly inaccessible cardinal, Solovay showed in [210] that the theory $\mathbf{ZF}$ and the theory

$$\mathbf{ZF} \ \& \ \mathbf{DC} \ \& \ (every \ subset \ of \ \mathbf{R} \ is \ Lebesgue \ measurable)$$

are equiconsistent. In this connection, he posed the question whether the existence of a large cardinal is essential for this equiconsistency. Shelah [192] gave a positive answer to the question. Moreover, he even showed (by means of complicated metamathematical techniques) that assuming

$$\mathbf{ZF} \ \& \ \mathbf{DC} \ \& \ (every \ \Sigma_3^1 \ set \ of \ reals \ is \ Lebesgue \ measurable),$$

one can conclude that ω_1 is an inaccessible cardinal in the Gödel Constructible Universe. On the other hand, Shelah proved in the same work [192] that the theory

$$\mathbf{ZF} \ \& \ \mathbf{DC} \ \& \ (every \ subset \ of \ \mathbf{R} \ has \ the \ Baire \ property)$$

is equiconsistent with $\mathbf{ZF}$, without any assumption concerning the existence of large cardinals.

Some further extensions of these results are discussed in [176].

EXERCISES

1. Let $\{X_n : n < \omega\}$ be a countable family of subsets of ω, such that

$$(\forall n < \omega)(card(X_n) = \omega),$$

and let

$$\phi : \omega \to \omega$$

be an arbitrary mapping. Prove that there exists a strictly increasing mapping

$$\psi : \omega \to \omega$$

satisfying the following two relations:
(a) $\phi(n) \leq \psi(n)$ for all $n < \omega$;
(b) $\psi(n) \in X_n$ for all $n < \omega$.
In particular, denoting

$$X = ran(\psi) = \{\psi(n) \ : \ n < \omega\},$$

we have

$$(\forall n < \omega)(X_n \cap X \neq \emptyset).$$

Moreover, slightly modify this argument and show that there exists a strictly increasing mapping $\psi : \omega \to \omega$ satisfying (a) and the following relation:
(c) $card(ran(\psi) \cap X_n) = \omega$ for each $n < \omega$.
Start with this fact and prove, assuming the Continuum Hypothesis and applying the method of transfinite induction, that there exists a rapid filter on the set ω.

Generalize the above argument and obtain the existence of rapid filters on ω by assuming Martin's Axiom.

2. Let $\mathcal{F}$ be a filter on ω, considered as a subset of the standard Cantor space $C = \{0, 1\}^\omega$ which is equipped with the Lebesgue probability measure λ (the completion of the Haar probability measure on C). Suppose also that $\mathcal{F}$ contains all co-finite sets in ω.

Demonstrate that the following two assertions are true.

(a) If $\mathcal{F}$ is λ-measurable, then it is of λ-measure zero. (Apply the metrical transitivity of λ and its invariance with respect to the symmetry in the group C.)

(b) If $\mathcal{F}$ has the Baire property, then it is of first category. (Use an analogous method.)

3. Show that the equivalence relation R_ϕ in Lemma 5 is Borel and has countably many equivalence classes.

4. Define a function
$$g \ : \ [0, 1[\ \to \mathbf{R}$$
by putting
$$g(x) = x + ln(1 - x) \quad (0 \le x < 1).$$
Prove that
$$(\forall x \in dom(g))(g(x) \le 0).$$
In other words, prove that the inequality
$$x \le -ln(1 - x)$$
is valid for any $x \in [0, 1[$.

5. We preserve the notation of the proof of Lemma 5. Demonstrate that
$$card(N_\phi \cap [0, \phi(k)[) \le 2^{4k} k (3k + 3)^2$$
for all $k < \omega$. In order to do this, first take into account that
$$N_\phi \cap [0, \phi(k)[\ = \{n(a, b) < \phi(k) : a \in X, b \in X, a \ne b, R_\phi(a, b)\}.$$

If $a \in X$, $b \in X$, $a \ne b$ and $R_\phi(a, b)$, then there exist some $j < \omega$ and $s \in 2^{<\omega}$ for which
$$F(a) = F(b) = \theta(j, s),$$
$$a|[0, \phi(\theta(j, s))[\ = b|[0, \phi(\theta(j, s))[.$$
Therefore, by definition of $n(a, b)$, we must have
$$n(a, b) \ge \phi(\theta(j, s)).$$

Since ϕ is increasing, we also have
$$n(a, b) < \phi(k) \Rightarrow \theta(j, s) < k.$$

In addition to this, the inequalities

$$k > \theta(j,s) \geq j$$

imply $j < k$. In view of $(*)$, we can write

$$B_\phi \cap C_s \neq \emptyset,$$

$$B_\phi \cap C_s \cap A(a|[0,\phi(k)[,k,j) = \emptyset,$$

$$B_\phi \cap C_s \cap A(b|[0,\phi(k)[,k,j) = \emptyset.$$

Let us denote

$$\Delta(k,j,s) = \{t \in 2^{\phi(k)} : B_\phi \cap C_s \cap A(t,k,j) = \emptyset\},$$

$$\delta(k,j,s) = card(\Delta(k,j,s)).$$

Show that

$$card(N_\phi \cap [0,\phi(k)[) \leq \sum_{\theta(j,s)<k, B_\phi \cap C_s \neq \emptyset} (\delta(k,j,s))^2.$$

Further, verify the inclusion

$$B_\phi \cap C_s \subset \bigcap_{t \in \Delta(k,j,s)} (2^\omega \setminus A(t,k,j))$$

and, utilizing the stochastical independence of the sets $A(t,k,j)$ with respect to λ, infer

$$\lambda(B_\phi \cap C_s) \leq (1 - 2^{-(k+j)})^{\delta(k,j,s)}.$$

Since $B_\phi \cap C_s \neq \emptyset$, claim by Lemma 3 that

$$\lambda(B_\phi \cap C_s) \geq (1/8)^{|s|+1}$$

and, applying Exercise 4, deduce from the above-mentioned inequalities that

$$\delta(k,j,s) \leq d \cdot 2^{k+j}(3|s| + 3),$$

where $d = ln(2) < 1$. Consequently,

$$card(N_\phi \cap [0,\phi(k)[) \leq \sum_{\theta(j,s)<k} (2^{k+j}(3|s| + 3))^2.$$

Finally, take into account the implication

$$\theta(j, s) < k \Rightarrow (j < k \ \& \ |s| < k)$$

with the injectivity of θ and establish the required inequality

$$card(N_\phi \cap [0, \phi(k)[) \leq 2^{4k} k(3k + 3)^2.$$

These nontrivial purely technical details enable the proof of Lemma 5 to be completed.

6. Demonstrate, in the theory **ZF** & **DC**, that if there exists a finitely additive diffused probability measure ν with $dom(\nu) = \mathcal{P}(\omega)$, then there exists a subset of **R** without the Baire property.

Appendix 1
Logical aspects of the existence of nonmeasurable sets

The main part of this book has been devoted to various constructions of nonmeasurable sets and their role in certain questions of analysis, measure theory and point set theory. We also have touched upon several purely logical problems connected with the existence of such sets. (See especially Chapters 1, 7 and 15.)

Here we want to discuss some analogous logical and set-theoretical questions concerning nonmeasurable sets or sets without the Baire property.

Briefly, the aim of this Appendix is to consider some statements of set theory, which are usually applied in constructions of non Lebesgue-measurable sets on the real line.

First, let us recall the standard formulation of the Axiom of Choice (**AC**):

If $(X_i)_{i \in I}$ is an arbitrary family of nonempty sets, then there exists a family $(x_i)_{i \in I}$ such that $x_i \in X_i$ for all $i \in I$.

In other words, every family of nonempty sets admits at least one selector.

The necessity of **AC** for various fields of mathematics was observed by several authors at the end of the nineteenth century. In particular, Cantor repeatedly utilized this axiom in his original set-theoretical investigations. Moreover, the importance of **AC** was recognized by those mathematicians of the nineteenth century whose research interests were concerned with basic notions of classical mathematical analysis and the theory of differential equations. For example, it was discovered that **AC** plays a linking role in two different approaches to the concept of the continuity of a partial function $f : \mathbf{R} \to \mathbf{R}$ at a point $t \in dom(f)$. The first approach is based on the widely known $(\varepsilon - \delta)$-definition and the second one appeals to convergent sequences of points. Assuming **AC** (or some weaker form of it), these two approaches become equivalent.

Later, in the beginning of the twentieth century, it was demonstrated by Zermelo that **AC** is equivalent to the theorem stating that every set can be well ordered. (See his works [237] and [238].)

Also, it turned out that **AC** is equivalent to the Tychonoff theorem on products of quasicompact topological spaces. This result was obtained by Kelley [76]. In order to formulate it, let us recall two standard definitions from general topology.

Let $(E, \mathcal{T})$ be a topological space.

E is said to be a T_1-space if all singletons in E are closed.

E is said to be quasicompact if every open covering of E contains a finite subcovering.

In other words, E is quasicompact if every centered family of closed subsets of E has nonempty intersection.

E is said to be compact if E is a T_2-space (Hausdorff) and quasicompact.

In the theory **ZF**, the following three assertions are equivalent:

(1) the Axiom of Choice;

(2) the topological product of any family $((E_i, \mathcal{T}_i))_{i \in I}$, where $card(\mathcal{T}_i) \leq 3$ for all $i \in I$, is quasicompact;

(3) the topological product of any family of quasicompact T_1-spaces is qusicompact;

(4) the topological product of any family of quasicompact spaces is quasicompact.

Note that the implication (1) $\Rightarrow$ (4) is precisely the Tychonoff theorem (see, for instance, [37], [77] or [127]).

The implications (4) $\Rightarrow$ (2) and (4) $\Rightarrow$ (3) are trivial.

Let us prove the implication (3) $\Rightarrow$ (1). Indeed, suppose that (3) is valid and take an arbitrary family $(X_i)_{i \in I}$ of nonempty sets. Pick an element x such that $x \notin \cup\{X_i \ : \ i \in I\}$. For each index $i \in I$, define:

$X_i' = \{x\} \cup X_i$;

$\mathcal{T}_i = \{\emptyset\} \cup \{Z \subset X_i : card(X_i \setminus Z) < \omega\}$;

$\mathcal{T}_i'$ = the topological sum of $\mathcal{T}_i$ and $\mathcal{T}_x$, where $\mathcal{T}_x$ is the unique topology on the one-element set $\{x\}$.

In this way, each pair $(X_i', \mathcal{T}_i')$ becomes a quasicompact T_1-space. Consequently, the product space

$$X' = \prod_{i \in I} X_i'$$

is quasicompact. Consider the family $\mathcal{F}'$ of all those sets in X' which are representable in the form $\prod_{i \in I} Y_i$, where

$$(\forall i \in I)(Y_i = X_i' \ \lor \ Y_i = X_i),$$

$$card(\{i \in I \ : \ Y_i \neq X_i'\}) < \omega.$$

Clearly, $\mathcal{F}'$ is a centered family of closed subsets of the space X'. Therefore, there exists a point $(x_i)_{i \in I}$ belonging to all members of $\mathcal{F}'$. It can easily be seen that $x_i \in X_i$ for each $i \in I$. This establishes the implication $(3) \Rightarrow (1)$.

By applying an analogous argument, the implication $(2) \Rightarrow (1)$ can be proved. Indeed, suppose that (2) is valid and, for each index $i \in I$, define the topology $\mathcal{T}_i'$ on X_i' by putting

$$\mathcal{T}_i' = \{X_i', \{x\}, \emptyset\}.$$

Then, obviously, $card(\mathcal{T}_i') = 3$ for all $i \in I$, the product space X' is again quasicompact and the family $\mathcal{F}'$ of closed subsets of X' is again centered. Thus, there exists a point $(x_i)_{i \in I}$ belonging to all members of $\mathcal{F}'$ and, evidently, this point is the required selector for $(X_i)_{i \in I}$.

At the same time, the Tychonoff theorem for compact topological spaces does not imply the Axiom of Choice. (In this connection, see [74].)

Numerous applications of **AC** are known in the mathematical literature (see especially [13], [64], [74], [127] and [194]). In this book, we have demonstrated many times that **AC** plays the key role in different constructions of subsets of **R** having very bad descriptive properties.

Now, let us consider some weak forms of **AC**, which are not able to guarantee the existence of non Lebesgue-measurable subsets of **R**. (It should be mentioned, in this context, that various forms of **AC** are thoroughly discussed in [184] and [194].)

One of the weakest versions of the Axiom of Choice can be formulated as follows:

If $(X_n)_{n<\omega}$ is an arbitrary countable family of nonempty finite sets, then there exists a family $(x_n)_{n<\omega}$ such that $x_n \in X_n$ for all $n < \omega$.

It is not hard to show that in the theory **ZF**, this statement is equivalent to the well-known König lemma on ω-trees with finite levels (compare Exercise 5 from Chapter 7).

The countable form of **AC** is the following set-theoretical assertion:

If $(X_n)_{n<\omega}$ is an arbitrary countable family of nonempty sets, then there exists a family $(x_n)_{n<\omega}$ such that $x_n \in X_n$ for each $n < \omega$.

The Axiom of Dependent Choice (in short, **DC**) is the following set-theoretical statement:

If $S(x,y)$ is a binary relation on a nonempty set X, satisfying

$$(\forall x \in X)(\exists y \in X)S(x,y),$$

then there exists a sequence $(x_n)_{n<\omega}$ of elements of X such that

$$(\forall n < \omega)S(x_n, x_{n+1}).$$

In some sense, **DC** is a safe form of **AC** completely sufficient for most domains of classical mathematics: elementary point set theory, geometry of euclidean spaces, mathematical analysis on the real line **R**, and so on. On the other hand, it is worth noticing that many branches of contemporary mathematics need various uncountable forms of **AC** as a necessary tool in proving unexpected and strong results. For example, it is reasonable to recall the famous Banach-Tarski paradox or the Hahn-Banach theorem on extensions of continuous linear functionals.

It can easily be demonstrated, in the theory **ZF**, that:

(a) **DC** implies the countable form of **AC**;

(b) for a function f acting from a metric space into another one and for a point $t \in dom(f)$, the countable form of **AC** is sufficient to prove the equivalence of two classical definitions of the continuity of f at t (we mean here the $(\varepsilon - \delta)$-definition due to Cauchy and the Heine definition in terms of convergent sequences of points);

(c) the countable form of **AC** implies that the union of a countable family of countable sets is also countable;

(d) the countable form of **AC** implies that every infinite set contains an infinite countable subset.

Note also that traditionally the statement **DC** is more preferred by specialists than the countable form of the Axiom of Choice. For example, all usual properties of the classical Lebesgue measure can be established within the theory **ZF** & **DC** but, as we know, **DC** is not strong enough to prove the existence of subsets of **R** nonmeasurable in the Lebesgue sense.

Another formulation of **DC** was found by Blair [12]. Namely, he discovered that **DC** is closely connected with the classical Baire theorem on category of complete metric spaces. More precisely, in the theory **ZF**, the Axiom of Dependent Choice is equivalent to the following statement (Baire theorem):

Every nonempty complete metric space is of second category.

Indeed, the standard proofs of the Baire theorem belong to **ZF** & **DC**. Suppose now that the statement formulated above is valid in **ZF** and consider an arbitrary nonempty set X with a binary relation $S(x,y)$ on it such that

$$(\forall x \in X)(\exists y \in X)S(x,y).$$

Equip X with the discrete metric and introduce the product space

$$E = X^{\omega} = X \times X \times ... \times X \times ... \, .$$

A straightforward verification (within **ZF**) shows that E is a complete metric space. The elements of E will be denoted by $e = (e_n)_{n<\omega}$. Further, for any two natural numbers n and m such that $n < m$, define

$$U_{n,m} = \{e \in E \; : \; S(e_n, e_m)\}$$

and, for each $n < \omega$, put

$$V_n = \cup\{U_{n,m} : n < m < \omega\}.$$

Then all sets V_n are open and everywhere dense in E. Consequently, according to the Baire theorem, we have

$$\cap\{V_n \; : \; n < \omega\} \neq \emptyset.$$

Let $v = (v_n)_{n<\omega}$ be an arbitrary element from $\cap\{V_n \; : \; n < \omega\}$. Then we can easily construct (by recursion) a strictly increasing sequence

$$\{n(k) \; : \; k < \omega\} \subset \omega$$

such that $S(v_{n(k)}, v_{n(k+1)})$ for all $k < \omega$. Thus, denoting $x_k = v_{n(k)}$, we get

$$(\forall k < \omega)S(x_k, x_{k+1}).$$

In this way, we have shown the equivalence of **DC** and the Baire theorem.

It has already been mentioned that the weak forms of the Axiom of Choice considered above are not sufficient for the existence of pathological subsets of the real line $\mathbf{R}$. Freaks and monsters appear when various uncountable forms of the Axiom of Choice are utilized. In particular, the proof of the classical Vitali theorem (see Chapter 1) shows directly that the following version of **AC** implies the existence of subsets of $\mathbf{R}$ nonmeasurable in the Lebesgue sense and not possessing the Baire property:

If $(X_i)_{i \in I}$ is a family of nonempty sets, such that there exists a surjection from $\mathbf{R}$ onto I, then there exists a family $(x_i)_{i \in I}$ such that $x_i \in X_i$ for all $i \in I$.

Indeed, if $\{V_i : i \in I\}$ denotes the Vitali partition of $\mathbf{R}$, then it is obvious that there exists a surjection from $\mathbf{R}$ onto I. Therefore, we may apply the statement just formulated to $\{V_i : i \in I\}$ in order to obtain a Vitali set on $\mathbf{R}$.

In connection with the last version of **AC**, let us also underline that it readily implies the inequality $\omega_1 \leq \mathbf{c}$. Indeed, there are various partitions $\{X_\xi : \xi < \omega_1\}$ of $\mathbf{R}$ constructed within the theory **ZF**. The best known example of such a partition is due to Lebesgue (see, for instance, [127] or [194]). Thus, we claim that there are surjections from $\mathbf{R}$ onto ω_1. Now, applying the statement formulated above, we come to a subset of $\mathbf{R}$ whose cardinality is ω_1.

In Chapter 15, the remarkable result of Shelah and Raisonnier was discussed which states that the inequality $\omega_1 \leq \mathbf{c}$ is already sufficient for the existence of subsets of $\mathbf{R}$ nonmeasurable in the Lebesgue sense.

In Chapter 7, it was demonstrated that the Tychonoff theorem for compact topological spaces implies the existence of functions which act from $\mathbf{R}$ into $\{0, 1\}$ and are not measurable in the Lebesgue sense.

Let us also recall an old result of Kolmogorov [114] stating that the existence (in the theory **ZF** & **DC**) of a universal integral for the class of all Lebesgue measurable functions on $[0, 1]$ implies the existence of a non Lebesgue-measurable set on $\mathbf{R}$. (In this connection, see also Chapter 13 from [102].)

Another interesting result, due to Pincus, states in the same theory **ZF** & **DC** that if there exists a finitely additive probability measure which is defined on a σ-algebra of sets but is not countably additive, then there exists a subset of $\mathbf{R}$ without the Baire property. (See, for instance, [226]; compare also Exercise 6 from Chapter 15.)

It was shown in the present book that some additional set-theoretical axioms enable us to establish the existence of sets which are extremely bad from the point of view of measurability. For example, the Continuum Hypothesis implies the existence of a Sierpiński set in $\mathbf{R}$ which is uncountable and has the property that all its uncountable subsets are nonmeasurable in the Lebesgue sense. See Chapter 5 where many other interesting properties of Sierpiński sets are observed. Moreover, under the same **CH**, there exists a function $f : \mathbf{R} \to \mathbf{R}$ which is nonmeasurable with respect to every nonzero σ-finite diffused measure on $\mathbf{R}$.

The corresponding examples (generalized Sierpiński sets, absolutely nonmeasurable functions) can also be constructed under Martin's Axiom. (See again Chapter 5.) We would like to finish this Appendix with a brief discussion of Martin's Axiom and its useful consequences in classical Lebesgue measure theory. For a more detailed account, see [47], [64] or [122].

Let us recall some definitions and concepts from the theory of partially ordered sets.

Let $(P, \preceq)$ be an arbitrary partially ordered set and let D be a subset of P. We shall say that D is coinitial in P if for every element $p \in P$, there exists an element $q \in D$ satisfying the relation $q \preceq p$. In other words, $D \subset P$ is coinitial in P if it is cofinal in the partially ordered set $(P, \succeq)$.

A nonempty set $G \subset P$ is called a filter in $(P, \preceq)$ if:

$$(\forall p \in G)(\forall q \in P)(p \preceq q \Rightarrow q \in G),$$

$$(\forall p \in G)(\forall q \in G)(\exists r \in G)(r \preceq p \,\&\, r \preceq q).$$

Note that this definition resembles the definition of a filter in a Boolean algebra (see [127]).

Two elements p and q of P are called inconsistent (incompatible) if there is no $r \in P$ such that $r \preceq p$ and $r \preceq q$.

We say that a set $A \subset P$ is totally inconsistent (totally incompatible) if any two distinct elements of A are inconsistent (incompatible).

Finally, we say that $(P, \preceq)$ satisfies the countable chain condition if each totally inconsistent subset of P is at most countable. Sometimes (especially, in topological applications), the countable chain condition is called the Suslin condition.

Martin's Axiom, denoted usually by **MA**, is the following statement:

If $(P, \preceq)$ is a partially ordered set satisfying the countable chain condition and $\mathcal{D}$ is a family of coinitial subsets of P with $\mathrm{card}(\mathcal{D}) < 2^{\omega}$, then there *exists a filter $G \subset P$ which intersects every element of $\mathcal{D}$; in other words,*

$$(\forall D \in \mathcal{D})(D \cap G \neq \emptyset).$$

The next statement (similar to the Baire theorem on category) is a topological equivalent of **MA** (for more details, see [122]):

If E is an arbitrary nonempty compact topological space satisfying the Suslin condition, then E cannot be covered by a family of nowhere dense subsets, whose cardinality is strictly less than 2^ω.

The Continuum Hypothesis implies Martin's Axiom. Indeed, let $(P, \preceq)$ be any partially ordered set and let $(D_n)_{n \in \omega}$ be any sequence of coinitial subsets of P. Then we can easily construct (by recursion) a decreasing sequence $(p_n)_{n \in \omega}$ of elements of P, such that $p_n \in D_n$ for each $n \in \omega$. Now, we put

$$G = \{p \in P : (\exists n \in \omega)(p_n \preceq p)\}.$$

Evidently, G is a filter in P which intersects every D_n $(n \in \omega)$. In particular, we see that **MA** follows from **CH**.

Martin and Solovay proved that the statement **MA** & $\neg$**CH** is consistent with the theory **ZFC**. The size of $\mathbf{c} = 2^\omega$ is not precisely determined by **MA**. For example, it was proved that each of the statements

$$\mathbf{MA} \ \& \ (2^\omega = \omega_2), \qquad \mathbf{MA} \ \& \ (2^\omega = \omega_3)$$

is consistent with **ZFC**, as well as many other analogous statements. (For more detailed information, see [122].)

In the formulation of Martin's Axiom the restriction to a family $\mathcal{D}$ of coinitial subsets with $card(\mathcal{D}) < 2^\omega$ is not accidental. To see this, let us consider the complete binary tree $P = \{0,1\}^{<\omega}$ ordered by the reverse inclusion. Let $\mathcal{D}$ denote the family consisting of all sets of the form

$$A_n = \{p \in P : n \in dom(p)\}, \qquad D_f = \{p \in P : \neg(p \subset f)\}$$

for any $n \in \omega$ and for any $f \in 2^\omega$. Observe that $card(\mathcal{D}) = 2^\omega$ and each set from $\mathcal{D}$ is coinitial in P. Suppose that G is a filter in P which intersects every set $D \in \mathcal{D}$. Then we define

$$g = \cup\{p : p \in G\}$$

and easily verify that g is a function. Since $G \cap A_n \neq \emptyset$ for each $n \in \omega$, we see that $dom(g) = \omega$. Thus, we have

$$g : \omega \to \{0, 1\}.$$

But we also have $G \cap D_f \neq \emptyset$ for any $f \in 2^\omega$. Finally, we get $g \neq f$ for every $f \in 2^\omega$, and this is a contradiction.

The restriction to a partial ordering satisfying the countable chain condition is also matured in the formulation of Martin's Axiom. In fact, the following two sentences are equivalent (in **ZFC**):

(i) the Continuum Hypothesis;

(ii) for any partially ordered set $(P, \preceq)$ and for every family $\mathcal{D}$ of coinitial subsets of P with $card(\mathcal{D}) < \mathbf{c}$, there exists a filter $G \subset P$ which intersects each set from $\mathcal{D}$.

The proof of the equivalence of (i) and (ii) is left to the reader.

We thus conclude that if one wishes to have an additional set-theoretical statement srongly weaker than the Continuum Hypothesis, then the countable chain condition is essential in the formulation of Martin's Axiom.

For classical theory of Lebesgue measure, Martin's Axiom yields effects very similar to those ones which can be obtained by using the Continuum Hypothesis (compare Chapter 5 of this book). Let us prove, assuming **MA**, two theorems which establish the $\mathbf{c}$-additivity of the standard σ-ideals on **R**. (We mean the σ-ideal of all first category sets and the σ-ideal of all Lebesgue measure zero sets.)

Let κ be an infinite cardinal. We recall that a family of sets $\mathcal{T}$ is κ-additive if, for any subfamily $\mathcal{T}'$ of $\mathcal{T}$ with $card(\mathcal{T}') < \kappa$, we have $\cup \mathcal{T}' \in \mathcal{T}$.

Theorem 1. *If Martin's Axiom holds, then the σ-ideal $\mathcal{K}(\mathbf{R})$ of all first category subsets of $\mathbf{R}$ is $\mathbf{c}$-additive.*

Proof. Assume that Martin's Axiom is valid. Take any infinite cardinal $\kappa < \mathbf{c}$ and a family $(N_\alpha)_{\alpha < \kappa}$ of nowhere dense subsets of $\mathbf{R}$. It suffices to show that

$$\bigcup_{\alpha < \kappa} N_\alpha \in \mathcal{K}(\mathbf{R}).$$

Let $\mathcal{J}_r$ denote the family of all finite sequences of nonempty open intervals in $\mathbf{R}$ with rational end–points. Let

$$P = \{(f, U) \ : \ f \in \mathcal{J}_r \ \& \ U \text{ is a dense open subset of } \mathbf{R}\}.$$

We define a partial ordering $\preceq$ on the set P. Namely, we put

$$(f, U) \preceq (g, V)$$

if and only if the relation

$$(U \subset V) \ \& \ (f \supset g) \ \& \ (\forall i \in dom(f) \setminus dom(g))(f(i) \subset V)$$

is valid.

Now, we must check that the partially ordered set $(P, \preceq)$ satisfies the countable chain condition. For this purpose, take an uncountable family $((f_\alpha, V_\alpha))_{\alpha < \omega_1}$ of elements from P. Since the range of the family $(f_\alpha)_{\alpha < \omega_1}$ is countable, there are two distinct ordinals $\alpha < \omega_1$ and $\beta < \omega_1$ such that $f_\alpha = f_\beta$. Let us put

$$V = V_\alpha \cap V_\beta, \quad f = f_\alpha = f_\beta.$$

Then V is a dense open subset of $\mathbf{R}$ and

$$(f, V) \preceq (f_\alpha, V_\alpha), \quad (f, V) \preceq (f_\beta, V_\beta).$$

The last two relations show us that $(P, \preceq)$ satisfies the countable chain condition. Thus, we may apply Martin's Axiom to this partially ordered set.

For every $\alpha < \kappa$, every $n \in \mathbf{N}$ and any two numbers $p \in \mathbf{Q}$ and $q \in \mathbf{Q}$ such that $p < q$, we denote

$$D_\alpha = \{(f, U) \in P \ : \ N_\alpha \cap U = \emptyset\},$$

$$E_{p,q}^n = \{(f, U) \in P \ : \ (\exists m > n)(m \in dom(f) \ \& \ f(m) \cap \ ]p, q[\ \neq \emptyset)\}.$$

It can easily be checked that:

a) for each $\alpha < \kappa$, the set D_α is coinitial in $(P, \preceq)$;

b) for each $n \in \mathbf{N}$ and for all p and q from $\mathbf{Q}$ such that $p < q$, the set $E_{p,q}^n$ is coinitial in $(P, \preceq)$.

Let us define

$$\mathcal{S} = \{D_\alpha \ : \ \alpha < \kappa\} \cup \{E_{p,q}^n \ : \ n \in \mathbf{N}, \ p \in \mathbf{Q}, \ q \in \mathbf{Q}, \ p < q\}.$$

Clearly, we have

$$card(\mathcal{S}) \leq \kappa + \omega = \kappa < \mathbf{c}.$$

Consequently, there exists a filter F in $(P, \preceq)$ which intersects all sets from the family $\mathcal{S}$. We now put

$$h = \cup\{f \ : \ (\exists U)((f, U) \in F)\}.$$

Since F is a filter, h is a function. Moreover, since the relation $F \cap E^n_{p,q} \neq \emptyset$ holds for each natural number n and for any two rational numbers p and q such that $p < q$, we see that

$$dom(h) = \mathbf{N} = \omega.$$

Further, for every natural number n, we define

$$U_n = \bigcup_{n<m<\omega} h(m). \qquad H = \bigcap_{n<\omega} U_n.$$

It is obvious that all sets U_n $(n \in \mathbf{N})$ are open in $\mathbf{R}$.

If $n \in \mathbf{N}$, p and q are rational numbers and $p < q$, then $F \cap E^n_{p,q} \neq \emptyset$. Consequently, there exists a natural number $m > n$ such that

$$h(m) \ \cap \ ]p, q[\ \neq \emptyset.$$

Thus, for every $n \in \mathbf{N}$, the set U_n is dense and open in $\mathbf{R}$, and we conclude that H is a dense G_δ-subset of $\mathbf{R}$.

Notice, at the end, that if $\alpha < \kappa$, then $F \cap D_\alpha \neq \emptyset$, so there exists an element (f, U) of F for which $N_\alpha \cap U = \emptyset$. But, since F is a filter, it is not difficult to check that $H \subset U$, so $N_\alpha \cap H = \emptyset$, too. Therefore, we can write

$$\bigcup_{\alpha<\kappa} N_\alpha \subset \mathbf{R} \setminus H.$$

In particular, $\cup\{N_\alpha : \alpha < \kappa\} \in \mathcal{K}(\mathbf{R})$. Theorem 1 has thus been proved.

Remark 1. Evidently, the same method works for proving a more general theorem which states that under Martin's Axiom, the σ-ideal of all first category subsets of a topological space with a countable base is **c**-additive.

In order to formulate and prove the next statement, let us recall that the symbol λ denotes the standard Lebesgue measure on the real line $\mathbf{R}$.

Theorem 2. *If Martin's Axiom holds, then the σ-ideal $\mathcal{I}(\lambda)$ of all λ-measure zero sets is* **c**-*additive.*

Proof. Take again an arbitrary infinite cardinal $\kappa < \mathbf{c}$ and a family $(L_\alpha)_{\alpha < \kappa}$ of λ-measure zero subsets of $\mathbf{R}$. Fix a real $\varepsilon > 0$. It suffices to show that there exists an open set $U \subset \mathbf{R}$ such that

$$\lambda(U) \leq \varepsilon, \qquad \bigcup_{\alpha < \kappa} L_\alpha \subset U.$$

Let us put

$$P = \{V \subset \mathbf{R} \ : \ V \ is \ open \ in \ \mathbf{R} \ and \ \lambda(V) < \varepsilon\}$$

and consider a partial ordering $\preceq$ on P defined by the formula

$$U \preceq V \Leftrightarrow U \supset V.$$

First, we need to establish that $(P, \preceq)$ satisfies the countable chain condition. To demonstrate this, suppose that $(V_\alpha)_{\alpha < \omega_1}$ is an uncountable subfamily of P. Then there exist a strictly positive number $\varepsilon_1 < \varepsilon$ and an uncountable subset A of ω_1 such that, for each $\alpha \in A$, we have $\lambda(V_\alpha) < \varepsilon_1$. Now, for every $\alpha \in A$, let J_α be a finite union of intervals in $\mathbf{R}$ with rational end–points, such that

$$J_\alpha \subset V_\alpha, \qquad \lambda(V_\alpha \setminus J_\alpha) < \varepsilon - \varepsilon_1.$$

Then there are two distinct α and β in A such that $J_\alpha = J_\beta$. Obviously, we have

$$V_\alpha \cup V_\beta = V_\alpha \cup (V_\beta \setminus J_\beta).$$

Therefore, we may write

$$\lambda(V_\alpha \cup V_\beta) \leq \lambda(V_\alpha) + \lambda(V_\beta \setminus J_\beta) < \varepsilon,$$

$$V_\alpha \cup V_\beta \in P, \quad V_\alpha \cup V_\beta \preceq V_\alpha, \quad V_\alpha \cup V_\beta \preceq V_\beta.$$

Thus, $(P, \preceq)$ satisfies the countable chain condition and we may apply Martin's Axiom to this partially ordered set.

Note that for every $\alpha < \kappa$, the set

$$G_\alpha = \{V \in P \ : \ L_\alpha \subset V\}$$

is a coinitial subset of P and

$$card(\{G_\alpha : \alpha < \kappa\}) \leq \kappa < \mathbf{c}.$$

Consequently, there exists a filter G in P such that

$$(\forall \alpha < \kappa)(G_\alpha \cap G \neq \emptyset).$$

Let us consider the open set $U = \cup G$. It is easy to see that

$$(\forall \alpha < \kappa)(L_\alpha \subset U).$$

Notice now that each element of G is an open set in $\mathbf{R}$, and we can find a countable family $(U_n)_{n \in \omega}$ of elements from G, such that

$$U = \bigcup_{n \in \omega} U_n.$$

Moreover, G is a directed family of sets (with respect to the inclusion relation). Therefore,

$$\lambda(U_1 \cup U_2 \cup ... \cup U_n) < \varepsilon$$

for any natural number n. This fact immediately implies that

$$\lambda(U) = \lambda(\bigcup_{n \in \omega} U_n) \leq \varepsilon.$$

Remembering that ε is an arbitrary strictly positive number, we conclude that the set $\cup\{L_\alpha : \alpha < \kappa\}$ is of λ-measure zero. Theorem 2 has thus been proved.

Remark 2. Using the same method, we can establish a more general result. Namely, if E is a topological space with a countable base and μ is a σ-finite Borel measure on E, then, under Martin's Axiom, the σ-ideal $\mathcal{I}(\mu')$ is **c**-additive, where μ' denotes the completion of μ.

Remark 3. Assuming Martin's Axiom, we easily deduce from Theorems 1 and 2 that:

1) the σ-algebra of all subsets of $\mathbf{R}$ having the Baire property is **c**-additive;

2) the σ-algebra of all Lebesgue measurable subsets of $\mathbf{R}$ is **c**-additive.

This remark leads to a nontrivial application of Martin's Axiom in descriptive set theory. Let us consider the projective class $\Sigma_2^1(\mathbf{R})$ of subsets of the real line $\mathbf{R}$. We recall that a set X belongs to this class if and only if there exists an analytic set $Y \subset \mathbf{R}$ such that X is the image of $\mathbf{R} \setminus Y$ under some continuous mapping. According to the classical construction of Luzin (see [125] or [135]), every analytic (co-analytic) set can be represented as the union of an ω_1-sequence of Borel sets. Therefore, we claim that any set from the class $\Sigma_2^1(\mathbf{R})$ can be expressed in the same form. Now, if Martin's Axiom with the negation of the Continuum Hypothesis are valid, then we obtain that each set from $\Sigma_2^1(\mathbf{R})$ is measurable in the Lebesgue sense (and possesses the Baire property). This fact is of interest since it is well known that in the Constructible Universe of Gödel there are sets from $\Sigma_2^1(\mathbf{R})$ non-measurable with respect to λ (see [64], [65]).

Remark 4. In Chapter 10 the question of measurability of selectors associated with various uncountable subgroups of $\mathbf{R}$ was considered. We have demonstrated there that Martin's Axiom is an efficient tool for solving this question.

In particular, one can assert that if Martin's Axiom holds and Γ is any uncountable subgroup of $\mathbf{R}$ with $card(\Gamma) < \mathbf{c}$, then all Γ-selectors are nonmeasurable in the Lebesgue sense and do not possess the Baire property.

Indeed, choose any Γ-selector X and suppose for a while that X is in $dom(\lambda)$. Then, on the one hand, we must have $\lambda(X) = 0$ since there are uncountably many pairwise disjoint Γ-translates of X. On the other hand, taking into account the equality

$$\mathbf{R} = \cup\{g + X \ : \ g \in \Gamma\}$$

and Theorem 2 proved above, we deduce that $\lambda(\mathbf{R}) = 0$ which is impossible. This contradiction shows that X is not λ-measurable. A dual argument (based on Theorem 1) works for the Baire property.

Let us also recall that, under Martin's Axiom, there exists a group $\Gamma \subset \mathbf{R}$ with $card(\Gamma) = \mathbf{c}$ for which all Γ-selectors are nonmeasurable in the Lebesgue sense (see Theorem 1 of Chapter 10).

Many other interesting consequences of Martin's Axiom in various fields of mathematics are discussed in the monograph by Fremlin [47].

Appendix 2
Some facts from the theory of commutative groups

Here we wish to recall some material from the general theory of infinite commutative groups. This theory is thoroughly considered, for example, in well-known monographs [53], [54] and [129].

We shall present only those statements about commutative groups which were essentially utilized in preceding chapters of this book. For instance, the reader could see throughout the text of Chapters 11, 13, 14 how specific features of a given transformation group G imply the corresponding properties of measures which are invariant (or quasi-invariant) with respect to G.

Here we do not touch upon deep relationships between commutative groups and modules. Obviously, any commutative group may be regarded as a module over the ring $\mathbf{Z}$ of all integers. Numerous facts of the theory of commutative groups can be treated from a more general viewpoint, as facts of the theory of modules. Such an approach is fruitful for both of these theories and very often leads to important results.

Concerning finite commutative groups, we may state that there is no problem connected with their description. Indeed, the classical theorem says that any finitely generated (in particular, finite) commutative group G is representable as a direct sum of cyclic groups. The proof can be carried out by induction on $card(X)$, where X is a finite set of generators of G. (See [53] or [129].)

For infinite commutative groups, we do not have such a nice result and the corresponding theory becomes more complicated. Moreover, there are many infinite commutative groups which do not admit a nontrivial representation in the form of a direct sum. (Some simple examples will be given below; it is known that there exist commutative groups with the above-mentioned property, whose cardinalities are arbitrarily large.)

Let G be a commutative group. We recall that G is free if it contains at least one set X satisfying the following condition: for any commutative group H and for any mapping

$$\phi : X \to H,$$

there exists a unique homomorphism $\phi' : G \to H$ extending ϕ.

Actually, this is a particular case of the definition of a free object for a given algebraic structure. A set X in the definition above is usually called a basis of a given group G.

A finite commutative group G is free if and only if G is trivial (in other words, $G = \{0\}$).

The group $(\mathbf{Z}, +)$ of all integers is a natural example of an infinite free commutative group. The one-element set $\{1\}$ is a basis of this group. It can easily be verified that $\mathbf{Z}$ is not decomposable; that is, $\mathbf{Z}$ does not admit a representation in the form of a direct sum of its two proper subgroups.

Evidently, the direct sum of free commutative groups is also a free commutative group. More precisely, if $\{G_i : i \in I\}$ is a family of free commutative groups and X_i is a basis of G_i for each $i \in I$, then the set

$$X = \cup\{X_i : i \in I\}$$

is a basis of the free commutative group

$$G = \sum_{i \in I} G_i.$$

Of course, we identify here each set X_j $(j \in I)$ with its image under the canonical monomorphism from G_j into $\sum_{i \in I} G_i$.

We say that a commutative group G is projective if for any two commutative groups H and H' satisfying the relation $H \subset H'$ and for an arbitrary homomorphism

$$\phi : G \to H'/H,$$

there exists a homomorphism $\phi' : G \to H'$ such that

$$\phi = \psi \circ \phi',$$

where $\psi : H' \to H'/H$ is the canonical epimorphism.

The following statement yields a characterization of free commutative groups.

Theorem 1. *Let G be a commutative group. Then these three assertions are equivalent:*

1) G is a free group;

2) G can be represented as a direct sum of groups isomorphic to $\mathbf{Z}$;

3) G is a projective group.

For the proof of this statement, see [53] and [129].

Thus, if G is a free commutative group, then we may write

$$G = \sum_{i \in I} Z_i,$$

where I is some set of indices and, for each $i \in I$, the group Z_i is isomorphic to $\mathbf{Z}$. In other words, we have

$$Z_i = \{ke_i \ : \ k \in \mathbf{Z}\},$$

where e_i denotes a certain element of Z_i generating Z_i and such that

$$ke_i \neq le_i \quad (k \in \mathbf{Z}, \ l \in \mathbf{Z}, \ k \neq l).$$

Clearly, the set $\{e_i : i \in I\}$ may be regarded as a basis of our group G.

Let us mention an important statement saying that any subgroup of a free commutative group is free, too. This statement remains true for noncommutative groups as well (see, for instance, [129]; a beautiful topological proof of this result can be found in [142]).

We say that a commutative group $(G, +)$ is divisible if for every natural number $n > 0$ and for any element $g \in G$, the equation $nx = g$ has at least one solution in G.

Evidently, the product of a family of divisible commutative groups is a divisible commutative group; however, the standard proof of this simple fact is based on the Axiom of Choice.

Example 1. The additive group $(\mathbf{Q}, +)$ of all rational numbers is divisible. All nonzero elements in $\mathbf{Q}$ are of infinite order. Similarly to $\mathbf{Z}$, the group $\mathbf{Q}$ is not decomposable.

The commutative group $(\mathbf{S}_1, \cdot)$ is divisible, too. (Here $\mathbf{S}_1$ denotes the unit circumference in $\mathbf{R}^2$ equipped with the natural group operation which is the restriction to $\mathbf{S}_1$ of the usual multiplication operation of complex numbers.) Moreover, for every nonzero $n \in \mathbf{N}$, there are elements in $\mathbf{S}_1$ of order n. Clearly, there are also uncountably many elements in $\mathbf{S}_1$ of infinite order.

Example 2. Let p be an arbitrary prime natural number. For each natural number k, let us denote by G_k the subgroup of $\mathbf{S}_1$ consisting of all those elements $z \in \mathbf{S}_1$ for which the equality

$$z^{p^k} = 1$$

is valid. In this way, we obtain a strictly increasing (by inclusion) sequence of finite commutative groups:

$$G_0 \subset G_1 \subset ... \subset G_k \subset ... .$$

Let us define

$$\Gamma_p = \cup\{G_k \ : \ k < \omega\}.$$

Then Γ_p is also a subgroup of $\mathbf{S}_1$ and

$$card(\Gamma_p) = \omega.$$

Any group isomorphic to Γ_p is usually called a quasi-cyclic group of type p^∞.

Actually, the group Γ_p may be considered as the inductive limit of the family of groups $(G_k)_{k<\omega}$ with respect to their canonical embeddings

$$\phi_{k,k+1} \ : \ G_k \to G_{k+1} \quad (k < \omega).$$

Let us verify that Γ_p is divisible. First, let us observe that it suffices to establish the following fact:

For any prime natural number q and for any element $s \in \Gamma_p$, there exists an element $z \in \Gamma_p$, such that $z^q = s$.

Indeed, assume that this fact is true. If n is an arbitrary nonzero natural number, then we can represent n in the form

$$n = q_1 q_2 ... q_k$$

for some finite sequence $(q_1, q_2, ..., q_k)$ of prime numbers. According to our assumption, the equalities

$$s = z_1^{q_1}, \quad z_1 = z_2^{q_2}, \quad ..., \quad z_{k-2} = z_{k-1}^{q_{k-1}}, \quad z_{k-1} = z_k^{q_k}$$

are satisfied for certain elements $z_1, z_2, ..., z_{k-1}, z_k$ of Γ_p, from which it immediately follows that

$$s = z_k^{q_1 q_2 \cdots q_k} = z_k^n.$$

Now, let q be a prime natural number and let s be an arbitrary element from Γ_p. Then there exists a natural number k, such that

$$s^{p^k} = 1.$$

Only two cases are possible.

1. $q \neq p$. In this case, the numbers p^k and q are co-prime. According to a well-known theorem, we may write

$$mp^k + lq = 1$$

for some two integers m and l. Therefore, putting $z = s^l$, we obtain

$$z^q = s^{lq} = s^{1-mp^k} = s.$$

Thus, the element z belongs to Γ_p and is a solution of the equation $x^q = s$.

2. $q = p$. In this case, consider an element $z \in \mathbf{S}_1$ such that $z^p = s$. Since

$$z^{p^{k+1}} = s^{p^k} = 1,$$

we claim that z belongs to Γ_p and is a solution of the equation $x^q = s$.

Finally, let us point out the following interesting property of the group Γ_p:

Each proper subgroup of Γ_p is necessarily finite.

We leave to the reader the checking this fact (which also yields that Γ_p is not decomposable).

A commutative group G is called injective if, for any two commutative groups H and H' satisfying the relation $H \subset H'$ and for an arbitrary homomorphism

$$\phi : H \to G,$$

there exists a homomorphism $\phi' : H' \to G$ extending ϕ.

The next statement yields a characterization of divisible commutative groups.

Theorem 2. *Let G be a commutative group. Then the following assertions are equivalent:*

1) G is a divisible group;

2) G can be represented as a direct sum of groups each of which is isomorphic either to $\mathbf{Q}$ *or to a group of type* p^∞, *where p is a prime number;*

3) G is an injective group.

The proof of this theorem can be found in [53] and [129].

From assertion 2) we infer at once that if $(G, +)$ is a divisible commutative group whose all nonzero elements are of infinite order, then G is a vector space over $\mathbf{Q}$.

Notice that assertion 3) easily implies the next useful fact: if G is a divisible subgroup of a commutative group H, then G is a direct summand in H. In other words, we have the representation

$$H = G + G' \qquad (G \cap G' = \{0\})$$

for some subgroup G' of H. Indeed, let us consider the identity mapping

$$Id_G \; : \; G \to G$$

which obviously is an isomorphism of G onto itself. Applying the injectivity of G, we see that there exists a homomorphism

$$\phi \; : \; H \to G$$

extending Id_G. Let us put

$$G' = ker(\phi) = \phi^{-1}(0).$$

Then G' is a subgroup of H, and it can readily be verified that H is a direct sum of G and G'.

This property (be a direct summand in any larger commutative group) characterizes divisible commutative groups.

Let us mention, in addition to the said above, that any direct summand in a divisible commutative group is divisible itself. Indeed, suppose that G is a divisible commutative group and let

$$G = H + H' \qquad (H \cap H' = \{0\})$$

be a representation of G in the form of the direct sum of its two subgroups H and H'. Let us demonstrate that H is divisible, too. Take any natural number $n > 0$ and an arbitrary element $h \in H$. Since G is divisible, there exists an element $x \in G$ such that $nx = h$. Clearly, we may write $x = y + z$ where $y \in H$ and $z \in H'$. Further, we have

$$nx = n(y + z) = ny + nz = h, \quad h - ny = nz, \quad h - ny \in H, \quad nz \in H'.$$

Consequently,

$$h - ny = nz = 0, \quad h = ny,$$

which shows that H is divisible.

Theorem 2 also implies that, for any commutative group G, there are sufficiently many homomorphisms from G into $\mathbf{S}_1$. Indeed, take any nonzero element $g \in G$ and consider the group $[g]$ generated by g. In view of Example 1, the group $\mathbf{S}_1$ contains a subgroup T isomorphic to $[g]$. Let

$$\phi \; : \; [g] \to T$$

be an isomorphism between these two subgroups and let

$$\phi' \; : \; G \to \mathbf{S}_1$$

be a homomorphism extending ϕ. Obviously, we have $\phi'(g) \neq e$ where e stands for the neutral element of $\mathbf{S}_1$.

It immediately follows from the said above that if $x \in G$, $y \in G$ and $x \neq y$, then there always exists a homomorphism

$$\phi_{x,y} : G \to \mathbf{S}_1$$

such that $\phi(x) \neq \phi(y)$.

In other words, the family of all homomorphisms from G into $\mathbf{S}_1$ separates the elements of G. Taking into account this circumstance and applying the standard argument, we see that every commutative group G can be embedded in the corresponding product group $\mathbf{S}_1^\kappa$, where κ is some cardinal number.

From this fact we readily infer that for any commutative group G, there exists a divisible commutative group G' containing G as a subgroup. Moreover, we may assume without loss of generality that

$$card(G') \leq card(G) + \omega.$$

Let us point out another important consequence of the above-mentioned fact. Namely, every infinite commutative group G can be endowed with a nondiscrete Hausdorff topology compatible with the group operation in G. This result does not hold for infinite noncommutative groups. For instance, assuming **CH**, Shelah constructed a group of cardinality continuum which does not admit a nondiscrete Hausdorff group topology (see [191]). Later, Ol'shanskii gave an example (without using extra axioms) of an infinite group which also does not admit a nontrivial Hausdorff group topology.

We have the following important statement.

Theorem 3. *Let G be a commutative group and let H be a subgroup of G. If G can be represented in the form of a direct sum of cyclic groups, then H can be represented in the same form.*

A detailed proof of Theorem 3 is given in [129]. It is useful to compare this theorem with the result mentioned earlier and stating that any subgroup of a free commutative group is free, too.

Example 3. It can easily be verified that the additive group $\mathbf{Q}$ is not a direct sum of cyclic groups.

However, there exists an increasing (by the inclusion relation) countable family $\{H_n : n < \omega\}$ of subgroups of $\mathbf{Q}$, such that each H_n is representable as a direct sum of cyclic groups and

$$\mathbf{Q} = \cup\{H_n : n < \omega\}.$$

Indeed, it suffices to put $H_n = [\{1/n!\}]$ for all $n \in \omega$.

The same property is trivially valid for any group of type p^∞, where p is an arbitrary prime natural number (see Example 2).

Finally, we are able to establish the following statement which was essentially employed in this book.

Theorem 4. *Let G be an arbitrary commutative group. Then there exists an increasing (by inclusion) countable family $\{G_n : n < \omega\}$ of subgroups of G, such that:*

1) $\cup\{G_n : n < \omega\} = G$;

2) each group G_n $(n < \omega)$ is a direct sum of cyclic groups.

Proof. We know that G can be embedded in a divisible commutative group G'. In view of Theorem 2, G' is a direct sum of groups isomorphic either to $\mathbf{Q}$ or to a group of type p^∞, where p is a prime natural number. Taking into account Example 3, we see that G' can be represented in the form

$$G' = \cup\{G'_n : n < \omega\},$$

where $\{G'_n : n < \omega\}$ is an increasing (by inclusion) countable family of subgroups of G' and each G'_n is a direct sum of cyclic groups. Now, let us put

$$G_n = G \cap G'_n \qquad (n < \omega).$$

Clearly, G_n is a subgroup of G for any $n < \omega$. Moreover, we have

$$(\forall n < \omega)(G_n \subset G_{n+1}),$$

$$\cup\{G_n : n < \omega\} = G.$$

Finally, since G_n is a subgroup of G'_n, we may apply Theorem 3 and claim that G_n is a direct sum of cyclic groups.

Theorem 4 has thus been proved.

The last theorem played a significant role in our considerations connected with nonmeasurable sets and measure extension problems. (See, for instance, Chapters 11 and 13.) This theorem also implies that the family of all subgroups of an uncountable commutative group is sufficiently large from the measure-theoretical viewpoint. For example, we know that if G is a commutative group of cardinality ω_1, then there always exists a subgroup of G nonmeasurable with respect to a given nonzero σ-finite diffused measure on G. For noncommutative groups of the same cardinality, an analogous fact does not hold because there are groups G with $card(G) = \omega_1$ but without proper uncountable subgroups (see [191]).

Let us underline once more that a rich material concerning the structure of infinite commutative groups is presented in widely known monographs [53], [54], [129].

Bibliography

1. Armstrong T., Prikry K., *κ-additivity and κ-finiteness of measures on sets and left invariant measures on discrete groups*, Proc. Amer. Math. Soc., vol. 80, 1980, 105 - 112.

2. Ascherl A., Lehn J., *Two principles of extending probability measures*, Man. Math., vol. 21, 1977, 43 - 50.

3. Baker R., *"Lebesgue measure" on $\mathbf{R}^\infty$*, Proc. Amer. Math. Soc., vol. 113, no. 4, 1991, 1023 - 1029.

4. Balcerzak M., *A classification of σ-ideals in Polish groups*, Demonstratio Mathematica, vol. 20, no. 1 - 2, 1987, 77 - 88.

5. Balcerzak M., *Another nonmeasurable set with property (s^0)* , Real Analysis Exchange, vol. 17, no. 2, 1991 - 1992, 781 - 784.

6. Balcerzak M., Kharazishvili A., *On uncountable unions and intersections of measurable sets*, Georgian Mathematical Journal, vol. 6, no. 3, 1999, 201 - 212.

7. Banach S., *Sur les suites d'ensembles excluant l'existence d'une mesure* (with comments by E. Marczewski), Coll. Math., vol. 1, 1948, 103 - 108.

8. Banach S., Kuratowski K., *Sur une généralisation du probléme de la mesure*, Fund. Math., vol. 14, 1929, 127 - 131.

9. Banach S., Tarski A., *Sur la décomposition des ensembles de points en parties respectivement congruentes*, Fund. Math., vol. 6, 1924, 244 - 277.

10. Bernstein F., *Zur Theorie der trigonometrischen Reihen*, Sitzungsber. Sächs. Akad. Wiss. Leipzig. Math.-Natur. Kl. 60, 1908, 325 - 338.

11. Bing R.H., Bledsoe W.W., Mauldin R.D., *Sets generated by rectangles*, Pacific Journal Math., vol. 51, 1974, 27 - 36.

12. Blair C.E., *The Baire category theorem implies the principle of dependent choices*, Bull. Acad. Polon. Sci., Ser. Math., vol. 25, 1977, 933 - 934.

13. Bourbaki N., *Set Theory*, Izd. Mir, Moscow, 1965 (in Russian, translation from French).

14. Bourbaki N., *Integration*, Izd. Nauka, Moscow, 1977 (in Russian, translation from French).

15. Brown J.B., Cox G.V., *Classical theory of totally imperfect spaces*, Real Analysis Exchange, vol. 7, 1982, 1 - 39.

16. Brzuchowski J., Cichoń J., Grzegorek E., Ryll-Nardzewski C., *On the existence of nonmeasurable unions*, Bull. de L'Acad. Pol. des Sci., vol. 27, no. 6, 1979, 447 - 448.

17. Bukovský L., *Any partition into Lebesgue measure zero sets produces a non-measurable set*, Bull. de L'Acad. Pol. des Sci., vol. 27, no. 6, 1979, 431 - 435.

18. Cichoń J., Kharazishvili A., Węglorz B., *On sets of Vitali's type*, Proc. Amer. Math. Soc., vol. 118, no. 4, 1993, 1221 - 1228.

19. Cichoń J., Kharazishvili A., Węglorz B., *Subsets of the Real Line*, Part 1, Łódź University Press, Łódź, 1995.

20. Cichoń J., Kharazishvili A., *On ideals with projective bases*, Georgian Mathematical Journal, vol. 9, no. 3, 2002, 461 - 472.

21. Ciesielski K., *Algebraically invariant extensions of σ-finite measures on Euclidean spaces*, Trans. Amer. Math. Soc., vol. 318, 1990, 261 - 273.

22. Ciesielski K., *Set-theoretic real analysis*, Journal of Applied Analysis, vol. 3, no. 2, 1997, 143 - 190.

23. Ciesielski K., *Measure zero sets whose algebraic sum is not measurable*, Real Analysis Exchange, vol. 26, no. 2, 2000-2001, 919 - 922.

24. Ciesielski K., Pelc A., *Extensions of invariant measures on Euclidean spaces*, Fund. Math., vol. 125, 1985, 1 - 10.

25. Comfort W.W., *Topological groups*, in: Handbook of Set-Theoretic Topology, edited by K.Kunen and J.E.Vaughan, Elsevier Science Publishers B.V., 1984.

26. Corazza P., *Ramsey sets, the Ramsey ideal and other classes over* **R**, J. Symbolic Logic, vol. 57, 1992, 1441 - 1467.

27. Cornfeld I.P., Sinaj J.G., Fomin S.V., *Ergodic Theory*, Izd. Nauka, Moscow, 1980 (in Russian).

28. Davies R.O., *Equivalence to the Continuum Hypothesis of a certain proposition of elementary plane geometry*, Zeitschr. f. math. Logik und Grundlagen d. Math., vol. 8, 1962, 109 - 111.

29. Davies R.O., *The power of the continuum and some propositions of plane geometry*, Fund. Math., vol. LII, 1963, 277 - 281.

30. Davies R.O., *On a problem of Erdös concerning decompositions of the plane*, Proc. Camb. Phil. Soc., vol. 59, 1963, 33 - 36.

31. Davies R.O., *Covering the plane with denumerably many curves*, J. London Math. Soc., vol. 38, 1963, 433 - 438.

32. Davies R.O., *Covering space with denumerably many curves*, Bull. London Math. Soc., vol. 6, 1974, 189 - 190.

33. Doob J.L., *On a problem of Marczewski*, Coll. Math., vol. 1, no. 3, 1948, 216 - 217.

34. Doob J.L., *Stochastic Processes*, Wiley and Sons, Inc., New York, 1953.

35. Dougherty R., Foreman M., *Banach-Tarski decompositions using sets with the property of Baire*, Journal Amer. Math. Soc., vol. 7, 1994, 75 - 124.

36. Dye H.A., *On groups of measure preserving transformations*, Amer. Journ. Math., vol. 81, 1959, 119 - 159; vol. 85, 1963, 551 - 576.

37. Engelking R., *General Topology*, PWN, Warszawa, 1985.

38. Erdös P., Mauldin R.D., *The nonexistence of certain invariant measures*, Proc. Amer. Math. Soc., vol. 59, 1976, 321 - 322.

39. Erdös P., Jackson S., Mauldin R.D., *On partitions of lines and space*, Fund. Math., vol. 145, 1994, 101 - 119.

40. Erdös P., Jackson S., Mauldin R.D., *On infinite partitions of lines and space*, Fund. Math., vol. 152, 1997, 75 - 95.

41. Erdös P., Kunen K., Mauldin R.D., *Some additive properties of sets of real numbers*, Fund. Math., vol. CXIII, no. 3, 1981, 187 - 199.

42. Erdös P., Rado R., *A partition calculus in set theory*, Bull. Amer. Math. Soc., vol. 62, 1956, 427 - 489.

43. Erdös P., Stone A.H., *On the sum of two Borel sets*, Proc. Amer. Math. Soc., vol. 25, no. 2, 1970, 304 - 306.

44. Erdös P., Szekeres G., *A combinatorial problem in geometry*, Compositio Math., vol. 2, 1935, 463 - 470.

45. Ershov M.P., *Measure extensions and stochastic equations*, Probability Theory and its Applications, vol. 19, no. 3, 1974 (in Russian).

46. Foreman M., Wehrung F., *The Hahn-Banach theorem implies the existence of a non-Lebesgue measurable set*, Fund. Math., vol. 138, 1991, 13 - 19.

47. Fremlin D.H., *Consequences of Martin's Axiom*, Cambridge University Press, Cambridge, 1984.

48. Fremlin D.H., *Measurable functions and almost continuous functions*, Man. Math., vol. 33, 1981, 387 - 405.

49. Fremlin D.H., *Measure-additive coverings and measurable selectors*, Dissertationes Math. (Rozprawy Mat.), vol. 260, 1987, 1 - 116.

50. Fremlin D.H., *Real-valued-measurable cardinals*, Israel Math. Conf. Proc., vol. 6, 1993, 151 - 304.

51. Friedman H., *A consistent Fubini-Tonelli theorem for nonmeasurable functions*, Illinois J. Math., vol. 24, 1980, 390 - 395.

52. Frolik Z., *A measurable map with analytic domain and metrizable range is quotient*, Bull. Amer. Math. Soc., vol. 76, 1970, 1112 - 1117.

53. Fuchs L., *Infinite Abelian Groups*, vol. I, Academic Press, New York - London, 1970.

54. Fuchs L., *Infinite Abelian Groups*, vol. II, Academic Press, New York - London, 1973.

55. Gelbaum B.R., Olmsted J.M.H., *Counterexamples in Analysis*, San Francisco, Holden-Day, 1964.

56. Gnedenko B.V., Kolmogorov A.N., *Limit Distributions for Sums of Independent Random Variables*, Gos. Tech. Izdat., Moscow - Leningrad, 1949 (in Russian).

57. Grzegorek E., *Remarks on σ-fields without continuous measures*, Coll. Math., vol. 39, 1978, 73 - 75.

58. Grzegorek E., *Solution of a problem of Banach on σ-fields without continuous measures*, Bull. Acad. Polon. Sci., Ser. Sci. Math., vol. 28, 1980, 7 - 10.

59. Hajian A., Ito Y., *Weakly wandering sets and invariant measures for a group of transformations*, J. Math. Mec. vol. 18, 1969, 1203 - 1216.

60. Hall Ph., *On representatives of subsets*, J. London Math. Soc., v. 10, 1935, 26 - 30.

61. Halmos P.R., *Lectures on Ergodic Theory*, The Mathematical Society of Japan, Tokyo, 1956.

62. Halmos P.R., *Measure Theory*, D. Van Nostrand, Princeton, New York, 1950.

63. Hamel H., *Eine Basis aller Zahlen und die unstetigen Lösungen der Funktionalgleichung:* $f(x + y) = f(x) + f(y)$, Math. Ann., vol. 60, 1905, 459 - 462.

64. *Handbook of Mathematical Logic*, edited by J.Barwise, North-Holland Publishing Company, Amsterdam, 1977.

65. *Handbook of Mathematical Logic*, Part 2, Izd. Nauka, Moskva, 1982 (in Russian, translation from English with comments).

66. Hausdorff F., *Grundzüge der Mengenlehre*, Veit und Co., Leipzig, 1914.

67. Hejduk J., Kharazishvili A., *On density points with respect to von Neumann's topology*, Real Analysis Exchange, vol. 21, no. 1, 1995 - 1996, 278 - 291.

68. Hewitt E., Ross K.A., *Abstract Harmonic Analysis*, vol. I, Springer-Verlag, Berlin, 1963.

69. Hewitt E., Stromberg K., *Some examples of nonmeasurable sets*, Journ. Austral. Math. Soc., vol. 18, 1974, 236 - 238.

70. Hopf E., *Theory of measures and invariant integrals*, Trans. Amer. Math. Soc., vol. 34, 1932, 373 - 393.

71. Hulanicki A., *Invariant extensions of the Lebesgue measure*, Fund. Math., vol. 51, 1962, 111 - 115.

72. Hulanicki A., Ryll-Nardzewski C., *Invariant extensions of the Haar measure*, Coll. Math., vol. 42, 1979, 223 - 227.

73. Jacab Zs., Laczkovich M., *A characterization of Jordan and Lebesgue measures*, Coll. Math., vol. XL, no. 1, 1978, 39 - 52.

74. Jech, T., *The Axiom of Choice*, North-Holland Publishing Company, Amsterdam, 1973.

75. Kakutani S., Oxtoby J., *Construction of a nonseparable invariant extension of the Lebesgue measure space*, Ann. Math., vol. 52, 1950, 580 - 590.

76. Kelley J.L., *The Tychonoff product theorem implies the Axiom of Choice*, Fund. Math., vol. 37, 1950, 75 - 76.

77. Kelley J.L., *General Topology*, D. Van Nostrand, New York, 1955.

78. Kharazishvili A.B., *Absolutely nonmeasurable sets in Abelian groups*, Bull. Acad. Sci. Georgian SSR, vol. 97, no. 3, 1980 (in Russian).

79. Kharazishvili A.B., *Applications of Set Theory*, Izd. Tbil. Gos. Univ., Tbilisi, 1989 (in Russian).

80. Kharazishvili A.B., *Certain types of invariant measures*, Dokl. Akad. Nauk SSSR, vol. 222, no. 3, 1975, 538 - 540 (in Russian).

81. Kharazishvili A.B., *Γ-topology and the Baire property*, Bull. Acad. Sci. of Georgia, vol. 148, no. 1, 1993 (in Russian).

82. Kharazishvili A.B., *Invariant Extensions of the Lebesgue Measure*, Izd. Tbil. Gos. Univ., Tbilisi, 1983 (in Russian).

83. Kharazishvili A.B., *Martin's Axiom and Γ-selectors*, Bull. Acad. Sci. Georgian SSR, vol. 137, no. 2, 1990 (in Russian).

84. Kharazishvili A.B., *On nonmeasurable subgroups of the real line*, Journal of Applied Analysis, vol. 2, no. 2, 1996, 171 - 181.

85. Kharazishvili A.B., *On the absolute nonmeasurability of the unit ball in an infinite-dimensional separable Hilbert space*, Bull. Acad. Sci. GSSR, vol. 107, no. 1, 1982 (in Russian).

86. Kharazishvili A.B., *On translations of sets and functions*, Journal of Applied Analysis, vol. 1, no. 2, 1995, 145 -158.

87. Kharazishvili A.B., *Selected Topics in Geometry of Euclidean Space*, Izd. Tbil. Gos. Univ., Tbilisi, 1978 (in Russian).

88. Kharazishvili A., *Subsets of the plane with small linear sections and invariant extensions of the two-dimensional Lebesgue measure*, Georgian Mathematical Journal, vol. 6, no. 5, 1999, 441 - 446.

89. Kharazishvili A., *Small sets in uncountable Abelian groups*, Acta Universitatis Lodziensis, Folia Mathematica, vol. 7, 1995, 31 - 39.

90. Kharazishvili A.B., *Some applications of Hamel bases*, Bull. Acad. Sci. Georgian SSR, vol. 85, no. 1, 1977, 17 - 20 (in Russian).

91. Kharazishvili A.B., *Some remarks on density points and the uniqueness property for invariant extensions of the Lebesgue measure*, Acta Universitatis Carolinae, Mathematica et Physica, vol. 35, no. 2, 1994, 33 - 39.

92. Kharazishvili A.B., *The Baire property and its applications*, Proceedings of I.N. Vekua Institute of Applied Mathematics, vol. 43, 1992, 5 - 113 (in Russian).

93. Kharazishvili A.B., *Some remarks on nonmeasurable almost invariant sets*, Acta Universitatis Lodziensis, Folia Mathematica, vol. 7, 1995, 41 - 50.

94. Kharazishvili A.B., *Some questions concerning invariant extensions of Lebesgue measure*, Real Analysis Exchange, vol. 20, no. 2, 1994 - 1995, 580 - 592.

95. Kharazishvili A.B., *Some remarks on additive properties of invariant σ-ideals on the real line*, Real Analysis Exchange, vol. 21, no. 2, 1995 - 1996.

96. Kharazishvili A.B., *Some Questions of Set Theory and Measure Theory*, Izd. Tbil. Gos. Univ., Tbilisi, 1978 (in Russian).

97. Kharazishvili A.B., *The absolute negligibility of compacta in infinite dimensional Banach spaces*, Bull. Acad. Sci. GSSR, vol. 104, no. 3, 1981 (in Russian).

98. Kharazishvili A.B., *Topological Aspects of Measure Theory*, Izd. Naukova Dumka, Kiev, 1984 (in Russian).

99. Kharazishvili A.B., *Vitali's Theorem and its Generalizations*, Izd. Tbil. Gos. Univ., Tbilisi, 1991 (in Russian).

100. Kharazishvili A.B. *Transformation Groups and Invariant Measures*, World Scientific Publ. Co., Singapore, 1998.

101. Kharazishvili A.B., *Applications of Point Set Theory in Real Analysis*, Kluwer Academic Publishers, Dordrecht, 1998.

102. Kharazishvili A.B., *Strange Functions in Real Analysis*, Marcel Dekker, Inc., New York - Basel, 2000.

103. Kharazishvili A.B., *On Vitali's theorem for groups of motions of Euclidean space*, Georgian Mathematical Journal, vol. 6, no. 4, 1999, 323 - 334.

104. Kharazishvili A.B., *On countably generated invariant σ-algebras which do not admit measure type functionals*, Real Analysis Exchange, vol. 23, no. 1, 1997 - 1998, 287 - 294.

105. Kharazishvili A.B., *Some remarks on quasi-invariant and invariant measures*, Real Analysis Exchange, vol. 24, no. 1, 1998 - 1999, 427 - 434.

106. Kharazishvili A.B., *On measurability properties of subgroups of a given group*, Real Analysis Exchange, 24-th Summer Symposium Conference Reports, May 2000, 143 - 146.

107. Kharazishvili A.B., *On vector sums of measure zero sets*, Georgian Mathematical Journal, vol. 8, no. 3, 2001, 493 - 498.

108. Kharazishvili A.B., *On the existence of nonmeasurable subgroups of commutative groups*, Real Analysis Exchange, vol. 27, no. 1, 2001 - 2002, 71 - 76.

109. Kharazishvili A.B., *On negligible and absolutely nonmeasurable subsets of the Euclidean plane*, Georgian Mathematical Journal, vol. 11, 2004 (to appear).

110. King B., *Some remarks on difference sets of Bernstein sets*, Real Analysis Exchange, vol. 19, no. 2, 1993 - 1994, 478 - 490.

111. Kirtadze A.P., Pantsulaia G.R., *On invariant measures in the space $\mathbf{R}^N$*, Bull. Acad. Sci. GSSR, vol. 141, no. 2, 1991 (in Russian).

112. Kleinberg E.M., *Infinitary Combinatorics and the Axiom of Determinateness*, Lecture Notes in Mathematics, vol. 612, Springer-Verlag, Berlin, 1977.

113. Kodaira K., Kakutani S., *A nonseparable translation-invariant extension of the Lebesgue measure space*, Ann. Math., vol. 52, 1950, 574 - 579.

114. Kolmogoroff A.N., *Sur la possibilité de la définition générale de la dérivée, de l'intégrale et de la sommation des séries divergentes*, C. R. Acad. Sci. Paris, vol. 180, 1925, 362 - 364.

115. Krawczyk A., Zakrzewski P., *Extensions of measures invariant under countable groups of transformations*, Trans. Amer. Math. Soc., vol. 326, 1991, 211 - 226.

116. Krieger W., *On ergodic flows and the isomorphism of factors*, Math. Ann., vol. 223, 1976, 19 - 70.

117. Krieger W. *On nonsingular transformations of a measure space*, Z. Wahrsch. Verw. Gebiete, vol. 11, 1969, 83 - 97 and 98 - 119.

118. Kuczma M., *An Introduction to the Theory of Functional Equations and Inequalities: Cauchy's Equation and Jensen's Inequality*, PWN, Katowice, 1985.

119. Kunen K., *Inaccessibility Properties of Cardinals*, Ph. D. Thesis, Department of Mathematics, Stanford University, 1968.

120. Kunen K., *Combinatorics*, in: Handbook of Mathematical Logic, North-Holland Publishing Company, Amsterdam, 1977.

121. Kunen K., *Random and Cohen reals*, in: Handbook of Set-Theoretic Topology, edited by K.Kunen and J.E.Vaughan, North-Holland Publishing Company, Amsterdam, 1984.

122. Kunen K., *Set Theory*, North–Holland Publishing Company, Amsterdam, 1980.

123. Kupka J., *Strong liftings with application to measurable cross sections in locally compact groups*, Israel J. Math., vol. 44, 1983, 243 - 261.

124. Kupka J., Prikry K., *The measurability of uncountable unions*, Amer. Math. Monthly, vol. 91, no. 2, 1984, 85 - 97.

125. Kuratowski K., *Topology*, vol. 1, Academic Press, New York and London, 1966.

126. Kuratowski K., *Applications of the Baire-category method to the problem of independent sets*, Fund. Math., vol. LXXXI, 1973, 65 - 72.

127. Kuratowski K., Mostowski A., *Set Theory*, North-Holland Publishing Company, Amsterdam, 1967.

128. Kuratowski K., Ryll-Nardzewski C., *A general theorem on selectors*, Bull. Acad. Polon. Sci., Ser. Sci. Math., vol. 13, no. 6, 1965, 397 - 402.

129. Kurosh A.G., *The Theory of Groups*, Izd. Nauka, Moscow, 1967 (in Russian).

130. Labuda I., Mauldin R.D., *Problem 24 of the "Scottish Book" concerning additive functionals*, Coll. Math., vol. XLVIII, no. 1, 1984, 89 - 91.

131. Lebesgue H., *Contribution a l'étude des correspondances de M. Zermelo*, Bull. de la Soc. Math. France, vol. 35, 1907, 202 - 214.

132. Lipecki Z., *On continuity of group homomorphisms*, Coll. Math., vol. XLVIII, no. 1, 1984, 93 - 94.

133. Lubotzky A., *Discrete Groups, Expanding Graphs and Invariant Measures*, Birkhäuser, 1994.

134. Lusin N., *Sur une probléme de M. Baire*, C. R. Acad. Sci. Paris, vol. 158, 1914, 1259 - 1261.

135. Lusin N., *Lecons sur les Ensembles Analytiques et leurs Applications*, Gauthier-Villars, Paris, 1930.

136. Luzin N.N., *Integral and Trigonometric Series*, Izd. GITTL, Moscow, 1956 (in Russian).

137. Mabry R.D., *Sets which are well-distributed and invariant relative to all isometry invariant total extensions of Lebesgue measure*, Real Analysis Exchange, vol. 16, 1990 - 1991, 425 - 459.

138. Mabry R.D., *Some remarks concerning the uniformly gray sets of G. Jacopini*, Rend. Accad. Naz. Sci., Mem. Mat. Appl., vol. 22, no. 5, 1998, 43 - 49.

139. Mackey G.W., *Borel structures in groups and their duals*, Trans. Amer. Math. Soc., vol. 85, 1957, 134 - 169.

140. Maharam D., *On measure in abstract sets*, Trans. Amer. Math. Soc., vol. 51, 1942, 413 - 433.

141. Maharam D., *On a theorem of von Neumann*, Proc. Amer. Math. Soc., vol. 9, 1958, 987 - 994.

142. Massey W.S., *Algebraic Topology: An Introduction*, Harcourt, Brace and World, Inc., New York, 1967.

143. Marczewski E., Ryll-Nardzewski C., *Sur la mesurabilité des fonctions de plusieurs variables*, Ann. de la Societé Polonaise de Mathématique, vol. 25, 1952, 145 - 155.

144. Mauldin R.D., Schlee G.A., *Borel measurable selections and applications of the boundedness principle*, Real Analysis Exchange, vol. 15, no. 1, 1989 - 1990, 70 - 92.

145. Michael E., *Continuous selections I*, Ann. Math., vol. 63 (2), 1956, 361 - 382.

146. Michael E., *Selected selection theorems*, Amer. Math. Monthly, vol. 63, 1956, 233 - 238.

147. Miller A.W., *Special subsets of the real line*, in Handbook of Set-Theoretic Topology, North-Holland Publishing Company, Amsterdam, 1984, 201 - 234.

148. Miller A.W., Popvassilev S.G., *Vitali sets and Hamel bases that are Marczewski measurable*, Fund. Math., vol. 166, no. 3, 2000, 269 - 279.

149. Miller D.E., *On the measurability of orbits in Borel actions*, Proc. Amer. Math. Soc., vol. 63, no. 1, 1977, 165 - 170.

150. Mokobodzki G., *Ensembles a coupes denomerables et capacities dominee par une mesure*, Universite de Strasbourg, Seminaire de Probabilites, 1977 - 1978.

151. Mokobodzki G., *Ultrafiltres rapides sur* **N**, Sémin. Brelot-Choquet-Deny, Théorie de Potentiel, no. 12, 1967 - 1968.

152. Montgomery M.D., *Nonseparable metric spaces*, Fund. Math., vol. 25, 1935, 527 - 533.

153. Morayne M., *On differentiability of Peano type functions*, Coll. Math., vol. 53, no. 1, 1987, 129 - 132.

154. Morayne M., Ryll-Nardzewski C., *Functions equivalent to Lebesgue measurable ones*, Bull. Polish Acad. Sci., Mathematics, vol. 47, no. 3, 1999, 263 - 265.

155. Morgan II J.C., *Point Set Theory*, Marcel Dekker, Inc., New York and Basel, 1990.

156. Mycielski J., *On the paradox of the sphere*, Fund. Math., vol. 42, 1955, 348 - 355.

157. Mycielski J., *Independent sets in topological algebras*, Fund. Math., vol. 55, 1964, 139 - 147.

158. Natanson I.P., *The Theory of Functions of a Real Variable*, Izd. Nauka, Moscow, 1957 (in Russian).

159. Neumann (von) J., *Ein System algebraisch unabhängiger Zahlen*, Math. Ann., vol. 99, 1928, 134 - 141.

160. Neveu J., *Bases Mathématiques du Calcul des Probabilités*, Masson et Cie, Paris, 1964.

161. Nijaradze G.V., *On some properties of measures which are invariant with respect to discrete groups of motions*, Bull. Acad. Sci. GSSR, vol. 80, no. 2, 1975 (in Russian).

162. Nijaradze G.V., *On some geometric properties of G-measures*, Bull. Acad. Sci. GSSR, vol. 123, no. 2, 1986 (in Russian).

163. Nowik A., *On measure which measures at least one selector for every uncountable subgroup*, Real Analysis Exchange, vol. 22, no. 2, 1996 - 1997, 814 - 817.

164. Oxtoby J.C., *Invariant measures in groups which are not locally compact*, Trans. Amer. Math. Soc., vol. 60, 1946, 215 - 237.

165. Oxtoby J.C., *Measure and Category*, Springer-Verlag, Berlin, 1971.

166. Pantsulaia G.R., *Density points and invariant extensions of the Lebesgue measure*, Bull. Acad. Sci. of Georgia, vol. 151, no. 2, 1995 (in Russian).

167. Pawlikowski J., *The Hahn-Banach theorem implies the Banach-Tarski paradox*, Fund. Math., vol. 138, 1991, 21 - 22.

168. Pelc A., *Semiregular invariant measures on Abelian groups*, Proc. Amer. Math. Soc., vol. 86, 1982, 423 - 426.

169. Pelc A., *Invariant measures and ideals on discrete groups*, Dissertationes Math., vol. 255, 1986.

170. Pelc A., Prikry K., *On a problem of Banach*, Proc. Amer. Math. Soc., vol. 89, no. 4, 1983, 608 - 610.

171. Pelc A., Prikry K., *On the existence of measures on σ-algebras*, Fund. Math., vol. 125, no. 2, 1985, 115 - 123.

172. Pfeffer W.F., Prikry K., *Small spaces*, Proc. London Math. Soc., vol. 58, no. 3, 1989, 417 - 438.

173. Pkhakadze Sh., *The theory of Lebesgue measure*, Trudy Tbilis. Mat. Inst. im. Razmadze, Akad. Nauk Gruz. SSR, vol. 25, 1958, 3 - 272 (in Russian).

174. Plachky D., *A note on measurable subgroups of* $(\mathbf{R}, +)$, preprint.

175. Prikry K., *On images of the Lebesgue measure, I*, (unpublished manuscript dated 20 September, 1977).

176. Raisonnier J., *A mathematical proof of S. Shelah's theorem on the measure problem and related results*, Israel Journal of Mathematics, vol. 48, no. 1, 1984, 48 - 56.

177. Ramachandran D., *A new proof of Hopf's theorem on invariant measures*, Contemporary Mathematics, vol. 94, 1989, 263 - 270.

178. Ramachandran D., Misiurewicz M., *Hopf's theorem on invariant measures for a group of transformations*, Studia Math., vol. LXXIV, 1982, 183 - 189.

179. Ramsey F.P., *On a problem of formal logic*, Proc. London Math. Soc., vol. 30, 1930, 264 - 286.

180. Rao B.V., *On discrete Borel spaces and projective sets*, Bull. Amer. Math. Soc., vol. 75, 1969, 614 - 617.

181. Repovs D., Semenov P., *E.Michael's theory of continuous selections (development and applications)*, Usp. Mat. Nauk, vol. 49, no. 6, 1994, 151 - 190.

182. Rogers C.A., *A linear Borel set whose difference set is not a Borel set*, Bull. London Math. Soc., vol. 2, 1970, 41 - 42.

183. Rothberger F., *Eine Äquivalenz zwischen der Kontinuumhypothese und der Existenz der Lusinschen und Sierpińskischen Mengen*, Fund. Math., vol. 30, 1938, 215 - 217.

184. Rubin H., Rubin J., *Equivalents of the Axiom of Choice*, North-Holland Publishing Company, Amsterdam, 1963.

185. Ruziewicz S., *Sur une proprieté de la base hamelienne*, Fund. Math., vol. 26, 1936, 56 - 58.

186. Ruziewicz S., Sierpiński W., *Sur un ensemble parfait qui a avec toute sa translation au plus un point commun*, Fund. Math., vol. 19, 1932, 17 - 21.

187. Ryll-Nardzewski C., *On Borel measurability of orbits*, Fund. Math., vol. 56, 1964, 129 - 130.

188. Ryll-Nardzewski C., Telgársky R., *The nonexistence of universal invariant measures*, Proc. Amer. Math. Soc., vol. 69, 1978, 240 - 242.

189. Sazonov V.V., *On perfect measures*, Izv. Akad. Nauk SSSR (mathematics series), vol. 26, 1962, 391 - 414 (in Russian).

190. Scott D., *Invariant Borel sets*, Fund. Math., vol. 56, 1964, 117 - 128.

191. Shelah S., *On a problem of Kurosh, Jónsson groups, and applications*, in: S.I.Adian, W.W.Boone, and G.Higman, eds., *Word Problems, II*, North-Holland, Amsterdam, 1980, 373 - 394.

192. Shelah S., *Can you take Solovay's inaccessible away?*, Israel Journal of Mathematics, vol. 48, no. 1, 1984, 1 - 47.

193. Shipman J., *Cardinal conditions for strong Fubini theorems*, Trans. Amer. Math. Soc., vol. 321, 1990, 465 - 481.

194. Sierpiński W., *Cardinal and Ordinal Numbers*, PWN, Warszawa, 1958.

195. Sierpiński W., *L'hypothése du Continu*, Monogr. Mat., vol. 4, Warsaw, 1934.

196. Sierpiński W., *Sur la question de la mesurabilité de la base de M. Hamel*, Fund. Math., vol. 1, 1920, 105 - 111.

197. Sierpiński W., *Sur une fonction non mesurable partout presque symetrique*, Acta Litt. Scient., Szeged, vol. 8, 1936, 1 - 6.

198. Sierpiński W., *Sur le paradoxe de la sphére*, Fund. Math., vol. 33, 1945, 235 - 244.

199. Sierpiński W., *Sur les translations des ensembles linéaires*, Fund. Math., vol. 19, 1932, 22 - 28.

200. Sierpiński W., *Sur un théoréme équivalent á l'hypothése du continu*, Bull. Internat. Acad. Sci. Cracovie Ser. A, 1919, 1 - 3.

201. Sierpiński W., *Les correspondances multivoques et l'axiome du choix*, Fund. Math., vol. 34, 1947, 39 - 44.

202. Sierpiński W., Szpilrajn-Marczewski E., *Remarque sur le probléme de la mesure*, Fund. Math., vol. 26, 1936, 256 - 261.

203. Sierpiński W., Zygmund A., *Sur une fonction qui est discontinue sur tout ensemble de puissance du continu*, Fund. Math., vol. 4, 1923, 316 - 318.

204. Simms J.C., *Sierpiński's theorem*, Simon Stevin, vol. 65, no. 1 - 2, 1991, 69 - 163.

205. Skorokhod A.V., *Integration in Hilbert Space*, Izd. Nauka, Moscow, 1975 (in Russian).

206. Sodnomov B.S., *On arithmetical sums of sets*, Dokl. Akad. Nauk SSSR, vol. 80, no. 2, 1951, 173 - 175.

207. Sodnomov B.S., *An example of two G_δ-sets whose arithmetical sum is not Borel measurable*, Dokl. Akad. Nauk SSSR, vol. 99, 1954, 507 - 510.

208. Solecki S., *On sets nonmeasurable with respect to invariant measures*, Proc. Amer. Math. Soc., vol. 119, no. 1, 1993, 115 - 124.

209. Solecki S., *Measurability properties of sets of Vitali's type*, Proc. Amer. Math. Soc., vol. 119, no. 3, 1993, 897 - 902.

210. Solovay R.M., *A model of set theory in which every set of reals is Lebesgue measurable*, Ann. Math., vol. 92, 1970, 1 - 56.

211. Solovay R.M., *Real-valued measurable cardinals*, Proceedings of Symposia in Pure Mathematics, vol. 12, Axiomatic set theory, Part 1, Providence, 1971, 397 - 428.

212. Szpilrajn (Marczewski) E., *On problems of the theory of measure*, Uspekhi Mat. Nauk, vol. 1, no. 2 (12), 1946, 179 - 188 (in Russian).

213. Szpilrajn (Marczewski) E., *Sur une classe de fonctions de M.Sierpiński et la classe correspondante d'ensembles*, Fund. Math., vol. 24, 1935, 17 - 34.

214. Szpilrajn (Marczewski) E., *Sur l'extension de la mesure lebesgui-enne*, Fund. Math., vol. 25, 1935, 551 - 558.

215. Szpilrajn (Marczewski) E., *The characteristic function of a sequence of sets and some of its applications*, Fund. Math., vol. 31, 1938, 207 - 223.

216. Szpilrajn-Marczewski E., *Sur les ensembles et les fonctions absolument measurables*, Comp. rend Soc. Sci. Lett. Varsovie, vol. 30, 1937, 39 - 68 (in Polish).

217. Talagrand M., *Extension de la mesure de Lebesgue aux filtres*, C. R. Acad. Sci. Paris, Serie A, vol. 283, 1976, 95 - 98.

218. Talagrand M., *Filtres: mesurabilité, rapidité, propriété de Baire forte*, Studia Mathematica, vol. 74, 1982, 283 - 291.

219. Tall F., *The density topology*, Pacific Journal of Mathematics, vol. 62, 1976, 275 - 284.

220. Traynor T., *An elementary proof of the lifting theorem*, Pacific J. Math., vol. 53, 1974, 267 - 272.

221. Tsakadze R.A., *On totally homogeneous spaces*, Bull. Acad. Sci. GSSR, vol. 98, no. 1, 1980 (in Russian).

222. Ulam S., *Zur Masstheorie in der allgemeinen Mengenlehre*, Fund. Math., vol. 16, 1930, 140 - 150.

223. Van der Waerden B.L., *Algebra*, vol. I, Springer-Verlag, Berlin, 1971.

224. Vitali G., *Sul problema della misura dei gruppi di punti di una retta*, Bologna, Italy, 1905.

225. Vladimirov D.A., *Boolean Algebras*, Izd. Nauka, Moscow, 1969 (in Russian).

226. Wagon S., *The Banach-Tarski Paradox*, Cambridge University Press, Cambridge, 1985.

227. Walsh J.T., *Marczewski sets, measure and the Baire property II*, Proc. Amer. Math. Soc., vol. 106, no. 4, 1989, 1027 - 1030.

228. Węglorz B., *A note on automorphisms and partitions*, Coll. Math., vol. LV, 1988, 1 - 4.

229. Zakrzewski P., *Extending invariant measures on topological groups*, Annals New York Academy of Sciences, 1996, 218 - 222.

230. Zakrzewski P., *Extending isometrically invariant measures on $\mathbf{R}^n$ - a solution to Ciesielski's query*, Real Analysis Exchange, vol. 21, no. 2, 1995 - 1996, 582 - 589.

231. Zakrzewski P., *Extensions of isometrically invariant measures on Euclidean spaces*, Proc. Amer. Math. Soc., vol. 110, 1990, 325 - 331.

232. Zakrzewski P., *The existence of universal invariant measures on large sets*, Fund. Math., vol. 133, 1989, 113 - 124.

233. Zakrzewski P., *When do equidecomposable sets have equal measures?*, Proc. Amer. Math. Soc., vol. 113, 1991, 831 - 837.

234. Zakrzewski P., *When do sets admit congruent partitions?*, Quart. J. Math. Oxford, vol. 45, no. 2, 1994, 255 - 265.

235. Zakrzewski P., *The uniqueness of Haar measure and set theory*, Coll. Math., vol. 74, no. 1, 1997, 109 - 121.

236. Zakrzewski P., *The existence of universal invariant semiregular measures on groups*, Proc. Amer. Math. Soc., vol. 99, no. 3, 1987, 507 - 508.

237. Zermelo E., *Beweis dass jede Menge wohlgeordnet werden kann*, Math. Ann. vol. 59, 1904, 514 - 516.

238. Zermelo E., *Neuer Beweis für die Möglichkeit einer Wohlordnung*, Math. Ann., vol. 65, 1908, 107 - 128.

239. Zermelo E., *Untersuchungen über die Grundlagen der Mengenlehre, I*, Math. Ann., vol. 65, 1908, 261 - 281.

240. Zorn M., *A remark on method in transfinite algebra*, Bull. Amer. Math. Soc., vol. 41, 1935, 667 - 670.

241. Zygmund A., *Trigonometric Series*, vol. I, Cambridge University Press, Cambridge, 1959.